Desert Aeolian Processes

Desert Aeolian Processes

Edited by

Vatche P. Tchakerian
Department of Geography
Texas A&M University, USA

CHAPMAN & HALL
London · Glasgow · Weinheim · New York · Tokyo · Melbourne · Madras

Published by Chapman & Hall, 2–6 Boundary Row, London SE1 8HN, UK

Chapman & Hall, 2–6 Boundary Row, London SE1 8HN, UK

Blackie Academic & Professional, Wester Cleddens Road, Bishopbriggs, Glasgow G64 2NZ, UK

Chapman & Hall GmbH, Pappelallee 3, 69469 Weinheim, Germany

Chapman & Hall USA, 115 Fifth Avenue, New York, NY 10003, USA

Chapman & Hall Japan, ITP-Japan, Kyowa Building, 3F, 2-2-1 Hirakawacho, Chiyoda-ku, Tokyo 102, Japan

Chapman & Hall Australia, 102 Dodds Street, South Melbourne, Victoria 3205, Australia

Chapman & Hall India, R. Seshadri, 32 Second Main Road, CIT East, Madras 600 035, India

First edition 1995

Softcover reprint of the hardcover 1st edition 1995

Typeset in Times by Best-set Typesetter Limited

ISBN-13: 978-94-010-6519-1 e-ISBN-13: 978-94-009-0067-7
DOI: 10.1007/978-94-009-0067-7

A catalogue record for this book is available from the British Library

Library of Congress Catalog Card Number: 95-70865

♾ Printed on permanent acid free text paper, manufactured in accordance with ANSI/NISO Z39.48-1992 and ANSI/NISO Z39.48-1984 (Permanence of Paper).

Dedicated to my parents Panos and Marie.

CONTENTS

CONTRIBUTORS

Andrew J. Bach Department of Geography, Arizona State University, Tempe, AZ 85287, USA

Dan G. Blumberg Department of Geology, Arizona State University, Tempe, AZ 85287, USA

Charles A. Bush United States Geological Survey, Denver, CO 80225, USA

James Clark Department of Chemistry, Arizona State University, Tempe, AZ 85287, USA

Keith Clarke Department of Geology and Geography, Hunter College of the City, University of New York, New York City, NY 10021, USA

Scott D. Cowherd United States Geological Survey, Denver, CO 80225, USA

Anthony R. Dobrovolskis NASA-Ames Research Center, Moffett Field, CA 94035, USA

Ronald I. Dorn Department of Geography, Arizona State University, Tempe, AZ 85287, USA

Lisa R. Gaddis Astrogeology Branch, United States Geological, Survey, Flagstaff, AZ 86001, USA

Victor Goldsmith Department of Geology and Geography, Hunter College of the City, University of New York, New York City, NY 10021, USA

Ronald Greeley Department of Geology, Arizona State University, Tempe, AZ 85287, USA

James M. Gregory Department of Civil Engineering, Texas Tech University, Lubbock, TX 79409, USA

Peter K. Haff Department of Geology, Duke University, Durham, NC 27708, USA

James D. Iverson Department of Aerospace Engineering, Iowa State University, Ames, IA 50010, USA

Arnon Karnieli Remote Sensing Laboratory, Ben-Gurion University of the Negev, Sede-Boker 84993, Israel

Julie E. Laity Department of Geography, California State University, Northridge, CA 91330, USA

Nicholas Lancaster Quaternary Sciences Center, Desert Research Institute, Reno, NV 89506, USA

Jeffrey A. Lee Department of Economics and Geography, Texas Tech University, Lubbock, TX 79409, USA

Shannon Mahan United States Geological Survey, Denver, CO 80225, USA

Daniel R. Muhs United States Geological Survey, Denver, CO 80225, USA

Kevin R. Mulligan Department of Geography, Texas A&M University, College Station, TX 77843, USA

Steven L. Namikas Department of Geography, University of Southern California, Los Angeles, CA 90089, USA

David E. Presti Veterans Administration Medical Center, San Francisco, CA 94121, USA

Keld R. Rasmussen Institute of Geology, Aarhus University, DK 8000 Aarhus C, Denmark

R. Stephen Saunders Jet Propulsion Laboratory, Pasadena, CA 91109, USA

Steve Schoenhaus Department of Geology and Geography, Hunter College of the City, University of New York, New York City, NY 10021, USA

Douglas J. Sherman Department of Geography, University of Southern California, Los Angeles, CA 90089, USA

Udai B. Singh Department of Civil Engineering, Texas Tech University, Lubbock, TX 79409, USA

Vatche P. Tchakerian Department of Geography, Texas A&M University, College Station, TX 77843, USA

Haim Tsoar Department of Geography and Environmental Development, Ben-Gurion University of the Negev, Beer-Sheva 84105, Israel

Niccole Villa Department of Geography, Arizona State University, Tempe, AZ 85287, USA

Stephen D. Wall Jet Propulsion Laboratory, Pasadena, CA 91109, USA

Bruce R. White Department of Mechanical Engineering, University of California at Davis, Davis, CA 95616, USA

Steven H. Williams Department of Space Studies, University of North Dakota, Grand Forks, ND 58202, USA

Gregory R. Wilson Department of Plant and Soil Science, Texas Tech University, Lubbock, TX 79409, USA

James R. Zimbelman Center for Earth and Planetary Studies, Smithsonian Institute, Washington, DC 20560, USA

PREFACE

The idea for this volume came during the dryland sessions of the Association of American Geographers meeting in San Diego in April, 1992. The large number of papers devoted to aeolian processes and landforms indicated to me that aeolian geomorphology had come of age and the last 15 years or so had produced a plethora of papers, books, and edited volumes on all aspects of aeolian geomorphology. Chapter one is my tentative attempt to place developments in aeolian geomorphology in a historical perspective and to contemplate some thoughts about the future.

The fourteen papers selected address a wide range of issues ranging from micro-scale studies devoted to aeolian dust, sediment transport, and rock varnish in ventifacts to medium-scale studies of dunes and dune forms, reverse desertification, and macro-scale studies of ergs and sand transport pathways. The American Southwest, particularly the spectacular and unique Mojave Desert of California, is featured prominently in seven chapters. I hope this volume provides students and colleagues some new perspectives in aeolian geomorphology as well as pathways for future work.

I am grateful to many individuals who helped in the processing of the manuscripts for publication and volunteering their precious time for the review process. In particular, I especially acknowledge the assistance of the following in refereeing the papers: A. Bach, D. Blumberg, R. Balling, R. Dorn, A. Goudie, P. Haff, G. Kocurek, J. Laity, N. Lancaster, C. Mckenna-Neuman, T. Maxwell, K. Mulligan, W. Nickling, A. Orme, D. Sherman, H. Tsoar, K. White and J. Zimbelman. Special thanks to Dan Stupka and Lona Dearmont for editorial assistance and to Salwa Choucair for spending long hours assembling the manuscripts in PageMaker and providing much needed guidance and encouragement. I would also like to thank my colleague Robert Bednarz, editor of the *Journal of Geography*, for providing logistical assistance and editorial help and to Barbara Goldman (Chapman & Hall, New York) and Ian Francis (Chapman & Hall, London) for the successful start and end of this publication respectively.

V.P. Tchakerian

1 THE RESURGENCE OF AEOLIAN GEOMORPHOLOGY

Vatche P. Tchakerian
Department of Geography
Texas A&M University

ABSTRACT

The past decade has witnessed a major resurgence of interest in aeolian geomorphology. The first thirty years following publication of *The Physics of Blown Sand and Desert Dunes* by R. A. Bagnold in 1941 produced few significant works in aeolian geomorphology. More recently, however, the number of papers, edited volumes and conference proceedings has been phenomenal. This 'renaissance' in aeolian geomorphology is attributed to a number of interrelated factors including the increased designed number of studies to confirm or refine the theoretical foundations first proposed by Bagnold, the advent of satellite imagery, interest in planetary geomorphology, advances in instrumentation, heightened concern for land degradation and desertification, recognition of ergs (sand seas) as analogs for hydrocarbon sandstone reservoirs, advances in dating methods, interest in coastal dunes and the increased encouragement by university professors in the 1980s for postgraduate studies in aeolian geomorphology.

INTRODUCTION

The past two decades have seen a tremendous upsurge in research and publication dealing with aeolian processes and landforms. Numerous papers, edited volumes, special journal issues and conference proceedings have been published encompassing topics in both desert and coastal environments. This period has also witnessed the testing, fine-tuning and confirmation of many of the theoretical concepts in aeolian dynamics first outlined in 1941 by R. A. Bagnold in his now classic work on *The Physics of Blown Sand and Desert Dunes*.

This essay outlines the context within which this resurgence of aeolian geomorphology has occurred. The historical background is divided into three periods: pre-1940s, 1940-1980, and 1980 to the present. Rather than attempting to embrace all research in aeolian studies during these periods, this essay seeks to outline some of the more significant themes. It is hoped that this will generate further discussion about a discipline that is very alive, vigorous, exciting and challenging.

HISTORICAL FRAMEWORK

Pre-1940s (Pre-Bagnold)

Most aeolian geomorphological studies before 1940 were part of a larger framework of studies involving the early explorations of deserts. At first, wind

Desert Aeolian Processes. Edited by Vatche P. Tchakerian. Published in 1995 by Chapman & Hall, London.
ISBN 978-94-010-6519-1

was not considered a very effective geomorphic agent in deserts and was relegated to the background, relative to fluvial and weathering processes.

Towards the close of the nineteenth century, however, the importance of aeolian activity in deserts began to be recognized, especially as geoscientists such as Richthofen (1882) and Udden (1896) began to explore the deserts of interior Asia and realized the nature and extent of dust deposits. Some of the earliest contributions to the study of aeolian processes and dune morphology include the classic writings of Cornish (1897, 1914) and Beadnell (1910). For a while, wind acquired a more prominent role in shaping desert landscapes as a result of the travels and writings of a number of eminent scientists such as Passarge (1904). This period reached its zenith when Keyes (1912) proposed a major role for aeolian processes and coined the term 'aeoliation'. The 'aeolianists' were strong proponents of wind action as an effective denudational and deflational agent in desert basins. This interlude of prominence was put to rest by W. M. Davis and his disciples who wrote persuasively on the seminal role of fluvial processes, weathering and mass movement in the evolution of desert landscapes (Davis 1930). In summary, most aeolian geomorphology focused on describing the various aeolian landforms in deserts, both erosional and depositional, the role of wind in shaping desert landscapes and some early observations about dust deposits and dust storms in drylands (Cooke et al. 1993). Additionally, much early impetus for desert research and hence aeolian studies was driven by governments with imperial aspirations and colonial obligations who were keen to investigate deserts within their purview, although romance and reckless adventurism also played a crucial part in these endeavors (Livingstone 1993). The combination of scientific inquiry and high romance can be seen in a number of monographs and travel books from this period, including Bagnold's *Libyan Sands* (1935).

1940 - 1980 (Bagnold)

This period begins with the now classic work of R. A. Bagnold *The Physics of Blown Sand and Desert Dunes* (1941). Much has been said about Bagnold's contribution to aeolian geomorphology. Suffice to say, he laid the theoretical foundations for the discipline, even though he never held a regular university position nor, with a degree in engineering, was he a geomorphologist by academic training. He recognized the contributions to the understanding of fluid flow and particle motion of other scientists such as von Karman, Shields and Prandtl, and incorporated their work in his studies of aeolian transport and deposition (Pye and Tsoar 1990). Bagnold's autobiography, *Wind, War and Sand* (1990), written shortly before his death, is a wonderful odyssey into the mind of a genius who pioneered and strongly influenced subsequent studies in aeolian geomorphology.

The drastic consequences of the "Dust Bowl" years of the 1930s in the Great Plains of the United States, led a number of American scientists to investigate the mechanics of soil erosion by wind, spearheaded by W. S. Chepil and his colleagues, subsequently synthesized by Chepil and Woodruff (1963).

This applied work contributed immensely towards a better comprehension of aeolian entrainment under different land use activities.

A significant contribution to the study of dune morphology and dynamics came from French geomorphologists working in the Sahara Desert, whose work has gone unnoticed outside of Europe. Gautier (1935) and Aufrère (1931, 1935) and later Capot-Rey (1948) and Clos-Arceduc (1967) produced significant works that ought to be consulted more frequently.

Aeolian research and publications were few and far between during the 1960s and 1970s, but some notable accomplishments during this period included papers on the mechanics of aeolian transport by Belly (1964) and Williams (1964) as well as the very influential work of P. R. Owen (1964) in wind erosion. In dune morphology and dynamics, papers by Sharp (1966) on the Kelso Dunes of the Mojave Desert, California, Hastenrath (1967) and Lettau and Lettau (1969) on the barchans of southern Peru, Verstappen (1968), Folk (1971) on the longitudinal dunes in the Simpson Desert, Australia and Cooper (1958, 1967) and Inman et al. (1966), on Pacific coast dunes of North America, were significant. Howard et al. (1978) wrote one of the earliest papers on dune form and wind flow. One of the most original figures during this time was I. G. Wilson, who in a series of papers before his untimely death, advanced some controversial concepts involving grain size and dune spacing (the ripple/dune/draa classification) and a model for the development of ergs (1971, 1972, 1973). The culmination and the beginnings of the next phase in research was heralded by the publication in 1979 of *A Study of Global Sand Seas* edited by E. D. Mckee. This influential publication presented the first results of a global survey of all the major dune fields using Landsat imagery as its foundation (Breed et al. 1979). The dune classification system proposed by McKee (1979) has been subsequently adopted by most introductory texts in physical geography and geology. Given current developments in aeolian geomorphology, this classification warrants a substantial revision.

1980s to the present (Post-Bagnold)

During the past 15 years, several factors have led to a renaissance of interest in aeolian geomorphology.

First, scientists have made a sincere attempt to refine the theoretical foundations of aeolian geomorphology first outlined by Bagnold. A major focus has been in aeolian grain mechanics and the basic physics of aeolian sediment transport. Examples include a conference on 'The Physics of Blown Sand' in 1985, organized by the Department of Theoretical Statistics in Aarhus, Denmark (Barndorff-Nielsen et al. 1985). This meeting brought together various scientists interested in aeolian processes from a variety of disciplines. In 1986, the annual Binghamton Symposium in Geomorphology, devoted entirely to aeolian processes and landforms (Nickling 1986). Aarhus II was held in Denmark in 1990 (Barndorff-Nielsen and Willetts 1991a, b). During this period, significant monographs and symposia proceedings were also published (e.g. Brookfield and Ahlbrandt 1983, Greeley and Iverson 1985, Pye

1987, Kocurek 1988, Hesp and Fryberger 1988, Nordstrom et al. 1990, Pye and Tsoar 1990, Pye 1993, Pye and Lancaster 1993). Aeolian research was also published in desert geomorphology monographs (e.g. Thomas 1989, Cooke et al. 1993, Abrahams and Parsons 1994) as well as in a multitude of scientific journals. Certain research themes have attracted frequent attention including the mechanics of aeolian entrainment and wind erosion (Gillette and Stockton 1989, Raupach et al. 1993, among others), spatial and temporal variations in wind speed, shear stress and sediment transport (Lancaster 1985, Tsoar 1985, Livingstone 1986, Mulligan 1988, Wiggs 1993, among others), and mathematical models designed to simulate entrainment and sediment transport (Walmsey and Howard 1985, Anderson and Hallet 1986, Anderson and Haff 1988).

One reason for this renaissance in aeolian geomorphology has been the interaction of scientists unconstrained by traditional disciplinary barriers. The impetus for vigorous exchanges of ideas has come from collaboration between physical geographers, geologists, sedimentologists, engineers, soil scientists, atmospheric scientists and ecologists.

Second, since the first satellite-housed camera systems were sent aloft in the 1950s, the increasing availability of satellite imagery has made previously remote deserts relatively accessible, especially in Africa and Asia and provided for a more complete global perspective on aeolian landforms. This saw the publication of *A Study of Global Sand Seas* edited by Mckee (1979) and continues to the present. The arrival of geographic information systems (GIS) has re-generated renewed interest in remote sensing.

Third, concomitant with advances in remote sensing and air photo interpretation has been the rapid emergence of planetary geomorphology and the search for terrestrial analogs. Mars has been especially singled out owing to its plethora of apparent aeolian landforms (e.g. Tsoar et al. 1979). Laboratory studies for simulating sediment entrainment and transport in both the Martian and Venusian atmospheres have also been carried out with much success (e.g. Iverson et al. 1976, Iverson and White 1982, Greeley and Iverson 1985). The recent success of the 1990 Magellan mission to Venus has added still further more to our knowledge of planetary aeolian processes and landforms.

Fourth, our understanding of aeolian processes has also been improved by significant advances in field measurement techniques and field instrumentation. Whereas there has been little improvement in the basic design of sand traps and erosion stakes, our ability to measure wind flow has been markedly improved. Cup anemometers and wind vanes are used to study the complex relation between wind flow and dune morphology and robust hot wire anemometers have been developed for measuring wind close to the dune surface. Especially important is the use of dataloggers to record the wind measurements. Dataloggers in particular have revolutionized the gathering and analysis of wind data owing to the fact that they can record simultaneous measurements from a number of sensors and can be left in place for extended periods of time.

Fifth, renewed interest in aeolian processes and landforms has resulted from concerns for land degradation (desertification) in drylands, especially with respect to mitigating the effects of sand encroachment in agricultural and urban areas (Middleton and Thomas 1992, Thomas and Middleton 1994), and the increased frequency of dust storms (e.g. Middleton et al. 1986, Pye 1987, Middleton 1989). Additionally, recent concern in global environmental changes has led atmospheric scientists to evaluate critically the regional wind flow patterns and seek to include them in global circulation models (GCM's). This increased awareness has led to a more concentrated look at aeolian dust and dust deposits both on land and in the oceans. Data from ocean cores have provided scientists with clues to past aeolian inputs to the oceans, notably the movement of dust plumes from the Sahara onto and beyond the continental shelf of West Africa and the notion of increased aridity during the last glacial maximum (e.g. Sarnthein 1978). Mineral aerosols from desert source regions into the world's oceans have been the focus of increased attention owing to the complex role of aerosols in affecting biological productivity and atmospheric transparency (e.g. Prospero 1981, Prospero et al 1983, Duce and Tindale 1991). This period also witnessed an increased awareness and much work in applied urban geomorphology in drylands (e.g. Cooke et al. 1982).

Sixth, some of the current resurgence in aeolian geomorphology stems from the fact that aeolian sedimentary environments are good analogs for studying hydrocarbon reservoirs in the geologic record and can contribute towards the exploration and recovery of oil and natural gas (e.g. Fryberger et al. 1990). Ergs (sand seas) are studied as modern day analogs for ancient sandstone reservoirs (Glennie 1970). This has been made possible through advances in the electron microscope, both in the scanning (SEM) and backscatter (BSE) modes, and the publication of an atlas devoted to the SEM study of quartz grains (Krinsley and Doornkamp 1973). These tools are valuable for analyzing the textural characteristics of aeolian sediments and the micromorphological changes associated with early diagenesis (e.g. Smart and Tovey 1981, Fryberger et al. 1983, Marshall 1987, Krinsley and Manley 1989).

Seventh, an increased awareness of the fragility of coastal dune systems has led to a multitude of studies in dune dynamics and sediment transport in coastal environments (e.g. Hesp & Fryberger 1988, Nordstrom et al. 1990, Carter et al. 1992). Research has focused primarily in beach and dune replenishment projects, coastal engineering structures to control beach erosion and flooding (especially the threat of hurricanes along the Gulf Coast and the eastern seaboard of the United States), and planning and zoning in coastal communities (e.g. Hesp & Fryberger 1988, Bakker et al. 1990, Nordstrom et al. 1990, Carter et al. 1992). Research has also focussed on the evolution of coastal dune systems relative to changing Quaternary sea levels and sediment sources, for example by Orme and Tchakerian (1986) and Orme (1992) along the California coast.

Eight, within the past two decades there has been an explosion in the variety of dating methods available for temporal aeolian studies, although the accuracy and precision of some such methods is still open to debate. In addition to radiocarbon dating of organics in sands, two methods in particular, appear to offer a unique insight for establishing a time frame involving aeolian activity. These are the application of luminescence techniques to aeolian sands and the dating of rock varnish in ventifacts. Luminescence dating is ideal for aeolian sediments since the technique dates the quartz and feldspar grains which make up the majority of aeolian deposits (e.g. Wintle 1993, Rendell et al. 1994). Both cation-ratio and accelerator mass spectrometry (^{14}C TAMS) dating of rock varnish in ventifacts can also be used to provide a time frame for aeolian activities (e.g. Dorn et al. 1989).

Lastly, over the past two decades, many university professors have encouraged their students to pursue postgraduate studies in aeolian geomorphology, as opposed to fluvial geomorphology which has been traditionally well represented. Aeolian geomorphology was relatively uncrowded during the 1950s and 1960s, but subsequently many students have chosen to pursue aeolian interests. With technological improvements in field instrumentation and data loggers and the ready availability of remote sensing data, aeolian geomorphology took off in an unprecedented way around 1980 as witnessed today by the plethora of papers, monographs and conference proceedings which have inundated our visual and mental capabilities.

FUTURE TRENDS

The last two decades have witnessed major developments in aeolian geomorphology with some significant contributions in the fields of aeolian erosion and transport. Much work remains to be done, especially in understanding the relation between wind flow and dune form close to the dune surface and in correlating the data to simultaneous sediment transport. The understanding of spatial variations in shear stress and sand transport rates on single dunes and in dunefields should be a priority for the future. The continuation of numerical modeling studies for understanding sediment transport and especially its extension to dune dynamics at varying scales remains a challenge (Werner 1994). Our present knowledge in sediment transport and dune dynamics ought to take us next towards deciphering the complexities of draas and ergs. Some of the major ergs of the arid zone remain to be studied. Research will most likely continue in coastal aeolian processes, dating methods, Quaternary environments and climate change in deserts, and micrometeorological studies on dunes, among others. According to Pye and Lancaster (1993), understanding the 'dynamics of aeolian depositional systems at different temporal and spatial scales and their response to external changes in sea level, regional and global climates and tectonics', will be a major challenge for aeolian geomorphologists. These are indeed exciting times for aeolian geomorphology as we find ourselves in a renaissance begun by the late Brigadier Bagnold, and continued vigorously by a generation of scientists inspired by his ideas and works.

ACKNOWLEDGMENTS

I would like to thank A.R. Orme and K. R. Mulligan for their comments and suggestions on the manuscript.

REFERENCES

Abrahams, A. D. and Parsons, A. J. (1994) *Geomorphology of Desert Environments*. Chapman & Hall, New York.

Anderson, R. S. and Haff, P. K. (1988) Simulation of eolian saltation. *Science*, v. 241, p. 820-823.

Anderson, R. S. and Hallet B. (1986) Sediment transport by wind: toward a general model. *Geological Society America Bulletin*, v. 97, p. 523-535.

Aufrère, L. (1931) Le cycle morphologique de dunes. *Annales de Géographie*, v. 40, p. 362-385.

Aufrère, L. (1935) Essai sur les dunes du Sahara algérien. *Geografiska Annaler*, v. 18, p. 481-500.

Bagnold, R. A. (1935) *Libyan Sands*. Michael Haag, London.

Bagnold, R. A. (1941) *The Physics of Blown Sand and Desert Dunes*. Chapman & Hall, London.

Bagnold, R. A. (1990) *Sand, Wind, and War: Memoirs of a Desert Explorer*. University of Arizona Press, Tucson.

Bakker, Th. W., Jungerius P. D., and Klijn, J. A., eds., (1990) *Dunes of the European Coasts*. Catena Supplement No. 18.

Barndorff-Nielsen, O. E., Møller, J. T., Rasmussen, K. R., and Willetts, B. B. eds. (1985) *Proceedings of International Workshop on the Physics of Blown Sand*. Department of Theoretical Statistics, Institute of Mathematics, University of Aarhus, Memoir 8, v. 1-3.

Barndorff-Nielsen, O. E. and Willetts, B. B. eds. (1991a) *Aeolian Grain Transport 1. Mechanics*. Acta Mechanica Supplementum 1, Springer-Verlag, Vienna.

Barndorff-Nielsen, O. E. and Willetts, B. B. eds. (1991b) *Aeolian Grain Transport 2. The Erosional Environment*. Acta Mechanica Supplementum 2, Springer-Verlag, Vienna.

Beadnell, H.J.L. (1910) The sand dunes of the Libyan Desert. *Geographical Journal*, v. 35, p. 379-395.

Belly, P. Y. (1964) *Sand Movement by Wind*. US Army Corps of Engineers, Coastal Engineering Reasearch Center Technical Memoir 1.

Breed, C. S., Fryberger, S. G., Andrews, S., McCauley, C., Lennartz, F., Gebel, D. and Hortsman, K. (1979) Regional studies of sand seas, using Landsat (ERTS) imagery. In E. D. McKee (ed.) *A Study of Global Sand Seas*. United States Geological Survey Professional Paper No. 1052, p. 305-397.

Brookfield, M. E. and Ahlbrandt, T. S. eds. (1983) *Eolian Sediments and Processes*. Elsevier, Amsterdam.

Capot-Rey, R. (1948) Le déplacement des sables éolien et la formation des dunes désertiques, d'après Bagnold. *Traveaux Institut Recherches Sahariennes*, v. 5, p. 47-80.

Carter, R.W.G., Curtis, T.G.F. and Sheehy-Skeffington, M. J. eds. (1992) *Coastal Dunes: Geomorphology, Ecology and Management for Conservation*. Balkema, Rotterdam.

Chepil, W. S. and Woodruff, N. P. (1963) The physics of wind erosion and its control. *Advances in Agronomy*, v. 15, p. 211-302.

Clos-Arceduc, L. (1967) La direction des dunes et ses rapports avec celle du vent. *Comptes Rendus Academie des Sciences Paris Series*, D264, p. 1393-1396.

Cooke, R. U., Warren, A. and Goudie, A. S. (1993) *Geomorphology in Deserts*. UCL Press, London.

Cooke, R. U., Brunsden, D., Doornkamp, J. C., and Jones, D.K.C. (1982) *Urban Geomorphology in Drylands*. Oxford University Press, Oxford.

Cooper, W. S. (1967) *Coastal Dunes of California*. Geological Society of America, Memoir 104.

Cooper, W. S. (1958) *Coastal Sand Dunes of Oregon and Washington*. Geological Society of America, Memoir 72.

Cornish, V. (1897) On the formation of sand-dunes. *Geographical Journal*, v. 9, p. 278-309.

Cornish, V. (1914) *Waves of Sand and Snow*. Unwin-Fisher, London.

Davis, W. M. (1930) Rock floors in arid and in humid climates. *Journal of Geology*, v. 38, p. 1-27, 136-158.
Dorn, R. I., Jull, A.J.T., Donahue, D. J., Linick, T. W. and Toolin, L. J. (1989) Accelerator mass spectrometry radiocarbon dating of rock varnish. *Geological Society America Bulletin*, v. 101, p. 1363-1372.
Duce, R. A. and Tindale N. W. (1991) The atmospheric transport of iron and its deposition in the ocean. *Limnological Oceanography*, v. 36, p. 1715-1726.
Folk, R. L. (1971) Longitudinal dunes of the northwestern edge of the Simpson Desert, Northern Territory, Australia, 1: Geomorphology and grain size relationships. *Sedimentology*, v. 16, p. 5-54.
Fryberger, S. G., Al-Sari, A. M. and Clisham, T. J. (1983) Eolian dune, interdune, sand sheet and siliciclastic sabkha sediments of an offshore prograding sand see, Dhahran area, Saudi Arabia. *Bulletin American Association Petroleum Geologists*, v. 67, p. 280-312.
Fryberger, S. G., Krystinik, L. F. and Schenk, C. J. (1990) *Modern and Ancient Eolian Deposits: Petroleum Exploration and Production*. Rocky Mountain Section, Society of Economic Paleontologists and Mineralogists, Denver.
Gautier, E. F. (1935) *Sahara: The Great Desert*. Frank Cass, London (English Translation).
Gillette, D. A. and Stockton, P. H. (1989) The effect of nonerodible particles on the wind erosion of erodible surfaces. *Journal of Geophysical Research*, v. 93, p. 233-242.
Glennie, K. W. (1970) *Desert Sedimentary Environments*. Elsevier, Amsterdam.
Greeley, R. and Iversen, J. D. (1985) *Wind as a Geological Process*. Cambridge University Press, Cambridge.
Hastenrath, S. L. (1967) The barchans of the Arequipa region, southern Peru. *Zeitschrift für Geomorphologie*, v. 11, p. 300-331.
Hesp, P. A. and Fryberger, S. G. eds. (1988) *Special Issue on Eolian Sediments*. Sedimentary Geology, v. 55.
Howard, A. D., Morton, J. B., Gad-el-Hak, M. and Pierce, D. B. (1978) Sand transport model of barchan dune equilibrium. *Sedimentology*, v. 25, p. 307-338.
Inman, D. L., Ewing, G. C. and Corliss, J. B. (1966) Coastal sand dunes of Guerrero Negro, Baja California, Mexico. *Geological Society America Bulletin*, v. 77, p. 787-802.
Iverson, J. D. and White, B. R. (1982) Saltation threshold on Earth, Mars and Venus. *Sedimentology*, v. 29, p. 111-119.
Iverson, J. D., Greeley, R. and Pollack, J. B. (1976) Windblown dust on Earth, Mars and Venus. *Journal of Atmospheric Sciences*, v. 33. p. 2425-2429.
Keyes, C. R. (1912) Deflative scheme of the geographic cycle in an arid climate. *Geological Society America Bulletin*, v. 23, p. 537-562.
Kocurek, G. ed (1988) *Special Issue: Late Paleozoic and Mesozoic Eolian Deposits of the Western Interior of the United States*. Sedimentary Geology, v. 56.
Krinsley, D. H., and Manley, C. R. (1989) Backscattered electron microscopy as an advanced technique in petrography. *Journal of Geological Education*, v. 37, p. 202-209.
Krinsley, D. H. and Doornkamp J. C. (1973) *An Atlas of Quartz Sand Grain Surface Textures*. Cambridge University Press, Cambridge.
Lancaster, N. (1985) Variations in wind velocity and sand transport on the windward flanks of desert sand dunes. *Sedimentology*, v. 32. p. 581-593.
Lettau, H. and Lettau K. (1969) Bulk transport of sand by the barchans of Pampa La Joya in southern Peru. *Zeitschrift für Geomorphologie*, v. 13, p. 183-195.
Livingstone, D. N. (1993) *The Geographical Tradition*. Blackwell, London.
Livingstone, I. (1986) Geomorphological significance of wind flow patterns over a Namib linear dune. In W. G. Nickling (ed.) *Aeolian Geomorphology*. Allen and Unwin, p. 97-112.
Marshall, J. R. ed. (1987). *Clastic Particles*. Van Nostrand Reinhold, New York.
McKee, E. D. ed., (1979) *A Study of Global Sand Seas*. United States Geological Survey Professional Paper No. 1052.
Middleton, N. J. (1989) Desert dust. In D.S.G. Thomas (ed.) *Arid Zone Geomorphology*. Halsted Press, p. 262-283.
Middleton, N. J. and Thomas, D.S.G. (1992) *World Atlas of Desertification*. Edward Arnold, London.

Middleton, N. J., Goudie, A. S., and Wells, G. L. (1986) The frequency and source areas of dust storms. In W. G. Nickling (ed.) *Aeolian Geomorphology*. Allen and Unwin, p. 237-259.

Mulligan, K. R. (1988) Velocity profiles measured on the windward slope of a transverse dune. *Earth Surface Processes and Landforms*, v. 13, p. 573-582.

Nickling, W. G. ed. (1986) *Aeolian Geomorphology*. Binghampton Symposia in Geomorphology No. 17, Allen and Unwin, Boston.

Nordstrom, K. F., Psuty, N. P. and Carter, R.W.G., eds. (1990) *Coastal Dunes Form and Process*. John Wiley and Sons, New York.

Orme, A. R. (1992) Late Quaternary deposits near Point Sal, south-central California: a time frame for coastal dune emplacement. In C. H. Fletcher and J. F. Wehmiller (eds.), *Quaternary Coasts of the United States: Marine and Lacustrine Systems*, Society of Sedimentary Geology, p. 309-315.

Orme, A. R. and Tchakerian, V. P. (1986) Quaternary dunes of the Pacific Coast of the Californias. In W. G. Nickling (ed.) *Aeolian Geomorphology*. Allen and Unwin, p. 149-175.

Owen, P. R. (1964) Saltation of uniform grains in air. *Journal of Fluid Mechanics*, v. 20, p. 225-242.

Passarge, S. (1904) *Die Kalahari*. Reimer, Berlin.

Prospero, J. M. (1981) Arid regions as sources of mineral aerosols in the marine atmosphere. *Geological Society America Special Paper 186*, p. 71-86.

Prospero, J. M., Charlson, R. J., Mohnen, V., Jaenicke, R., Delaney, A. C., Moyers, J., Zoller, W., and Rahn, K. (1983) The atmospheric aerosol system: an overview. *Review Geophysics & Space Physics*, v. 21, p. 1607-1629.

Pye, K. (1987) *Aeolian Dust and Dust Deposits*. Academic Press, London.

Pye, K. and Tsoar, H. (1990) *Aeolian Sand and Sand Dunes*. Unwin Hyman, London.

Pye, K. ed. (1983) *The Dynamics and Environmental Context of Aeolian Sedimentary Systems*. Geological Society Special Publication No. 72, London.

Pye, K. and Lancaster, N. eds. (1983) *Aeolian Sediments Ancient and Modern*. International Association of Sedimentologists Special Publication No. 16, Blackwell, Oxford.

Raupach, M. R., Gillette, D. A. and Leys, J. F. (1993) The effect of roughness elements on wind erosion threshold. *Journal of Geophysical Research*, v. 98, p. 3023-3029.

Rendell, H. M., Lancaster, N. and Tchakerian, V. P. (1994) Luminescence dating of Late Quaternary aeolian deposits at Dale Lake and Cronese Mountains, Mojave Desert, California. *Quaternary Geochronology* (in press).

Richthofen, F. von (1882) On the mode of origin of the loess. *Geological Magazine*, v. 9, p. 293-305.

Sarnthein, M. (1978) Sand deserts during glacial maximum and climatic optimum. *Nature*, v. 272, p. 43-46.

Sharp. R. P. (1966) Kelso Dunes, Mohave Desert, California. *Geological Society America Bulletin*, v. 77, p. 1045-1074.

Smart, P. and Tovey, N. K. (1981) *Electron Microscopy of Soils and Sediments: Examples*. Oxford University Press, Oxford.

Thomas, D.S.G. ed., (1989) *Arid Zone Geomorphology*. Halsted Press, New York.

Thomas, D.S.G. and Middleton, N. J. (1994) *Desertification: Exploding the Myth*. John Wiley and Sons, London.

Tsoar, H. (1985) Profile analysis of sand dunes and their steady state significance. *Geografiska Annaler*, v. 67A, p. 47-59.

Tsoar, H., Greeley, R. and Peterfreund, A. R. (1979) Mars: the north polar sand sea and related wind patterns. *Journal of Geophysical Research*, v. 83 p. 8167-8180.

Udden, J. A. (1896) Dust and sand storms in the West. *Popular Science Monthly*, v. 49, p. 655-664.

Verstappen, H. T. (1968) On the origin of longitudinal (seif) dunes. *Zeitschrift für Geomorphologie*, v. 12, p. 200-220.

Walmsley, J. L. and Howard, A. D. (1985) Application of a boundary-layer model to flow over an eolian dune. *Journal of Geophysical Research*, v. 90, p. 10631-10640.

Werner, B. T. (1994) Computer simulation of eolian dunes. Response of Eolian Processes to Global Change, Desert Research Institute, *Quaternary Sciences Center Occasional Paper No. 2.*, p. 111.

Wiggs, G.F.S. (1993) Desert dune dynamics and the evaluation of shear velocity: an integrated approach. In K. Pye (ed.) *The Dynamics and Environmental Context of Aeolian Sedimentary Systems*. Geological Society Special Publication No. 72, p. 37-46.

Williams, G. (1964) Some aspects of the eolian saltation load. *Sedimentology*, v. 3, p. 257-287.

Wilson, I. G. (1971) Desert sandflow basins and a model for the development of ergs. *Geographical Journal*, v. 137, p. 180-199.

Wilson, I. G. (1972) Aeolian bedforms - their development and origins. *Sedimentology*, v. 19, p. 173-210.

Wilson, I. G. (1973) Ergs, *Sedimentary Geology*, v. 10, p. 77-106.

Wintle, A. G. (1993) Luminescence dating of aeolian sands: an overview. In K. Pye (ed.) *The Dynamics and Environmental Context of Aeolian Sedimentary Systems*. Geological Society Special Publication No. 72, p. 49-58.

2 ORIGIN OF THE GRAN DESIERTO SAND SEA, SONORA, MEXICO: EVIDENCE FROM DUNE MORPHOLOGY AND SEDIMENTOLOGY

Nicholas Lancaster
Quaternary Sciences Center
Desert Research Institute

ABSTRACT

Studies of dune morphology and sediments in the Gran Desierto sand sea show that it is comprised of a wide variety of dune types, each of which appears to have formed as a result of a genetically distinct episode of sediment input and/or reworking of pre-existing dunes. Formation of different dune generations was probably the result of fluctuations in sediment supply from source zones which were controlled by tectonic reorganizations of the Colorado River delta region and eustatic changes in sea levels. Climatic changes favored the stabilization and preservation of older generations of dunes.

INTRODUCTION

Sand seas are dynamic sedimentary bodies that form part of local-scale and regional-scale sand transport systems in which sand is moved by the wind from source zones to depositional sinks (e.g. Fryberger and Ahlbrandt 1979, Wilson 1971). The processes by which most sand seas have accumulated are, however, poorly known, although many appear to have accumulated episodically in response to Quaternary climatic and sea level changes and probably represent an amalgamation of different generations of dune and interdune deposits (e.g., Fryberger et al. 1983, Kocurek et al. 1991). The accumulation of ancient sand seas was also strongly influenced by changes in sea level (e.g., Chan and Kocurek 1988).

A full understanding of sand sea history and sedimentary processes should involve extensive surface and sub-surface investigations, but the logistics of these are often hampered by the large size and inaccessibility of many deserts. However, many sand seas exhibit well-developed spatial patterns of dune morphology and morphometry as well as near-surface sediments (e.g. Breed et al. 1979). These patterns can be regarded as the surface expression of the factors affecting sand sea accumulation and can therefore be used to infer the processes and sequences of events involved. A geomorphic approach to understanding aeolian accumulation has been shown to be widely applicable (e.g., Warren 1988, Wasson et al. 1988, Lancaster 1989a). This paper discusses

Desert Aeolian Processes. Edited by Vatche P. Tchakerian. Published in 1995 by Chapman & Hall, London.
ISBN 978-94-010-6519-1

application of this paradigm to understanding the origins of the Gran Desierto sand sea in northern Mexico.

Although small in size by comparison with sand seas in Africa and Asia, several features of the Gran Desierto sand sea and its location make it an ideal area to examine the effects of climate, tectonics, and sea level changes on sand sea development: (1) the sand sea contains a wide variety of dune types, as well as extensive areas of sand sheets; (2) multiple sources of dune sediments are evident; (3) the sand sea is located in an area of active tectonism; (4) records of late Pleistocene and Holocene climatic and vegetation changes are available from areas adjacent to the Gran Desierto; and (5) the late Quaternary history of sea level changes in the area is well documented (Ortlieb 1991).

REGIONAL SETTING

The Gran Desierto sand sea (Figure 1) covers an area of 5700 km^2 on the northeastern shores of the Gulf of California, east of the Colorado River delta and extends from a few meters above sea level to an elevation of approximately 120 m adjacent to the Pinacate shield volcanic complex and the piedmont of the Tinajas Altas and Cabeza Prieta mountains along the Arizona-Mexico frontier. To the west, the sand sea is separated from the Colorado delta by the Sonora Mesa at an elevation of 25-50 m. The Mesa Arenosa, at an elevation of up to 100 m, lies along the southwestern margin. The eastern areas of the sand sea extend inland from the sabkhas and salt marshes that fringe the Bahia del Adair.

The Gran Desierto sand sea and the Colorado River delta lie in the southern part of the Salton Trough, a rapidly subsiding basin bounded by two major strike-slip fault zones: the southeastern extension of the San Andreas fault and the San Jacinto fault (Figures 1 and 2). The Salton Trough formed by the detachment and oblique separation of Baja California from the mainland of North America. As a result, the region has been deformed extensively over the past 5.5 Ma in association with the opening of the Gulf of California, resulting in a 130 km westward migration of the lowermost part of the Colorado River since 2.8 Ma (Winkler and Kidwell 1986). Subsidence rates of as much as 1-1.5 mm per year are reported from the western part of the Salton Trough (Lonsdale 1989). The fault-bounded Mesa Arenosa along the western margin of the sand sea is composed of sands and silts of fluvial and deltaic origins with an age of 0.5-1.8 Ma, based on vertebrate fossils (Shaw and McDonald 1987). It was uplifted prior to 120,000 yr. ago (Colletta and Ortlieb 1984). Northeast of the sand sea is the Pinacate volcanic field, composed of numerous basaltic lava flows, cinder cones, and maar craters of Pleistocene and Holocene age (Lynch 1981).

The area of the sand sea is underlain by as much as a 1 km thickness of alluvial sediments that to the southwest overlie up to 4 km of siltstones in a series of fault-bounded basins (Sumner 1972). The sediments that immediately

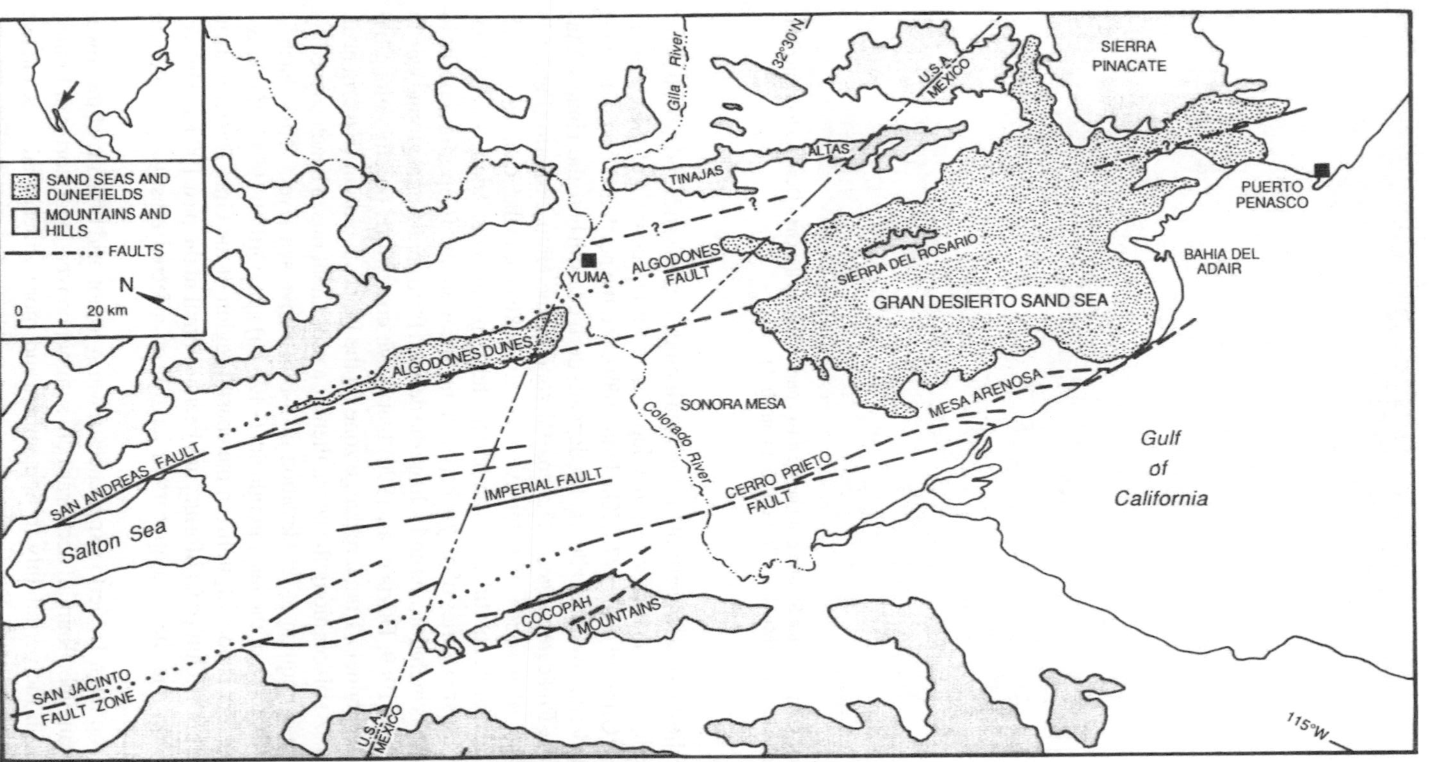

Figure 1. Location and regional setting of the Gran Desierto sand sea.

Figure 2. NASA Space Shuttle hand held camera photograph of the Gran Desierto (bottom right) and adjacent areas (Mission #37, Roll 75, Frame 071).

underlie the western parts of the sand sea include fluvial sands and gravels, silts, and silty clays that represent the deposits of former courses of the Colorado River (Olmsted et al. 1973, Kinsland 1989). To the north and east, the dunes are underlain by distal alluvial fan deposits derived from both the Tinajas Altas and the Pinacate ranges. The coastal zone of the Gran Desierto (Figure 1) consists of a series of extensive coastal sabkhas and large areas of high intertidal and supratidal flats around the head of Bahia del Adair. Inland from the coast are a series of playas between the linear dunes. Evaporite deposits in the playas include trona and schairerite derived from continental ground waters (Sinitiere 1989, Davis et al. 1990) that move coastwards possibly along old fluvial channels, from a recharge zone on the Pinacate margin (Ezcurra et al. 1988). At places along the coast, there is a prominent raised beach capped by a Chione coquina at an elevation of 5-8 m above sea level. This beach is believed to be of the last interglacial (~125 ka) age (Ortlieb 1991). If this age assessment is correct, uplift in the coastal region of the Gran Desierto, as in much of the Gulf of California, has been minimal in the past 100 ka (Muhs et al. 1992) and the area appears to have been affected mostly by strike-slip faulting.

The Gran Desierto experiences a warm to hot arid climate of the lower Sonoran type. Rainfall occurs as both summer convectional storms of limited extent in the period July to September, and more widespread, less intense winter rains when Pacific frontal storms penetrate the region (Ezcurra and Rodrigues 1986). Winter rainfall is very important for plant germination and

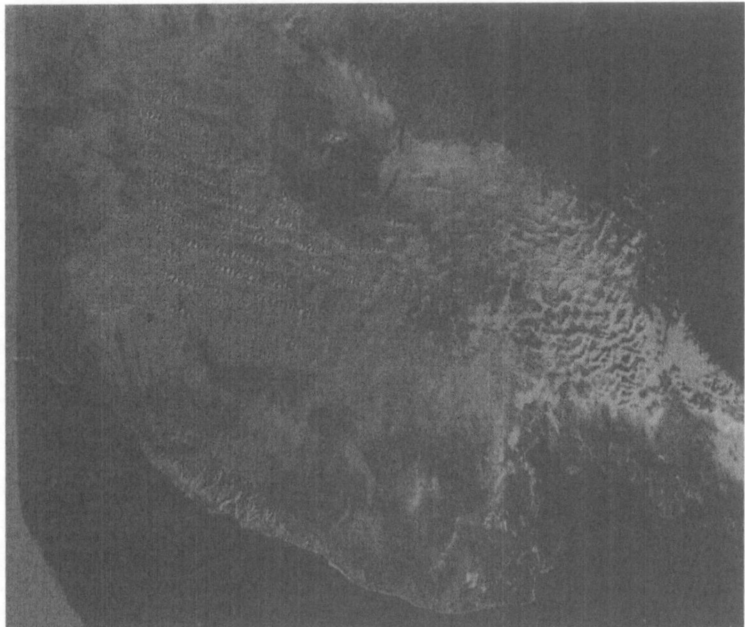

Figure 3. Landsat image (TM band 5) of the Gran Desierto sand sea, showing areas of different dune types.

growth (Felger 1980) and strongly affects aeolian sediment transport in the region (Mackinnon et al. 1990). Mean annual rainfall is 73 mm at Puerto Penasco and decreases northward toward Yuma, which receives 62 mm per year. Mean maximum temperatures similarly increase northwards and away from the coast. Winds in the area are controlled in part by the position and strength of the Sonoran low in summer, and by the Great Basin High in winter. This gives rise to southerly winds in the summer and northerly to northwesterly winds in winter and spring.

The Gran Desierto falls within the Lower Colorado Valley subdivision of the Sonoran Desert (Felger 1980). Large areas of the sand sea, especially the sand sheet areas, have a moderately dense cover of perennial plants, mostly low shrubs and herbs such as bursage (*Ambrosia dumosa*) and creosote bush (*Larrea tridentata*). Near the coast, increased salinity favors the growth of halophytes, including *Atriplex* spp. Overall vegetation cover on the dunes is as much as 9%-15% (Felger 1980). Although cattle do graze some areas of the dunes, their numbers appear to be very small and their effects are confined to a few areas (e.g., close to Puerto Penasco and the Pinacate).

PATTERNS OF DUNE MORPHOLOGY

The Gran Desierto sand sea can be divided into four main parts on the basis of dune morphology (Figures 3 and 4): (1) a northwestern area of sand sheets,

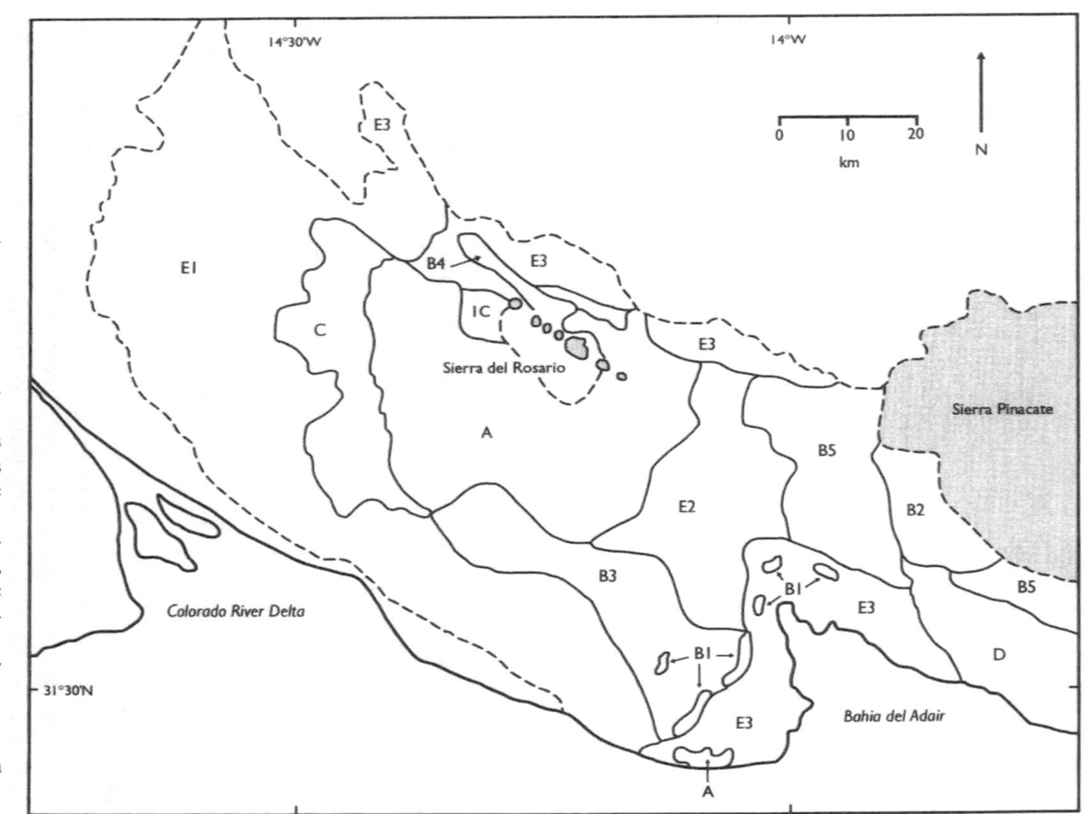

Figure 4. Areas of different dune morphology in the Gran Desierto sand sea.

(A) Chains and clusters of star dunes

(B) Crescentic dunes:
- (1) active simple crescentic dunes
- (2) active simple crescentic dunes
- (3) large relict crescentic dunes, stabilized by vegetation
- (4) coalescing multiple generations of largely relict crescentic dunes
- (5) compound/complex crescentic dunes

(C) Reversing dunes with active crescentic dunes on margins of the area, with stabilized and relict crescentic dunes in topographically lower areas

(D) Linear and parabolic dunes, largely vegetated

(E) Sand sheets
- (1) sparsely vegetated, low relief
- (2) moderately vegetated, 2- to 3-m local relief
- (3) undifferentiated

streaks, and low degraded crescentic dunes; (2) a western area of 80- to 150-m-high star dunes, with 5-20 m high largely vegetated crescentic dunes on their margins and extending to the coast west of the Bahia del Adair; (3) a central area of undulating to hummocky sand sheets and areas of 2- to 5-m-high active crescentic dunes that extends north from the Bahia del Adair and separates the western star dunes and the eastern crescentic dunes; and (4) an eastern area of mostly active compound and complex crescentic dunes 10-80 m high, with low vegetated linear and parabolic dunes and sand sheets towards the coast (Lancaster et al. 1987, Blount and Lancaster 1990).

Star Dunes

Star dunes (area A on Figure 4) are a prominent feature of the Gran Desierto sand sea and occupy some 10% of its area south and southwest of the Sierra del Rosario (Lancaster et al. 1987, Lancaster 1989b). A small area of star dunes also occurs north of these mountains (Breed et al. 1984). Many of the star dunes occur in linear clusters or chains up to 20 km long with WNW-ESE trends and a spacing of 2-3.5 km (Figure 3). Each star dune cluster or chain consists of a series of 80- to 100-m-high, sharp-crested, straight to slightly sinuous, near symmetrical ridges on a dominant NE-SW alignment and with a spacing of 300-400 m (Figure 5). The avalanche faces are oriented to the NW or NNW in summer, but are reversed seasonally to face southwest or south in the winter (Lancaster 1989c). Lower linear crests on NW-SE or N-S alignments form subsidiary arms on many star dunes, and may connect adjacent NE-SW-trending ridges to form a rectilinear pattern of peaks and ridges separated by deep hollows. The main NE-SW trending star dune arms lie on a wide plinth, on the lower parts of which are multiple transverse and reversing ridges 3-10 m high on ENE-WSW or NE-SW alignments. These dunes increase in height to 10-20 m and become more symmetrical in cross section towards the main dune ridges. Interdune areas between the star dunes range from 2- to 4-m-high active crescentic dunes and very degraded 5- to 10-m-high crescentic ridges forming a reticulate pattern with deep enclosed hollows to gently undulating sand sheets.

Crescentic Dunes

Crescentic dunes of simple, compound, and locally complex varieties cover approximately 21% of the area of the sand sea (B1-5 on Figure 4). Simple crescentic ridges in the Gran Desierto exhibit a wide range of size, spacing, and morphology. They range from small active dunes north of the Bahia del Adair (B1) and adjacent to the westernmost star dunes to large, degraded crescentic ridges on the southern margins of the star dune zone. North of the Bahia del Adair, a series of 2- to 4-km-wide elongate tongues of active 2- to 5-m-high crescentic dunes extends inland. Dune crests strike N70°-90°E or transverse to S to SSE winds. The second major group of active simple crescentic dunes lies toward the eastern margin of the sand sea between areas of compound/ complex crescentic dunes (B2 on Figure 4). Here there are extensive areas of

Figure 5. Star dunes in the central part of the sand sea, with low crescentic and reversing dunes in interdune areas.

Figure 6. Stabilized, relict crescentic dunes on southern margins of star dune area.

small (2-4 m high) and closely spaced (150-200 m apart) straight-crested transverse dunes with alignments transverse to southerly winds.

Areas of degraded, stabilized, and largely vegetated crescentic dunes (B3-B4 on Figure 4) occur on the margins of the star dune area, north of the Rosario

Mountains, and west of the Bahia del Adair (Lancaster 1992). West of the Bahia del Adair these dunes are 15-20 m high with a spacing of 200-300 m and a crest strike of ~ N50°E (Figure 6). They typically have very broad (50-100 m wide) and sparsely vegetated crests, but stoss and lee slopes (with angles of 7°-10°) are more densely vegetated and exhibit a distinct surface crust. Additional areas of simple crescentic dunes with a very degraded form occur on the margins of the star dune area, as well as in some interdune areas between the star dunes (Figure 5). Crescentic dunes on the western and northern margins of the star dunes are vegetated and mostly stabilized with a height of 5-10 m and a spacing of 300-500 m. The strike of the crest lines of most of these dunes is N60°-65°E, indicating formation by winds from the NW to NNW. North of the Rosario Mountains, merging and superposition of four groups of crescentic dunes has created a sand accumulation up to 50 m thick that rests on distal alluvial fan deposits (area B4 on Figure 4, Figure 7). Group A and B dunes are typically 8- to 15-m-high, straight-crested crescentic dunes with a crest strike of N60°E and a spacing of 300-400 m. The crests of these dunes are active, whereas the lower slopes are stabilized by vegetation and surface crust with very degraded lee slopes (maximum 10°). Dunes of Groups C and D are distinctly different: their crest alignments indicate formation by winds from more westerly directions and they are generally much lower (5-12 m high) and more closely spaced (100-200 m) than the other groups of dunes.

Compound/complex crescentic dunes in the eastern part of the Gran Desierto (B5 on Figure 4) consist of straight or crescentic ridges that are 300-1000 m wide and 20-50 m high with a lee slope to the NW (Figure 8). Dune spacing increases from 1400 m near the coast to as much as 2000 m inland. Superimposed on the main ridges are multiple 2- to 5-m-high crescentic dunes with a spacing of 50-100 m. The main dune ridges are oriented at N50°-60°E or approximately transverse to southeast-south southeast winds. In winter, northerly winds reverse the lee face orientation of the superimposed dunes, and in the group of dunes north of the Bahia del Adair, the northernmost ridges have developed star dunes on their crests (Lancaster 1989c).

There are two groups of compound/complex crescentic dunes: an eastern group along the southeast margin of the Pinacate, characterized by short, wide ridges and narrow and discontinuous interdunes that probably result from merging or shingling of the main dune ridges against the Pinacate margins; and a western group north of the Bahia del Adair, characterized by long dune ridges and wide, open interdune areas (Lancaster et al. 1987). The upwind end of each dune area consists of a broad sparsely vegetated, gently sloping ramp of fine sand with local areas of granule ripples that taper to the south or southeast. Interdune areas between the compound crescentic dunes consist of undulating vegetated aeolian sand, with a discontinuous lag of basalt and cinder gravel and granules close to the Pinacate volcanic area. One interdune area west of the Pinacate is occupied by the distal reaches of an ephemeral stream.

Figure 7. Coalesced and superposed crescentic dunes north of the Rosario Mountains (area B4 on Figure 4).

Figure 8. Compound/complex crescentic dunes on eastern margins of sand sea, Pinacate volcanic field in background.

Reversing Dunes

Reversing dunes (C on Figure 4) in the Gran Desierto are characterized by near symmetrical, sharp-crested ridges 5-30 m high (Figure 9) with spacing of 300-500 m. They are located in areas adjacent to the main star dunes and appear to be a transitional form between crescentic and star dunes (Lancaster 1989c).

Linear and Parabolic Dunes

Vegetated linear and locally parabolic dunes 2-5 m high are widely distributed in the areas between the coast and the compound/complex crescentic dunes in the eastern parts of the Gran Desierto sand sea (area D on Figure 4). Typical linear ridges are straight to slightly sinuous with a spacing of 100-200 m. Ridge length varies from as little as 200 m to a maximum of 2 km. Shorter ridges tend to be associated with areas of closer and less regular dune spacing. At the inland end of the linear dunes north of First Salina, there is a zone where they appear to join by Y-junctions, open to the south. The trends of the linear dunes are S to N or SSE to NNE, curving slightly to the west and north. Interdune areas are undulating and sand covered.

Sand Sheets

Sand sheets cover 20%-25% of the sand sea area in the northwestern parts of the Gran Desierto and in the area between the star dunes and the eastern compound/complex crescentic dunes (areas E1-E3 on Figure 4). Sand sheets in the central parts of the sand sea (E2) are moderately vegetated and characterized by an irregular, hummocky topography with a local relief of as much as 2-3 m. They appear to be the degraded remnants of formerly active crescentic dunes.

Sand sheets in the northwestern parts of the sand sea (E1 on Figure 4) form a flat to gently undulating surface with a maximum local relief of 1-5 m (Figure 10). They consist of a series of coalescing streaks or stringers aligned at S30°E in central and western parts of the area, and S15°E in the east. Trenching of the sand sheets shows that they are composite features consisting of two or three successive generations of aeolian accumulations separated by erosional surfaces and/or weakly developed palaeosols (Lancaster 1993a).

To the north and west, the sand sheets thin out on the highest alluvial terraces of the Colorado River and the distal areas of alluvial fans that head in the Tinajas Altas Mountains. To the southeast, the sand sheets merge with crescentic and reversing dunes that form the western margin of the main star dune area. In many respects, the area appears to be a typical "back erg" (Porter 1986) or "trailing erg margin" (Sweet et al. 1988).

PATTERNS OF DUNE AND SAND SHEET SEDIMENTS

Potential sediment sources for the Gran Desierto sand sea include the Colorado River (Merriam 1969) and marine sands from the Gulf of California (Ives 1959). Blount and Lancaster (1990) identified three main sand populations on the basis of grain size and sorting, mineralogy, and spectral signature on Landsat TM image data. These sands were derived from (1) the ancestral Colorado Delta, (2) the present lower Colorado River valley, and (3) the coast of the Bahia del Adair. They concluded that there was no consistent regional scale pattern of grain size, sorting, and color of dune sediments and that the most important scale of variability was that between areas of different dune types.

Figure 9. Reversing dunes on the northern margin of the star dune area; Rosario Mountains in background. Note relict vegetated crescentic dunes in lower areas.

Figure 10. Sand sheets in the northwestern part of the sand sea. Star dunes on horizon.

New studies (e.g., Lancaster 1992) support this hypothesis and show that dunes of distinctly different colors, composition, and grain size and sorting occur next to each other in all parts of the sand sea. However, this scale of variability is superimposed on a general trend from coarse and moderately sorted sand to very fine and very well sorted sand from northwest to southeast across the sand

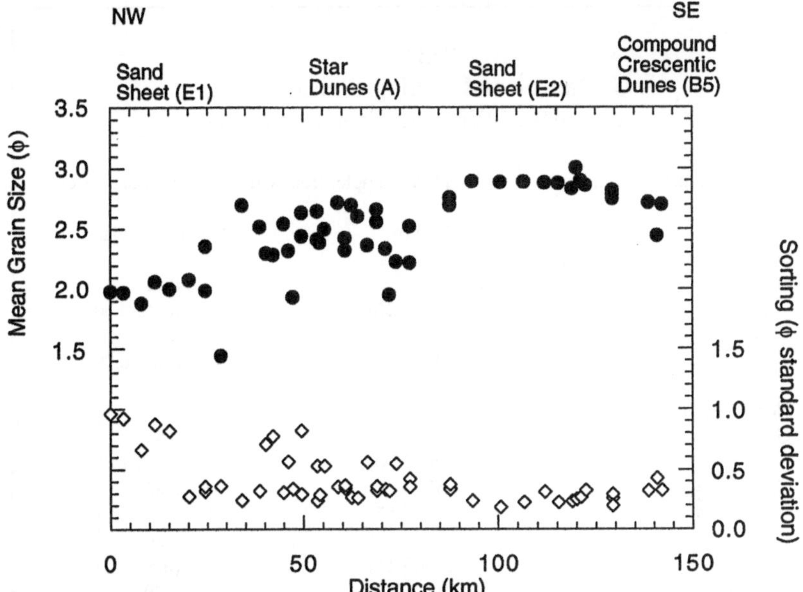

Figure 11. Variation in grain size and sorting from northwest to southeast across the sand sea. Transect runs through areas E1, C, A, E2, B5, and B2 on Figure 4.

sea (Figure 11). The eastern half of the sand sea is characterized by very fine and well sorted sand.

Star Dunes

Star dunes in the Gran Desierto are composed of very well sorted fine to very fine sands, with a range of mean grain sizes between 1.95 and 3.10 ϕ. Sand from dune crests is typically very well sorted (average s_I value = 0.30), and near symmetrical (average phi skewness = 0.03), with an average mean grain size of 2.49 ϕ. Dune plinths are slightly coarser (average mean grain size = 2.39 ϕ, and less well sorted (average s_I value = 0.49).

Crescentic Dunes

There is a wide range of composition between different areas of crescentic dunes in the Gran Desierto (Table 1). On the southeastern margin of the star dune zone, each group of crescentic dunes is composed of sediments that are distinctly different in grain-size composition and degree of post-depositional alteration. North of the Rosario Mountains, sand from dunes in Groups A, B, and D are typically bimodal, with modes at 1.5-2.0 and 3-3.5 ϕ. Group C dunes are distinctively fine, with a modal grain size of 3-3.5 ϕ. All sands are moderately to poorly sorted and sedimentary structures are weakly developed or not preserved except in the active crestal areas of these largely stabilized dunes.

Table 1
Grain Size and Sorting Characteristics of Areas of Different Cresentic Dunes

Range of values in parentheses (phi units, ϕ). All samples from wind ripple strata at dune crest

	Mean (ϕ)	Standard deviation (σ)
South east margin of star dunes		
Group I	1.95	0.51
	(1.71-2.26)	(0.40-0.60)
Group II	2.13	0.92
	(1.92-2.47)	(0.68-0.99)
North Rosario dunes		
Group A	2.33	0.84
	(1.55-2.80)	(0.55-1.05)
Group B	2.32	0.82
	(1.92-2.57)	(0.67-0.91)
Group C	2.79	0.80
	(2.24-3.14)	(0.57-1.08)
Group D	2.46	0.79
	(1.97-2.92)	(0.60-1.04)
West of star dunes	1.96	0.76
	(1.77-2.20)	(0.60-0.95)
Eastern sand sea		
Compound/complex crescentic	2.80	0.28
	(2.54-2.89)	(0.18-0.39)
Simple crescentic		
North of Bahia del Adair	2.70	0.35
	(2.56-2.85)	(0.26-0.47)
Between areas of compound/ complex crescentic dunes	2.89	0.28
	(2.86-3.00)	(0.24-0.32)

Simple and compound/complex crescentic dunes in the eastern parts of the Gran Desierto are composed of very fine, very well sorted quartz sand with up to 5% shell fragments near the coast. Mean grain size varies between 2.56 and 2.89 (average 2.72 ϕ) and ϕ standard deviations range from 0.20 to 0.42 with a mean of 0.32. The modal size is 3.0 ϕ.

Linear and Parabolic Dunes

Linear dunes in the eastern areas of the sand sea are composed of moderately well sorted to well sorted fine sands with a average mean grain size of 2.82 (range 2.60-2.90 ϕ) and a ϕ standard deviation of 0.43 (range 0.31-0.61). Modal grain size is 3.0 ϕ. Interdune sands are typically fine (average mean grain size 2.95 ϕ) and moderately sorted (average σ_1 value = 0.60).

Sand Sheets

Sand sheets on the northwestern margins of the Gran Desierto sand sea are composite features comprised of three main sedimentary units (Lancaster 1993a). Each unit is composed of moderately to poorly sorted (σ_1 = 0.70-1.10) sand with a mean grain size that varies between 1.95 and 2.25 ϕ. Older units tend to be finer than the surface wind ripple laminae. Samples from the eastern part of the sand sheet area are somewhat finer, with a mean grain size of 2.32 ϕ and moderate sorting (σ_1 = 0.81).

CONTROLS OF DUNE MORPHOLOGY

Analysis of the pattern of dune morphology and sediments in the Gran Desierto sand sea reveals two scales of variability: (1) From northwest to southeast, there is a progression from sand sheets through a zone of crescentic and reversing dunes to the central star dune area. The eastern part of the sand sea is characterized by large compound/complex crescentic dunes, as well as areas of vegetated sand sheets and linear and parabolic dunes. (2) A very characteristic feature of the Gran Desierto sand sea is the close juxtaposition of dunes of contrasting composition and morphology. In many areas, active star dunes occur next to stabilized crescentic dunes. Elsewhere, several sets of crescentic dunes, each with a different morphology and alignment, coalesce with each other.

Spatial changes in dune morphology at the first or regional scale can be related to regional changes in wind regimes, and are characterized by transitions from one dune type to another (e.g., from crescentic through reversing to star dunes) as described by Lancaster (1989c). There are no wind data for the immediate area of the sand sea, but winds at Yuma, 50 km from its northern margin, are from three major seasonally varying sectors (Figure 12): NNW-NNE (winter), W-WNW (spring), and S-SE (summer). In the vicinity of the Algodones Dunes, the first two sectors are dominant and give rise to net transport and dune migration toward the southeast (Havholm and Kocurek 1988). Archival wind data for Puerto Penasco, just east of the sand sea, indicate that winds there are dominated by spring westerly and year-round southerly winds, with a peak in south winds during the summer, and occasional strong north winds in the winter. Studies of the dynamics of star dunes in the Gran Desierto indicate that the southerly and northerly directional sectors are approximately balanced in the central part of the sand sea (Lancaster 1989c).

The regional-scale pattern of dune morphology in the Gran Desierto can therefore be explained by changes in regional wind regimes from one dominated by northerly and westerly winds to the west and north of the sand sea (sand sheets and crescentic dunes migrating toward the southeast) and south-

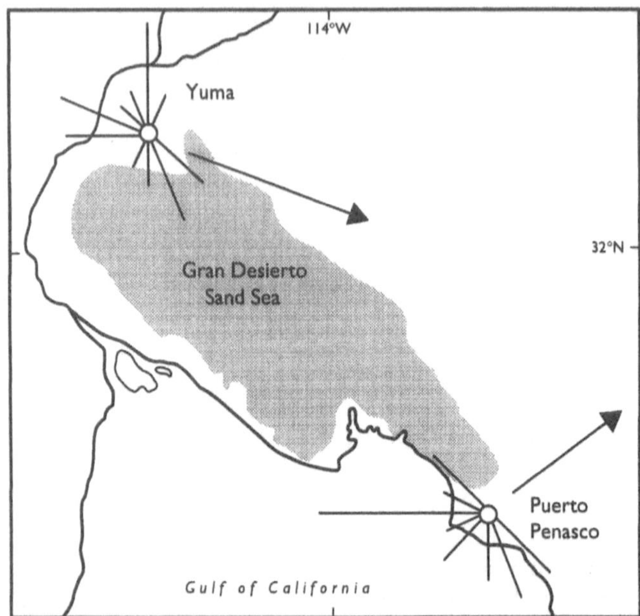

Figure 12. Regional changes in wind regimes and sand transport potential in the area of the Gran Desierto sand sea. Length of rose arms is proportional to potential sand transport calculated using formula of Fryberger (1979). Wind data cover the period 1948-1971 for Yuma and 1940-1941 for Puerto Penasco.

erly winds to the east and south (crescentic dunes migrating to the north and northwest). The star dunes occur in the zone where these two wind regimes interact with each other. The association of star dunes with areas of complex and seasonally reversing wind regimes has been widely documented (e.g., (Fryberger 1979, Lancaster 1989b). Complex crescentic and reversing dunes are found in the transitional areas around the zone of interaction between different wind regimes. These trends have apparently persisted throughout the period of sand sea development, as older stabilized crescentic dunes reflect the same basic pattern of winds.

The effects of regional changes in wind regime are modified because of the effect of increasing dune size on bedform reconstitution time in a seasonally varying wind regime: small dunes can be reformed in a single wind season, whereas larger dunes exhibit a morphology that is controlled by several wind directions (e.g., Lancaster 1989c). Thus, small crescentic ridges can co-exist with large star dunes in many areas of the central sand sea.

The northwestern sand sheets represent a response to limited sediment supply and sediment bypassing. Compared to many unvegetated sand sheets, they lack the prominent coarse or very coarse sand fraction and bimodal grain size distribution (Kocurek and Nielson 1986, Lancaster 1993a) and are composed of sand that is in the 2.0 or 2.5 ϕ range found in dunes of all types in the western part of the Gran Desierto. Vegetation cover in the area is sparse, with a maximum cover of 10-15%. This is insufficient to prevent sand transport taking place, but sufficient to give rise to localized deposition of a poorly sorted mixture of fine and coarse sand. If sand supply is very high, then it is possible that these effects can lead to dune initiation (e.g. Kocurek et al. 1992). However, dunes are not being initiated by these processes in this part of the Gran Desierto. It appears, therefore, that sand supply is a limiting factor, and the sand sheets represent a zone of sediment bypassing with a neutral sediment budget. Any accumulation therefore represents a temporary storage of medium and fine sand that is en route to its final depositional site in the dunes of the sand sea.

The juxtaposition and/or superposition of dunes with different morphologic types, alignments, grain-size composition, and degree of post-depositional modification of sedimentary structures (e.g., cementation of laminae, carbonate accumulation, bioturbation) cannot be explained by regional changes in wind regimes. Detailed investigations of the geomorphic relations between adjacent areas of dunes of varying characteristics (Lancaster 1992) show that they represent genetically distinct episodes of dune formation and/or modification of existing bedforms, each of which forms a separate generation of dunes. Up to five generations of dunes and three periods of sand sheet formation can be recognized in the Gran Desierto sand sea. Dune and sand sheet generations are separated by weakly developed soils, erosional unconformities, and deflation lag surfaces (Lancaster 1992, Lancaster 1993a). These surfaces are comparable to regional scale or super bounding surfaces (Talbot 1985, Kocurek 1988). In the southeastern star dune area, the star dunes are superimposed on two older generations of crescentic dunes (Figure 13a). Elsewhere in the star dune area, the active star and reversing dunes overlie one main generation of crescentic dunes (Figure 13b), while north of the Rosario Mountains, four superposed and coalesced generations of crescentic dunes capped by star and reversing dunes can be recognized (Figure 13c). In the eastern part of the sand sea, the compound/complex crescentic dunes appear older than the vegetated linear and parabolic dunes closer to the coast. In turn, these dunes are overlain by active simple crescentic dunes near the coast of the Bahia del Adair (Figure 13d).

The Gran Desierto sand sea therefore represents an amalgamation of multiple generations of dunes that were deposited adjacent to or superposed on one another. Each generation of dunes was the product of distinct episodes of dune formation and/or reworking of older dunes separated by periods of varying duration when active dune migration was replaced by geomorphic stability and incipient soil formation.

DISCUSSION

Multiple dune generations represent the response of a sand sea to climatic, tectonic, or eustatic changes that have affected sediment supply and/or dune

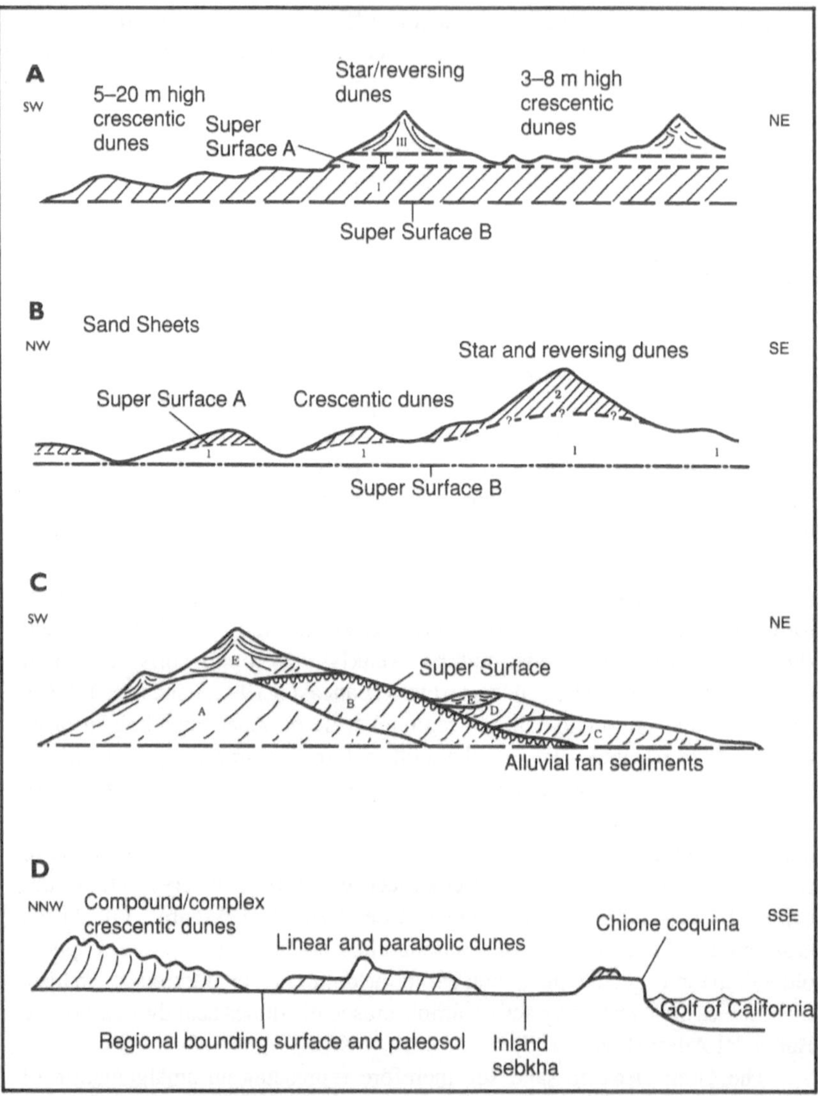

Figure 13. Examples of relations between dune generations in different parts of the Gran Desierto sand sea: (a) southeastern star dune area, (b) western star dunes, (c) north of Rosario Mountains, and (d) eastern part of sand sea. All illustrations are schematic.

mobility. Formation of areas of different dunes and their subsequent preservation also implies that later periods of sediment input and dune formation occurred in a manner that did not result in complete reworking of the older dunes.

A conceptual framework for assessing the effects of these changes on aeolian accumulation has been developed by Kocurek and Havholm (in press) based on principles of sediment conservation (Middleton and Southard 1984). In this model (Figure 14), the domains of sediment accumulation, bypassing, and erosion are defined in terms of (1) the saturation level of the sediment input (defined as the ratio between actual [q_{au}] and potential [q_{pu}] transport rates) and (2) the change in transport capacity with distance (defined as the ratio between the potential transport rate upwind and downwind of the area of concern). Assuming that the basic wind regime characteristics of an area do not change with time, it is clear from Figure 14 that changes in sediment availability from source zones and thus the saturation level of the input will determine the behavior of the system. Formation of different dune generations will therefore occur if there are major fluctuations in the availability of sediment to the system. The effects of these fluctuations will be manifested most clearly if there are changes in the location of sediment source areas over time, giving rise to spatially distinct dune generations.

Although there is local erosion in interdune areas and on stoss slopes of dunes, preservation of older generations of dunes is likely because sand seas and dune fields are areas of net accumulation or at least sediment bypassing (otherwise they would not exist). Preservation of older dune generations is favored by climatic changes that result in surface stabilization by vegetation and soil formation (e.g., Talbot 1985, Kocurek et al. 1991). However, surface stabilization is not only the result of climatic changes. Many coastal dune areas exhibit multiple dune generations that are the result of changes in sediment availability over time (e.g., Muhs 1992, Orme and Tchakerian 1986). Preservation of older dunes also occurs where different dunes are superposed on one another. This takes place where the accumulation is restricted by topographic obstacles, e.g., north of the Rosario Mountains, at Kelso Dunes, California (Lancaster 1993b), and Great Sand Dunes, Colorado (Andrews 1981). It is also favored by high sediment supply and rapid deposition, as well as subsidence of the accumulation zone. The widespread superposition of dunes in the central star dune zone of the Gran Desierto is a function of former high sediment supply and rapid accumulation in a zone of converging sand transport pathways, as well as possible tectonic subsidence in the rifted basin between the Rosario Mountains and the Mesa Arenosa (Figures 1 and 2).

The timing of periods of dune formation and the factors that determined episodic input of sediment to the Gran Desierto are poorly known at present. Eustatic changes in sea level have played a major role, especially the area of the Bahia del Adair, which is the source for three major areas of dunes: large compound/complex crescentic dunes to the north, large and small simple crescentic dunes north and west of the bay, and linear and parabolic dunes to

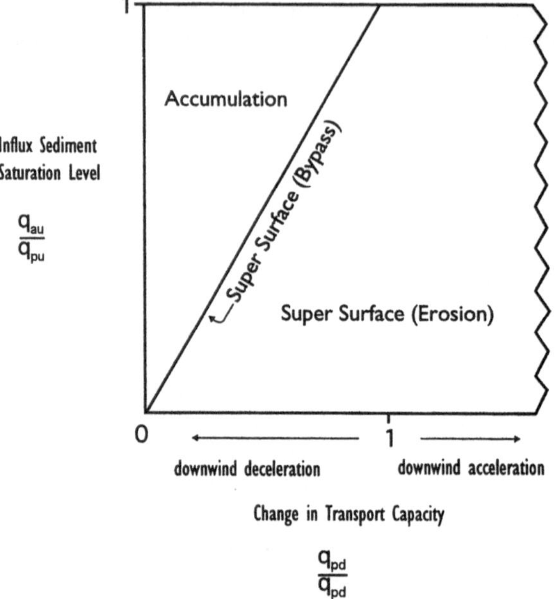

Figure 14. Domains of accumulation, bypass, and erosion in aeolian sedimentary systems. After Kocurek and Havholm (in press).

the east. With the exception of dunes west of the Bahia, all are composed of very fine and well-sorted sand. The source for the crescentic dunes is presently truncated by extensive sabkhas and salt marshes adjacent to the Bahia del Adair and an actively eroding sea cliff several meters high. Sediment input from this source and development of areas of crescentic dunes implies: (1) increased availability of sediment from this area; and (2) changes in the configuration of the coastline to permit onshore transport of sand by the wind. For example, during the last glacial maximum, when sea level was ~ 120 m lower than present 21,000 years ago (Bard et al. 1990), the coastline of the Bahia del Adair was located as much as 150 km south of its modern position and lay close to the position of the Colorado River delta front at this time (van Andel 1964), exposing a 50- to 100-km-wide zone of sandy sediments. These conditions would have provided a ready source of material for input to the sand sea by southerly winds.

The primary sand supply for dunes north of the Rosario Mountains and in the western part of the sand sea appears to have been from point bars and river terrace deposits of the lower Colorado River downstream from Yuma (Blount and Lancaster 1990). The bedload of the lower Colorado River contains 30%-40% sand-sized particles, with modal sizes prior to dam construction between 2.25 and 3.00 ϕ (Thompson 1968), or similar to those in most dune sands. The occurrence of coarse sand in the northwestern parts of the sand sea suggests derivation from deposits laid down when the river had a higher competence than in historical times.

Fluctuations in sand supply from the Colorado River source may have been the result of eustatic changes in sea level or tectonic events that altered the pattern of fluvial sedimentation in the delta system and/or climatic changes that restricted mobility of sand in transport corridors. For example, during the last glacial maximum the delta front was located 100 km southeast of its modern position (van Andel 1964), resulting in formation of an incised valley in the vicinity of the present delta region. In this case, eustatic changes in sea level that affected fluvial and deltaic sand sources may have resulted in a switch from northwestern to southerly sand input to the Gran Desierto and the formation of entirely different dune areas.

A further potential control on sand supply to the Gran Desierto from the Colorado River is the periodic diversion of the river to flow into the Salton depression for periods of decades to centuries to form Lake Cahuilla (Waters 1983). Such diversions had a major effect on sediment supply to the Algodones Dunes (Muhs et al., this volume). They likely cut off sand supply to the Gran Desierto sand sea. The avulsion of the Colorado River delta to the west following uplift of the Mesa Arenosa in the middle Pleistocene (Colletta and Ortlieb 1984) was also significant. Deflation of sediments from the abandoned flood plain could have provided the oldest population of sand in the sand sea (Blount and Lancaster 1990).

Periods of climatically controlled stabilization of dunes and sand sheets are suggested by stabilization surfaces in sand sheets on the northwestern margins of the Gran Desierto (Lancaster 1993a). Palaeobotanical evidence (Van Devender 1987, 1990) indicates that the Sonoran Desert experienced a number of periods of increased rainfall during the late Pleistocene and Holocene, although the area of the Gran Desierto sand sea was probably arid during this time (Cole 1986). For example, increased winter rainfall and cooler temperatures occurred coeval with the late Wisconsin 13,000-14,000 yr B.P. During the mid-Holocene, there is evidence from a variety of sources for increased summer rainfall and development of ephemeral grasses throughout the lower Sonoran Desert (Van Devender 1987). Such changes likely had a significant impact on sediment mobility in the Gran Desierto, by reducing actual transport rates during relatively wet periods.

CONCLUSIONS

Spatial variations in dune morphology and sediments in the Gran Desierto sand sea indicate that it is composed of multiple generations of dunes that have accumulated in response to episodic input of sediment to the area. Sediment supply to the sand sea appears to have been strongly influenced by eustatic changes in sea level and modulated by climatic changes which resulted in periods of dune stabilization. Understanding these relations is limited however by insufficient information on the ages of different dune generations, as well as the Late Quaternary history of the lower Colorado region. It is also clear that a geomorphic approach to understanding sand sea evolution has its limitations. Studies of sub-surface sediments and their stratigraphy show that many dunes and sand sheets are composite features, reflecting a number of periods of formation. This information cannot be gained from geomorphic studies alone.

The Gran Desierto sand sea is a good example of the complex response of aeolian geomorphic systems to external changes that affect sediment supply and mobility. Abrupt variations in dune morphology and sediments occur in many other sand seas, the Wahiba Sands of Oman (Warren 1988), the Namib Sand Sea (Lancaster 1983), and the Simpson Desert of Australia (Wasson 1983). Explanations for spatial variations in dune morphology in desert sand seas should therefore be re-examined in light of increasing knowledge of climatic and other changes in arid regions.

ACKNOWLEDGMENTS

Research in the Gran Desierto was supported by grants from NASA (NSG-7415) and the National Science Foundation (EAR 89-01457). I thank Judith Lancaster and Grady Blount for field assistance on many occasions over the years, and Joanne Hoffard and Jennifer Husek for sediment analyses. Thanks to Dan Muhs and an anonymous reviewer for their comments on the manuscript.

REFERENCES

Andrews, S. (1981) Sedimentology of Great Sand Dunes, Colorado. In F. P. Ethridge and R. M. Flores (eds.) *Recent and Ancient Non Marine Depositional Environments: Models for Exploration*. Society of Economic Paleontologists and Mineralogists, Special Publication, v. 31, p. 279-291.

Bard, E., Hamelin, B., Fairbanks, R. G., and Zindler, A. (1990) Calibration of the ^{14}C timescale over the past 80,000 years using mass spectrometric U-Th ages from Barbados corals. *Nature*, v. 345, p. 405-410.

Blount, G., and Lancaster, N. (1990) Development of the Gran Desierto sand sea. *Geology*, v. 18, p. 724-728.

Breed, C. S., Fryberger, S. G., Andrews, S., McCauley, C., Lennartz, F., Geber, D., and Horstman, K. (1979) Regional studies of sand seas using LANDSAT (ERTS) imagery. In E. D. McKee (ed.) *A Study of Global Sand Seas*. United States Geological Survey Professional Paper, v. 1052, p. 305-398.

Breed, C. S., Grolier, M. J., Breed, W. J., McCauley, C. K., and Cotera, A. S. (1984) Aeolian (wind formed) landscapes. In T. L. Smiley, J. D. Nations, T. L. Pewe, and J. P. Schafer (eds.) *Landscapes of Arizona, The Geological Story*. University Presses of America, p. 359-413.

Chan, M. A., and Kocurek, G. (1988) Complexities in aeolian and marine interactions: processes and eustatic controls on erg development. *Sedimentary Geology*, v. 56, p. 283-300.

Cole, K. L. (1986) The Lower Colorado River Valley: a Pleistocene Desert. *Quaternary Research*, v. 25, p. 392-400.

Colletta, B., and Ortlieb, L. (1984) Deformations of the middle and late Pleistocene deltaic deposits at the mouth of the Rio Colorado, northwestern Gulf of California. In V.E.A. Malpica-Cruz (ed.) *Neotectonics and Sea Level Variations in the Gulf of California Area: A Symposium*. University Nacional Auton Mexico, Institute Geolocia, Mexico City, p. 31-53.

Davis, O. K., Cutler, A. H., Meldahl, K., Palacios-Fest, M. R., Schreiber, J., Locke, B. W., Williams, L., Lancaster, N., Shaw, C., and Sinitiere, S. (1990) Quaternary and Environmental Geology of the Northeastern Gulf of California. In G. E. Gehrels and J. E. Spencer (ed.) *86th Annual Meeting Guidebook: Geological Excursions Throughout the Sonoran Desert Region, Arizona and Sonora*. Geological Society of America, Cordilleran Section, p. 136-154.

Ezcurra, E., Felger, R., Russell, A. D., and Equihua, M. (1988) Freshwater islands in a desert sand sea: the hydrology, flora and phytogeography of the Gran Desierto oases of northwestern Mexico. *Desert Plants*, v. 9, p. 35-44, 55-63.

Ezcurra, E., and Rodrigues, V. (1986) Rainfall patterns in the Gran Desierto, Sonora, Mexico. *Journal of Arid Environments*, v. 10, p. 13-28.

Felger, R. S. (1980) Vegetation and flora of the Gran Desierto, Sonora, Mexico. *Desert Plants*, v. 2, p. 87-114.

Fryberger, S. G. (1979) Dune forms and wind regimes. In E. D. McKee (ed.) *A Study of Global Sand Seas*. United States Geological Survey, Professional Paper, v. 1052, p 137-140.

Fryberger, S. G., and Ahlbrandt, T. S. (1979) Mechanisms for the formation of aeolian sand seas. *Zeitschrift für Geomorphologie*, v. 23, p. 440-460.

Fryberger, S. G., Al-Sari, A. M., and Clisham, T. J. (1983) Aeolian dune, interdune, sand sheet and siliciclastic sabkha sediments of an offshore prograding sand sea, Dhahran area, Saudi Arabia. *American Association of Petroleum Geologists Bulletin*, v. 67, p. 280-312.

Havholm, K. G., and Kocurek, G. (1988) A preliminary study of the dynamics of a modern draa, Algodones, southeastern California, USA. *Sedimentology*, v. 35, p. 649-669.

Ives, R. L. (1959) Shell dunes of the Sonora shore. *American Journal of Science*, v. 257, p. 449-457.

Kinsland, G. L. (1989) Proposed ancient Colorado River channel, Sonora, Mexico. *Geological Society of America, Abstracts with Programs*, v. 21, p. 101-102.

Kocurek, G. (1988) First order and super bounding surfaces in aeolian sequences — Bounding surfaces revisited. *Sedimentary Geology*, v. 56, p. 193-206.

Kocurek, G., and Havholm, K. G. (in press) Aeolian sequence stratigraphy — a conceptual framework. In H. Posamentier and P. Weimer (eds.) *Recent Advances in Siliciclastic Sequence Stratigraphy*. American Association of Petroleum Geologists, Tulsa, Oklahoma.

Kocurek, G., Havholm, K. G., Deynoux, M., and Blakey, R. C. (1991) Amalgamated accumulations resulting from climatic and eustatic changes, Akchar Erg, Mauritania. *Sedimentology*, v. 38, p. 751-772.

Kocurek, G. and Nielson, J. (1986) Conditions favourable for the formation of warm-climate aeolian sand sheets. *Sedimentology*, v. 33, p. 795-816.

Kocurek, G., Townsley, M., Yeh, E., Havholm, K., and Sweet, M. L. (1992) Dune and dunefield development on Padre Island, Texas, with implications for interdune deposition and water-table-controlled accumulation. *Journal of Sedimentary Petrology*, v. 62, p. 622-635.

Lancaster, N. (1983) Controls of dune morphology in the Namib sand sea. In T. S. Ahlbrandt and M. E. Brookfield (eds.) *Eolian Sediments and Processes*. Elsevier, Amsterdam, p. 261-289.

Lancaster, N. (1989a) *The Namib Sand Sea: Dune Forms, Processes, and Sediments*. A.A. Balkema, Rotterdam.

Lancaster, N. (1989b) Star dunes. *Progress in Physical Geography*, v. 13, p. 67-92.

Lancaster, N. (1989c) The dynamics of star dunes: an example from the Gran Desierto, Mexico. *Sedimentology*, v. 36, p. 273-289.

Lancaster, N. (1992) Relationships between dune generations in the Gran Desierto, Mexico. *Sedimentology*, v. 39, p. 631-644.

Lancaster, N. (1993a) Origins and sedimentary features of super surfaces in the northwestern Gran Desierto sand sea. In K. Pye and N. Lancaster (eds.) *Aeolian Sediments: Ancient and Modern*. Blackwell, Oxford, p. 71-83.

Lancaster, N. (1993b) Development of Kelso Dunes, Mojave Desert, California. *National Geographic Research and Exploration*, v. 9, p. 444-459.

Lancaster, N., Greeley, R., and Christensen, P. R. (1987) Dunes of the Gran Desierto Sand Sea, Sonora, Mexico. *Earth Surface Processes and Landforms*, v. 12, p. 277-288.

Lonsdale, P. (1989) Geology and tectonic history of the Gulf of California. In E. L. Winterer, D. M. Hussong, and R. W. Decker (eds.) *The Eastern Pacific Ocean and Hawaii*. Decade of North American Geology, volume N, Geological Society of America, Boulder, Colorado, p. 499-521.

Lynch, D. J. (1981) Genesis and geochronology of alkaline volcanism in the Pinacate volcanic field, northwestern Sonora, Mexico. Ph.D dissertation, University of Arizona.

Mackinnon, D. J., Elder, D. A., Helm, P. J., Tuesink, M. F., and Nist, C. A. (1990) A method of evaluating effects of antecedent precipitation on duststorms and its application to Yuma, Arizona, 1981-1988. *Climatic Change*, v. 17, p. 331-360.

Merriam, R. (1969) Source of sand dunes of southeastern California and northwestern Sonora, Mexico. *Geological Society of America Bulletin*, v. 80, p. 531-534.

Middleton, G. V. and Southard, J. B. (1984) *Mechanics of Sediment Movement* (2nd edition). Society of Economic Paleontologists and Mineralogists, Tulsa, Oklahoma.

Muhs, D. R. (1992) The last interglacial-glacial transition in North America: Evidence from uranium-series dating of coastal deposits. In P. U. Clark and P. D. Lea (eds.) *The Last Interglacial-Glacial Transition in North America. Geological Society of America.* Special Paper, v. 270, p. 31-51.

Muhs, D. R., Rockwell, T. K., and Kennedy, G. L. (1992) Late Quaternary uplift rates of marine terraces on the Pacific coast of North America, southern Oregon to Baja California Sur. *Quaternary Science Reviews*, v. 15/16, p. 121-133.

Olmsted, F. H., Loeltz, O. J., and Irelan, B. (1973). *Geohydrology of the Yuma area, Arizona and California;* United States Geological Survey, Professional Paper, v. 486-H.

Orme, A. R. and Tchakerian, V. P. (1986) Quaternary dunes of the Pacific Coast of the Californias. In W. G. Nickling (ed.) *Aeolian Geomorphology*. Allen & Unwin, Boston, p. 149-175.

Ortlieb, L. (1991) Quaternary shorelines along the northeastern Gulf of California: Geochronological data and neotectonic implications. In E. Pérez-Segura and C. Jacques-Ayala (eds.) *Studies of Sonoran Geology*. Geological Society of America, Special Paper, v. 254, p. 95-120.

Porter, M. L. (1986) Sedimentary record of erg migration. *Geology*, v. 14, p. 497-500.

Shaw, C. A., and McDonald, H. G. (1987) First record of Giant Anteater (*Xenarthra myrmecophagidae*) in North America. *Science*, v. 236, p. 186-188.

Sinitiere, S. M. (1989) The origin of a late Quaternary non-marine evaporite sequence in the Gran Desierto, Sonora, Mexico. M.S. thesis, University of Southwestern Louisiana.

Sumner, J. R. (1972) Tectonic significance of gravity and aeromagnetic investigations at the head of the Gulf of California. *Geological Society of America Bulletin*, v. 83, p. 3103-3120.

Sweet, M. L., Nielson, J., Havholm, K., and Farralley, J. (1988) Algodones dune field of southeastern California: case history of a migrating modern dune field. *Sedimentology*, v. 35, p. 939-952.

Talbot, M. R. (1985) Major bounding surfaces in aeolian sandstones: a climatic model. *Sedimentology*, v. 32, p. 257-266.

Thompson, R. W. (1968) Tidal flat sedimentation on the Colorado River Delta, northwestern Gulf of California. *Geological Society of America Memoir*, v. 125.

Van Andel, T. H. (1964) Recent marine sediments of the Gulf of California. In T. H. van Andel and G. C. Shor (eds.) *Marine geology of the Gulf of California*. American Association of Petroleum Geologists, Memoir, v. 3, p. 216-310.

Van Devender, T. R. (1987) Holocene vegetation and climate in the Puerto Blanco Mountains, Southwestern Arizona. *Quaternary Research*, v. 27, p. 51-72.

Van Devender, T. R. (1990) Late Quaternary vegetation and climate of the Sonoran Desert, United States and Mexico. In J. L. Betancourt, T. R. Van Devender, and P. S. Martin (eds.) *Packrat Middens: The Last 40,000 Years of Biotic Change*. The University of Arizona Press, Tucson, p. 134-166.

Warren, A. (1988) The dunes of the Wahiba Sands. In R. W. Dutton (ed.) *Scientific Results of the Royal Geographical Society's Oman Wahiba Sands Project 1985-1987. Journal of Oman Studies*, Special Report 3, Muscat, Oman, p. 131-160.

Wasson, R. J. (1983) Dune sediment types, sand colour, sediment provenance and hydrology in the

Strzelecki-Simpson Dunefield, Australia. In M. E. Brookfield and T. S. Ahlbrandt (eds.) *Eolian Sediments and Processes*. Elsevier, Amsterdam, p. 165-195.

Wasson, R. J., Fitchett, K., Mackey, B., and Hyde, R. (1988) Large-scale patterns of dune type, spacing, and orientation in the Australian continental dunefield. *Australian Geographer*, v. 19, p. 89-104.

Waters, M. R. (1983) Late Holocene lacustrine chronology and archaeology of ancient Lake Cahuilla, California. *Quaternary Research*, v. 19, p. 373-387.

Wilson, I. G. (1971) Desert sandflow basins and a model for the development of ergs. *Geographical Journal*, v. 137, p. 180-199.

Winkler, C. D., and Kidwell, S. M. (1986) Paleocurrent evidence for lateral displacement of the Pliocene Colorado River delta by the San Andreas fault system, southeastern California. *Geology*, v. 14, p. 788-791.

3 GEOMORPHIC AND GEOCHEMICAL EVIDENCE FOR THE SOURCE OF SAND IN THE ALGODONES DUNES, COLORADO DESERT, SOUTHEASTERN CALIFORNIA

Daniel R. Muhs, Charles A. Bush, Scott D. Cowherd, and Shannon Mahan
U.S. Geological Survey, Denver

ABSTRACT

The Algodones dunes of southeastern California comprise one of the largest active dune fields in the United States. The source of sand of the Algodones dunes is controversial, and the source of stabilized aeolian sand in the adjacent East Mesa area has not been investigated at all. We used mineralogical compositions and trace element concentrations to ascertain the most likely source of sand for these active and stabilized dunes. Results indicate that alluvium derived from the San Bernardino Mountains, which enters the Salton trough to the northwest of the dune fields, and alluvium derived from the Chocolate Mountains, which is deposited immediately to the northeast of the dunes, do not appear to be significant sources of sediment for the Algodones and East Mesa dunes. Both active aeolian sand from the Algodones dunes and stabilized aeolian sand on East Mesa are probably derived from sediments of ancient Lake Cahuilla, which formerly occupied part of the Salton Trough and left sandy shoreline sediments to the west and northwest of where the dune fields are now found. Lake Cahuilla sediments, in turn, were apparently derived from the Colorado River, when the river shifted its course and emptied into the Salton Trough, rather than the Gulf of California.

INTRODUCTION

The Algodones dunes are one of the largest active dune fields in the United States and are found in the Colorado Desert portion of southernmost California and northernmost Baja California Norte (Figures 1 and 2). Because the dunes are currently active, they have been studied by numerous researchers interested in understanding the processes of dune formation and sand sea evolution (Norris and Norris 1961, McCoy et al. 1967, Sharp 1979, Smith 1982, Nielson and Kocurek 1986, Kocurek and Nielson 1986, Havholm and Kocurek 1988, Sweet et al. 1988, 1991, Lee 1991). They have also been compared to dunes in the Hellespontus region of Mars, based on geomorphic similarities (Cutts and Smith 1973, Breed 1977). Study of active dune fields is important because sand seas that are presently inactive may have paleoclimatic significance (Fryberger and Ahlbrandt 1979, Lancaster 1990, Pye and Tsoar 1990, Thomas and Shaw 1991). In addition, stabilized dune fields may become active in the future with a change in climatic conditions, such as those predicted by models of a 21st century greenhouse climate (Muhs and Maat 1993). Thus, understanding of the dynamics of active dune fields is critical for interpretations of

Desert Aeolian Processes. Edited by Vatche P. Tchakerian. Published in 1995 by Chapman & Hall, London.
ISBN 978-94-010-6519-1

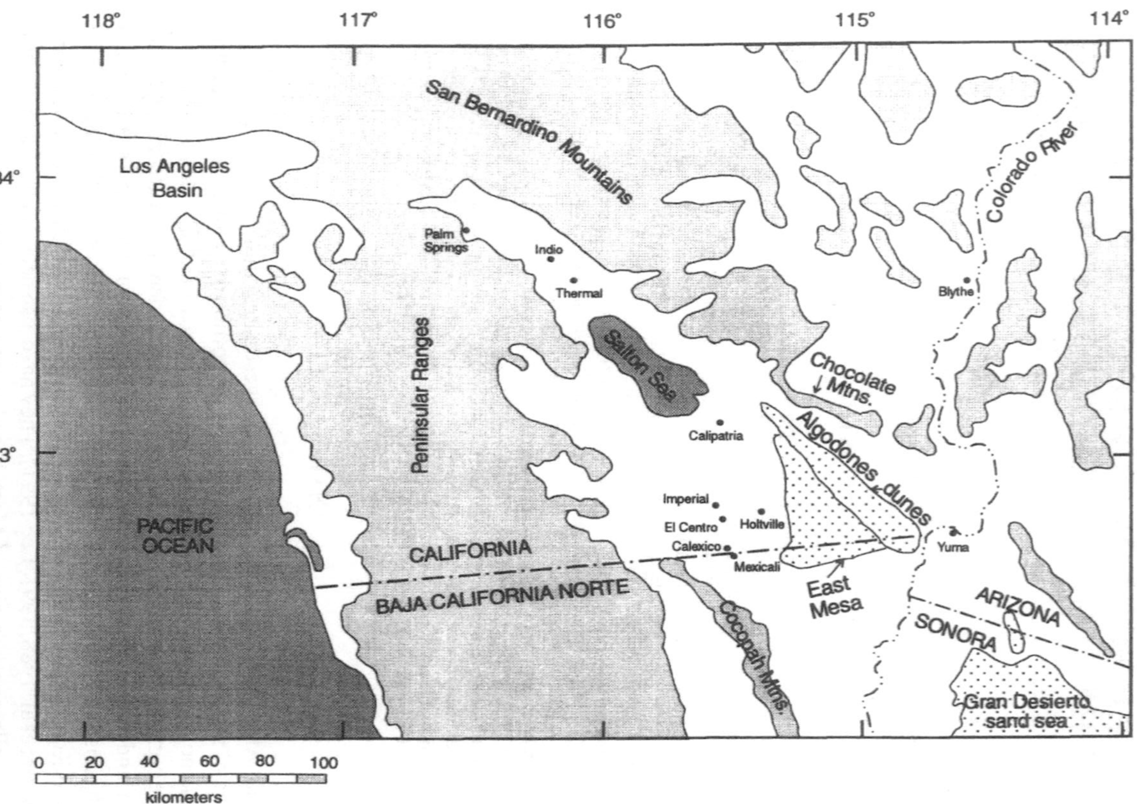

Figure 1. Map showing the localtion of the Algodones dunes and East Mesa in the Salton trough area and localities referred to in the text. Stippled areas are aeolian sand, light-shaded areas are mountain ranges, and dark-shaded areas are water bodies.

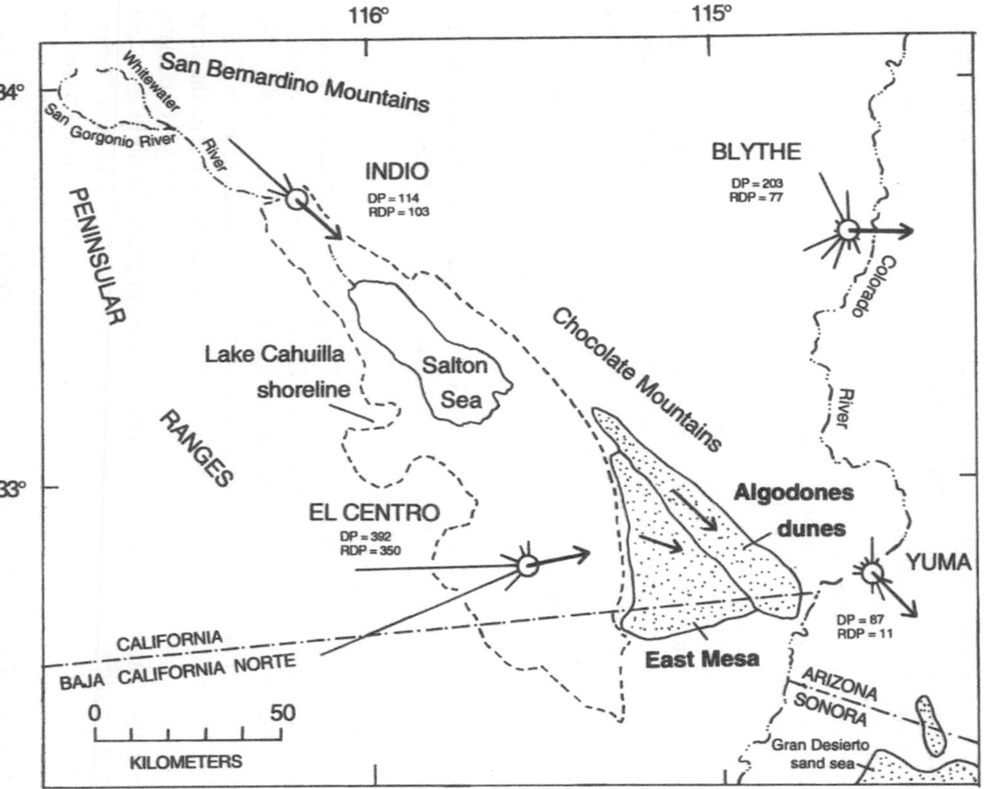

Figure 2. Location of Lake Cahuilla shoreline (dashed line) taken from Waters (1983), sand roses for various localities near the Algodones dunes and East Mesa, and arrows showing the mean direction of sand movement in the dune fields, inferred from dune orientations. Stippled areas are aeolian sand.

present-day processes, interpretation of extraterrestrial geologic processes, past events, and predictions of future activity.

An important part of understanding the dynamics of sand seas is identification of the source sediments for dunes. In some cases, this can be observed directly if source sands are actively feeding a dynamic dune field. In most cases, however, the source of aeolian sediments must be inferred either from the geologic and geomorphic setting, from the direction of prevailing strong winds, or from sedimentologic and mineralogic data which can effectively "fingerprint" the source. The source of aeolian sediments in the Algodones dunes is controversial. In this study, we report new geochemical and mineralogical data for the Algodones dunes, stabilized aeolian sediments of adjacent East Mesa, and the most reasonable candidate sediments that have been cited by previous workers as sources. These data allow us to test the hypothesized sources that previous workers have identified for this important dune field.

REGIONAL GEOLOGIC SETTING AND PREVIOUS STUDIES OF THE ORIGIN OF THE ALGODONES DUNES

Regional Geologic Setting

The Algodones dunes are in the Salton Trough, a fault-bounded, structural basin that is the landward extension of the Gulf of California. A good summary of the geology of the basin can be found in Dohrenwend and Smith (1991). The Salton Trough is bordered on the east by the Orocopia, Chocolate, and Cargo Muchacho Mountains, on the west by the Peninsular Ranges, and on the north by the San Bernardino Mountains (Figures 1 and 2). The Orocopia, Chocolate, and Cargo Muchacho Mountains are composed of Precambrian igneous and metamorphic rocks, Mesozoic granitic and metamorphic rocks, and Tertiary volcanic rocks (Jennings 1967, Haxel et al. 1987). The Peninsular Ranges are a batholith composed largely of Mesozoic granitic rocks and a variety of other pre-Cenozoic granitic and metamorphic rocks (Strand 1962). The San Bernardino Mountains in this area are composed mostly of Mesozoic plutonic rocks and Precambrian metamorphic rocks (Dibblee 1964, Morton et al. 1980).

The Salton Trough itself is filled with upper Cenozoic sediments that were derived from the Colorado River (Muffler and Doe 1968, Van de Kamp 1973, Dohrenwend and Smith 1991). These sediments were deposited in the basin whenever the Colorado River shifted its course away from the Gulf of California, where it now flows, to the Salton Trough. A modern analog to this change in course of the river occurred in the years 1905-1907, when a combination of floods and canal construction diverted the Colorado River away from its usual course to the Gulf of California into the Salton Trough, and created the present-day Salton Sea (Sykes 1937).

The most recent prehistoric episodes of Colorado River flow to the Salton Trough are recorded as abandoned shorelines and beach deposits of late Holocene lakes that occupied the basin (Waters 1983, Dohrenwend and Smith

1991). The late Holocene lacustrine shorelines are found at an elevation of about +12-14 m above sea level; collectively, the paleolakes represented by these shorelines are referred to as Lake Cahuilla (Figure 2). Radiocarbon dating of mollusks, charcoal, and peats associated with these shoreline deposits has shown that there have been several distinct high stands of Lake Cahuilla in the late Holocene at ~2300, ~1300-1200, ~900-600, and ~500-400 yr B.P. (Waters 1983, Rockwell and Gurrola 1993). With the exception of short-lived flood events such as the 1905-1907 diversion, however, the Colorado River has been flowing into the Gulf of California at least since the years 1539-1540 A.D., based on accounts of early Spanish explorers (Sykes 1937, Waters 1983).

Previous Studies of Algodones Dunes Source Sediments

Hypotheses of the origin of the Algodones dunes can be found as far back as the mid-19th century. Blake (1857) thought that Algodones dune sand was "derived from the surface of the upper gravelly plain of the Desert by the continued action of the northerly winds." Brown (1923) thought that beach sediments from ancient Lake Cahuilla (Figure 2) were the source of the Algodones dune sands. Norris and Norris (1961) proposed that most sediment entering the Salton Trough is derived from the San Bernardino Mountains via the San Gorgonio and Whitewater Rivers from the north, and from the Peninsular Ranges via San Felipe Creek from the west (Figure 2). They felt that little sediment was contributed to the basin by small streams draining the Chocolate Mountains to the east, or from the Colorado River (Figures 1 and 2). Mineralogical studies conducted by Norris and Norris (1961) did not allow them to differentiate between fluvial sediments, Lake Cahuilla sediments, and Algodones dunes sediments. They ultimately concluded, however, that the primary source of the dunes was Lake Cahuilla sediments that were in turn derived from fluvial sediments of the Whitewater River, with alluvial fan sediments derived from the Chocolate Mountains being a secondary source. McCoy et al. (1967) accepted the general model of Norris and Norris (1961) in their calculations of the volume of sediment moving through the Algodones dunes. Merriam and Bandy (1965) did not study the Algodones dunes, but conducted detailed mineralogical studies of other upper Cenozoic sediments in the Salton Trough and concluded that they were derived from the Colorado River. Similar conclusions were also reached by Muffler and Doe (1968), on the basis of mineralogical and Pb-isotope studies of late Cenozoic sediments in the Salton Trough, although they also did not specifically study the Algodones dunes. Merriam (1969) conducted mineralogical studies of the Algodones dunes and the Gran Desierto sand sea (Figures 1 and 2) and compared the data to the mineralogical composition of Colorado River sediments and sediments derived from the mountains bordering the Salton Trough. He concluded, in part on the basis of high quartz-to-feldspar ratios, that aeolian sediments in both the Algodones dunes and the Gran Desierto sand sea were derived primarily from Colorado River sediments; he did not, however, study the composition of Lake Cahuilla beach sediments. Van de Kamp (1973) reached conclusions similar to those of Merriam (1969) on the basis of mineralogical data and in addition analyzed Lake Cahuilla sediments. Van de Kamp (1973) reported that Lake Cahuilla sediments have lower quartz and higher volcanic rock fragments compared to Colorado River and Algodones dunes sediments. He inferred that deflation of Colorado River-derived sediments in the southern part of the Salton Trough was the source of the Algodones dunes and implied that Lake Cahuilla sediments were not an important source. Loeltz et al. (1975) thought that a higher shoreline, older than the Lake Cahuilla shorelines, was the source of sand for the dunes, but that these sediments were ultimately derived from the Colorado River, which originally deposited the sand on East Mesa. The Colorado River was also identified as the source of sediment for part of the nearby Gran Desierto sand sea (Figures 1 and 2) on the basis of textural, mineralogical, and Landsat TM spectral data by Blount and Lancaster (1990) and Lancaster (1993), although these workers did not study the Algodones dunes. Smith (1982) and Sweet et al. (1988, 1991) presented no new data on the source sediments for the Algodones dunes, but generally accepted the concept that Lake Cahuilla sediments were the source. Dohrenwend and Smith (1991) thought that the Colorado River was the ultimate source of the Algodones dunes and speculated that these sediments may have been reworked by Lake Cahuilla before accumulation into dunes; however, they pointed out that the exact mechanism for accumulation of aeolian sediments in the Algodones dunes is not known.

REGIONAL CLIMATIC SETTING

Synoptic Climatic Controls in the Colorado Desert

The Salton Trough is in the low-elevation (below ~300 m) portion of southeastern California called the Colorado Desert (also sometimes referred to as the "low desert" in contrast to the "high desert," which is the Mojave Desert to the north). It is a hot desert, classified as BWh in the Köppen system. Climatically, the Colorado Desert is an extension of the Sonoran Desert of Mexico and Arizona, if the climatic criteria of Schmidt (1989) are applied. Ezcurra and Rodrigues (1986) and Schmidt (1989) provide summaries of the synoptic climatic controls on the region. Pacific frontal systems derived from the Aleutian low pressure cell bring rain to the area in winter, and for most Colorado Desert localities, this is the time of maximum precipitation. In late winter, spring, and early summer, the region is under the influence of the eastern edge of the Pacific high pressure cell and there is little or no precipitation. In mid-summer, the Pacific high pressure cell migrates to the northwest, and the Colorado Desert comes under the influence of the western edge of the Bermuda high pressure cell. Combined with the thermally generated continental interior low pressure cell that develops in the southwestern United States in the summer, the presence of the Bermuda high sets up a monsoonal flow of air that results in summer thunderstorms, but precipitation during this season is much lower than during the winter. Overall precipitation ranges from 46 to 140 mm/yr, but most localities within the basin have mean annual totals of 60-80 mm/yr. January mean temperatures range from about 11°-13°C, and July mean temperatures range from about 32°-34°C. Potential evapotranspiration is thus very high, and ranges from about 1190-1275 mm/yr.

Sand-Moving Potential of Winds and Predicted Directions of Sand Drift
Particularly critical to a study of the Algodones dunes is an understanding of the climatic controls on wind regimes. Brazel and Nickling (1986) and MacKinnon et al. (1990) discuss the synoptic weather conditions that are capable of generating dust storms in Arizona, including the Yuma area, which is near the Algodones dunes (Figure 2). The two most important weather conditions for generating dust storms in the Yuma area are Pacific frontal systems that develop in winter, and convective storms (thunderstorms) that develop in the summer monsoon season. Pacific frontal systems in the Yuma area result in strong winds from the north and northwest, whereas convective storms generate strong winds with much more directional variability, and are of shorter duration. However, the majority of long-duration, high-intensity convective storms generate winds from the southwest and southeast (Brazel and Nickling 1986).

Wind regimes as they apply to the formation of sand dunes and sand sheets can be quantified in terms of sand-moving potential. Fryberger and Dean (1979) developed a method for graphic presentation of sand-moving potential called sand roses, which are circular histograms showing weighted magnitudes and directional variability of winds for a given station. The arms in a sand rose are weighted sums of the amount of time that the wind is above the threshold velocity for sand from a given direction; weights are applied to higher velocity winds because the sand-moving ability of wind is a function of the cube of wind speed. Fryberger and Dean (1979) define several parameters from sand rose data: drift potential (DP), which is the scalar sum of all sand-moving winds, regardless of direction; resultant drift potential (RDP), which is the vector sum of all sand-moving winds and will always be less than or equal to DP; and resultant drift direction (RDD), which is the net direction of sand movement. It should be noted, however, that some experimental data suggest that in wind regimes that are not unidirectional, bedform trends are not necessarily predicted correctly by RDD values (Rubin and Hunter 1987). Nevertheless, good agreement between RDD values and dune orientations from many parts of the world suggest that it is a valuable parameter in studying dune forms and their relations to winds (Fryberger and Dean 1979, Breed et al. 1979, Ahlbrandt and Fryberger 1980, Muhs 1985, Lancaster et al. 1987, Lancaster 1988, Wells et al. 1990). The ratio of RDP to DP is also a useful measure; this value decreases with greater directional variability. Calculation of DP, RDP, and RDD requires fairly detailed wind data, and we were able to find sufficient data for only four weather stations which surround the Algodones dunes (Figure 2), three of which have relatively short periods of record: Indio, California (1948-1950);

Blythe, California (1948-1954); El Centro, California (1950-1957); and Yuma, Arizona (1948-1988). In addition, Sweet et al. (1988) present sand rose data for a locality called Drop 1, at the junction of the All-American and Coachella Canals, immediately to the west of the southern edge of the Algodones dunes. In the classification scheme of Fryberger and Dean (1979), DP values for Yuma (87) and Indio (114) put them into "low-energy" wind regimes, and DP values for El Centro (392) and Blythe (203) put them into "intermediate-energy" wind regimes (Figure 2). RDP values for Yuma and Blythe are significantly lower than DP values because of directional variability (Figure 2). In contrast, Indio has winds that are from the northwest all year, and El Centro has winds mostly from the west all year (Figures 3a, 3b, 3c). The shift in wind direction at Yuma and Blythe is because of the change from a wintertime regime of predominantly Pacific-derived frontal systems to a summertime regime of convective storm systems owing to the summer monsoonal flow of air from south to north. The RDD values indicate a predominantly northwest-to-southeast drift of sand for Indio and Yuma, and an east-to-west drift of sand for El Centro and Blythe (Figure 2). Sweet et al. (1988) report that wind data derived from Drop 1 indicate an overall northwest to southeast drift of sand, in broad agreement with Indio and Yuma.

GEOMORPHOLOGY OF THE ALGODONES DUNES AND EAST MESA AEOLIAN DEPOSITS

Surficial Deposits

Using 1:80,000 black-and-white, stereo aerial photographs (Figure 4) as a base, we constructed a generalized surficial geologic map of the Algodones dunes, East Mesa, and the surrounding area (Figure 5). Our unit Qe consists of active sand in the Algodones dunes. Unit Qes includes all of the deposits of East Mesa, and includes sediments of stabilized linear dunes, coppice dunes, sand sheets and sand streaks; alluvial deposits that are apparently older based on the degree of soil development observed by us and by Zimmerman (1981) are included within this unit. Unit Qal includes alluvial and lacustrine deposits of the Imperial Valley; the youngest sediments of this unit were laid down by the last high stand of Lake Cahuilla (Waters 1983, Rockwell and Gurrola 1993). Alluvial fan deposits that are derived from the Chocolate and Cargo Muchacho Mountains were subdivided into two units: (1) Qafo, which consists of older fan deposits that are close to the mountain front, dissected, and covered with rock varnish, which gives them a dark tone on aerial photographs; and (2) Qaf, which consists of alluvial fan deposits found in modern washes and in younger fans that are inset against the older Qafo fan deposits. Small areas of pre-Quaternary bedrock, of too-limited extent to map separately, are included within unit Qafo. Although mapped independently, our Qafo and Qaf units correspond closely to Loeltz et al.'s (1975) "older alluvium" and "alluvium," respectively, that they mapped along the Chocolate Mountains range front. Bull (1991) has recently

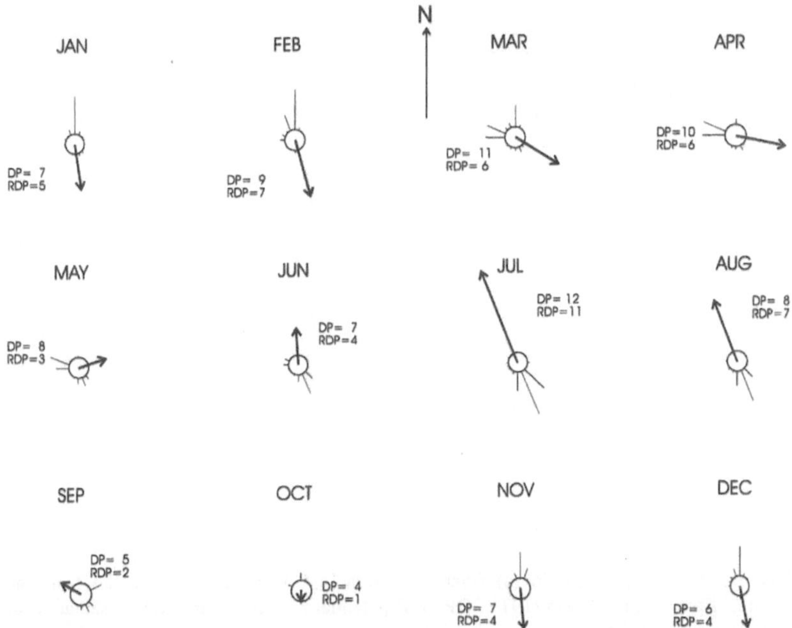

Figure 3a. Monthly sand roses for (a) Yuma, AZ, (b) Indio, CA, and (c) El Centro, CA, using the format of Fryberger and Dean (1979). DP, drift potential in vector units; RDP, resultant drift potential in vector units.

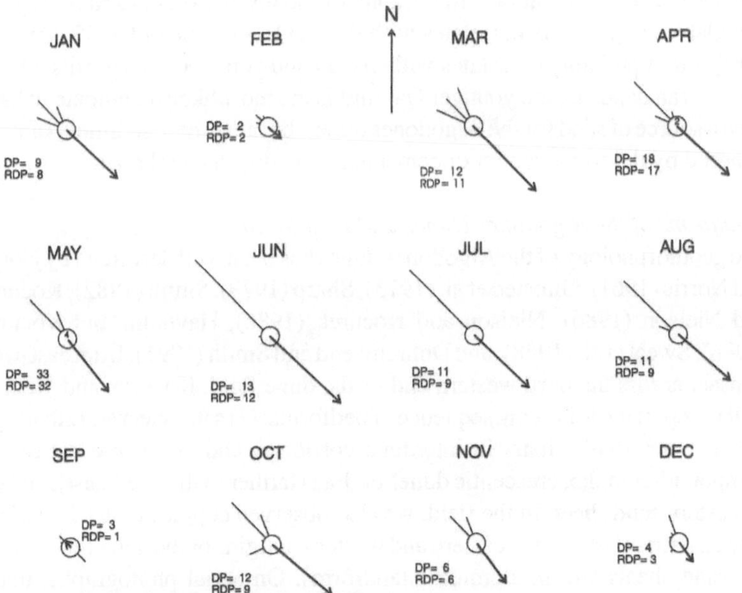

Figure 3b. Monthly sand roses for (a) Yuma, AZ, (b) Indio, CA, and (c) El Centro, CA, using the format of Fryberger and Dean (1979). DP, drift potential in vector units; RDP, resultant drift potential in vector units.

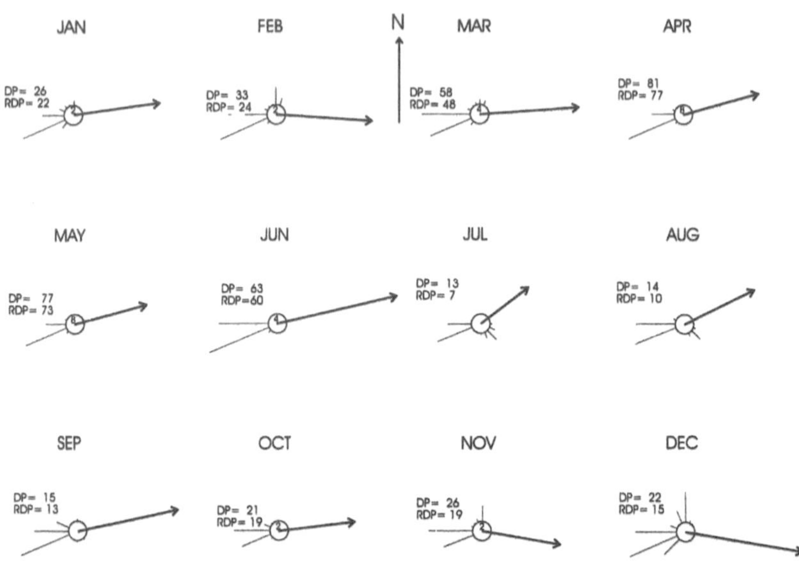

Figure 3c. Monthly sand roses for (a) Yuma, AZ, (b) Indio, CA, and (c) El Centro, CA, using the format of Fryberger and Dean (1979). DP, drift potential in vector units; RDP, resultant drift potential in vector units.

summarized the alluvial fan stratigraphy of the lower Colorado River region. Our Qafo unit probably correlates with Bull's Q1, Q2, and older Q3 deposits; our Qaf unit probably correlates with his Q4 and younger Q3 deposits. Of the alluvial fan deposits, the younger Qaf unit is the most likely candidate to have been a source of sand for the Algodones dunes, because these sediments are not armored by desert pavement or cemented by pedogenic carbonate.

Landforms of the Algodones Dunes and East Mesa

The geomorphology of the Algodones dunes has been well described by Norris and Norris (1961), Olmsted et al. (1973), Sharp (1979), Smith (1982), Kocurek and Nielson (1986), Nielson and Kocurek (1986), Havholm and Kocurek (1988), Sweet et al. (1988), and Dohrenwend and Smith (1991). In an east-west transect across the northwestern end of the dune field, Kocurek and Nielson (1986) report the following sequence of bedforms: (1) sand sheet and zibars, (2) linear dunes (with zibars in interdune corriders) and crescentic dunes, (3) compound-complex crescentic dunes or draa (farther to the southeast), and (4) an eastern sand sheet. In the field, we also observed coppice dunes in various places, particularly on the eastern and western margins of the dune field, where the sand sheets are the dominant landforms. On aerial photographs, linear dunes and transverse and crescentic ridges are prominent features in the Algodones dune field, and linear dunes and sand streaks (as defined by McKee,

1979) are visible on East Mesa (Figure 4). We mapped orientations of crests of linear dunes and sand streaks on East Mesa and linear dune crests, barchanoid ridge (draa) brinks, and transverse ridge (draa) brinks in the Algodones dunes on aerial photographs (Figure 5). Between the active linear and barchanoid/ transverse ridge dunes in the Algodones dunes there are smaller, transitional crescentic ridges whose orientations we did not measure. Results indicate that active linear dunes in the Algodones dune field have a mean orientation of N43°W ± 12° (n = 89), which approximates the dominant dune-forming wind direction. Transverse and barchanoid ridge draas in the Algodones dune field have a mean brink orientation of N45°E ± 16° (n = 57), also implying a dominant dune-forming wind direction from the northwest, about N45°W. Our measurements of the transverse and barchanoid ridges are in agreement with those reported by Havholm and Kocurek (1988). The inferred winds of formation based on dune orientations are nearly identical to the RDDs for Indio and Yuma (Figure 2), but imply a general drift of sand that is somewhat more easterly than the RDD for Drop 1 near the southern end of the dune field reported by Sweet et al. (1988). Havholm and Kocurek (1988) reported that superimposed features on the transverse ridge draas are the result of a secondary air flow that is the result of modification of the primary air flow by the draas themselves.

East Mesa dunes and sand sheets are presently stabilized by vegetation, mostly creosote (*Larrea divaricata*). The winds that formed the aeolian landforms on East Mesa had a more westerly component than the winds that formed the linear and transverse ridge dunes of the Algodones dune field. Linear dunes and sand streaks on East Mesa have a mean orientation of N71°W ± 9° (n = 255), although the southernmost dunes on East Mesa have an orientation that is closer to the mean orientation of the Algodones dunes. One explanation for the difference in mean orientations of East Mesa and Algodones dunes is that the East Mesa dunes were formed by winds different from those of today. However, the East Mesa dunes and sand sheets do not appear to be very old. We dug pits in aeolian deposits of East Mesa at several localities and found that there is no evidence of soil formation, consistent with the observations of Zimmerman (1981). Even minimal accumulations of organic matter to form simple A horizons, such as those that characterize the youngest (<2 ka) Holocene aeolian sands on the Great Plains and the Colorado Plateau (cf. Muhs 1985, Wells et al. 1990, Jorgensen 1992, Madole 1992) are absent in aeolian deposits of East Mesa. Thus, it appears that these deposits have only recently (i.e., within the past few hundred years or less) been stabilized by vegetation, and it is improbable that wind directions have been greatly different from the present in the past few hundred years. A more likely explanation is that there are significant differences in resultant drift directions in different parts of the Salton Trough, even over short distances. The resultant drift direction for El Centro, immediately west of East Mesa, is west-to-east, whereas that of Yuma is northwest-to-southeast (Figure 2).

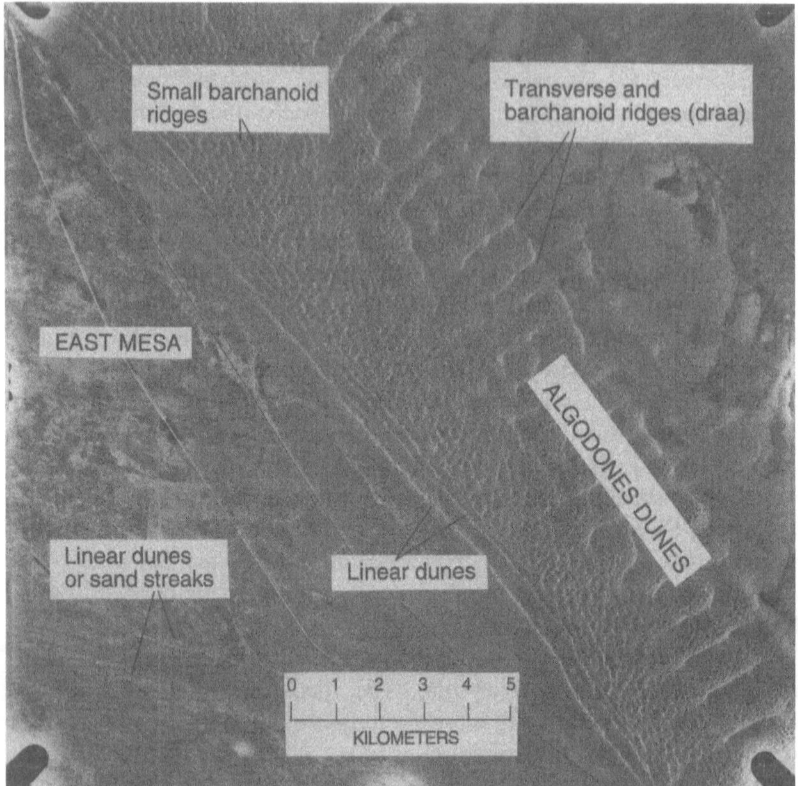

Figure 4. Vertical aerial photograph of the central portion of the Algodones dune field and a part of East Mesa, showing types of aeolian landforms.

Cause of Stability of East Mesa Aeolian Deposits

The relative stability of East Mesa aeolian sands contrasts sharply with the active Algodones dunes (Figure 4). It is possible that local climatic factors could be responsible for the relative stability of the East Mesa dunes. Lancaster (1988) has shown from studies in the Namib and Kalahari deserts in southern Africa that degree of dune activity is a function of the balance between the amount of time (W) that wind is above the threshold velocity for sand, and the precipitation-to-potential evapotranspiration ratio (P/PE). He generated an index of dune mobility (M) that is computed from these data as W/(P/PE). Higher values of M correspond to greater degrees of dune activity. Lancaster recognized, from field studies in southern Africa, four classes of dune activity: (1) inactive dunes, M values <50; (2) dunes with active crests only, M values between 50 and 100; (3) dunes that are fully active except for plinths and interdune areas, M values between 100 and 200; and (4) fully active dunes, M values >200. Muhs and Maat (1993) applied Lancaster's index to localities near dune fields in the Great Plains and found good agreement between predicted and actual degrees of dune activity.

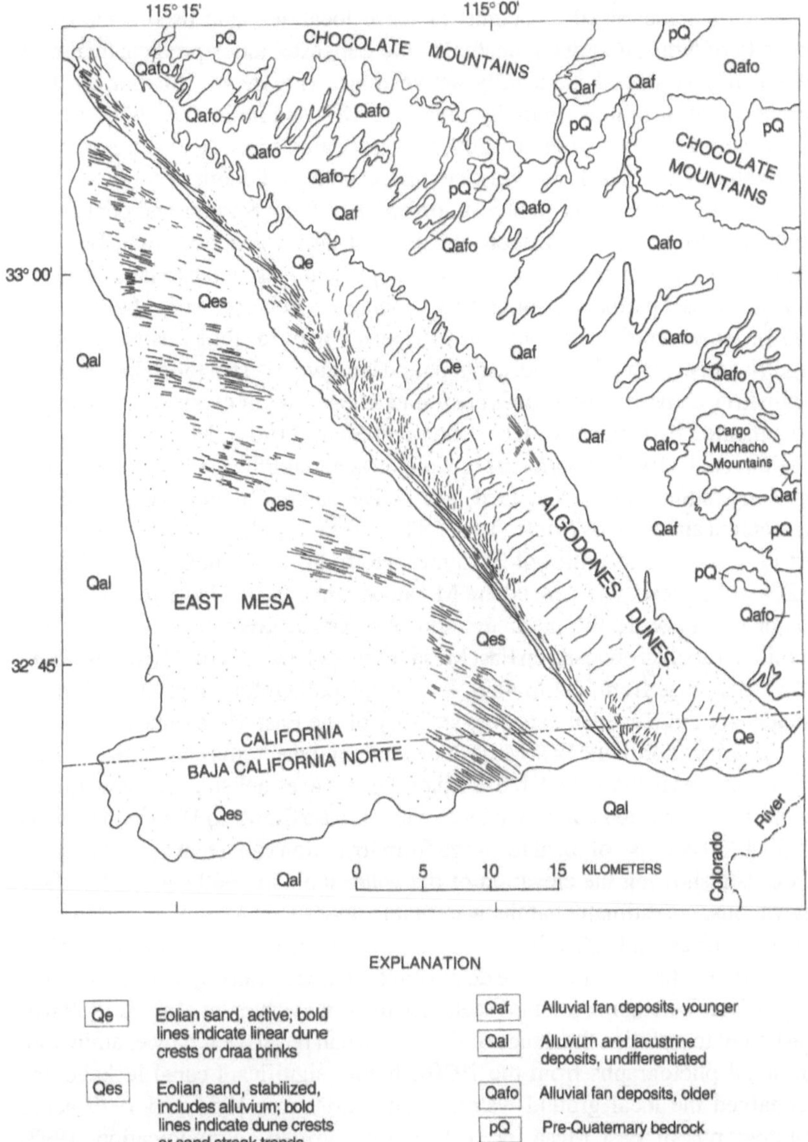

Figure 5. Generalized surficial geologic map of the Algodones dunes, East Mesa, and surrounding areas.

Using modern wind speed, P, and PE data, calculated by the method of Thornthwaite and Mather (1957), we computed M values for eleven localities in or adjacent to sand dunes or sand sheet areas in the Colorado Desert. Weather stations record wind data with anemometers at different heights; for most stations in the Colorado Desert, no anemometer heights are given and we assumed a standard 10 m height for these localities, and used a threshold velocity of 5 m/s. Results indicate that the Lancaster index predicts that most of the region should have fully active dunes (Figure 6). All but the two northernmost localities (Palm Springs, CA, 187; and Indio, CA, 202) have M values significantly greater than 200, and one (Calipatria, CA) has an extremely high value of 776. The field evidence supports the predictions of active sand by the mobility index. In the area just northwest of Palm Springs and continuing south to Indio (Figure 1), there are limited areas of active sand, but also significant tracts of aeolian sand sheets, coppice dunes, and barchanoid dunes that are at least partially stabilized by vegetation (Beheiry 1967, Shelton et al. 1978). However, farther south, in the areas where localities have M values significantly greater than 200, aeolian sand is active. West of the Salton Sea and Calipatria, active barchan dunes are found whose rates of movement have been documented by Rempel (1936), Long and Sharp (1964), Norris (1966), and Shelton et al. (1978). Southwest of Calipatria and northwest of Imperial there are longitudinal dunes that feed into active barchan dunes whose rates of movement are given by Smith (1982). The active Algodones dunes themselves are closest to the localities of El Centro, Holtville, and Calexico, CA, Yuma, AZ, and Mexicali, Mexico, where M values range from 274 to 469. The only major exception to the predicted and observed degrees of activity are the stabilized aeolian deposits on East Mesa (Figures 1 and 2), near localities where M values range from 274 to 469. We conclude, therefore, that local climatic factors are not responsible for the stability of the East Mesa dunes.

Stabilization of the East Mesa dunes and sand sheets may have been caused by recent rises in the water table that have helped to establish vegetation. Loeltz et al. (1975) reported that ground water levels rose from 0 to 18 m in the period 1939-1960 because of local recharge from irrigation canal leakages. Based on their data showing the elevation of the water table in 1960 and land surface elevations, we estimate that the water table under East Mesa was within 5-18 m of the surface in 1960. It is possible that the water table is closer to the surface at the present time, although we do not have, at present, any modern water table depth data. Whether or not the recent rise in ground water level since 1939 was sufficient to stabilize the dunes of East Mesa can be answered by examination of aerial photographs from the 1930s, before significant canal leakage had recharged the local ground water. A comparison of 1954 and 1979 aerial photographs of East Mesa by G. Kocurek (written communication, 1993) seems to indicate that there was less vegetation in the area in the 1950s, and a systematic study using older aerial photographs would be a worthwhile effort.

Possible Source Sediments

The observations of dune orientations and resultant drift directions from wind data help to constrain possible source sediments for the Algodones dunes and East Mesa aeolian sediments. Fluvial deposits of the San Gorgonio and

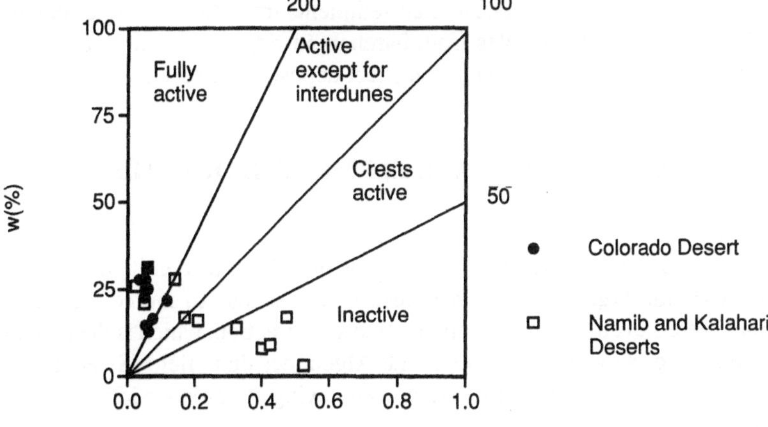

Figure 6. Plot of data from weather stations (localities shown in Figure 1) in or near bodies of aeolian sand in the Colorado Desert (solid circles) showing amount of time that wind is above the threshold velocity for sand (W) vs. ratio of mean annual precipitation to potential evapotranspiration (P/PE). Similar data for localities in the Namib and Kalahari Deserts (open squares), taken from Lancaster (1988) are also plotted for comparison. Bold numbers (50, 100, 200) are threshold values of M (W/[P/PE]) that separate dune activity classes.

Whitewater Rivers, which drain the San Bernardino Mountains, occur in dry, sandy washes to the northwest of the dune fields, and are thus potential sources of sand (Figure 2). We witnessed dramatic aeolian movement of sand from these river beds to the southeast after passage of a Pacific frontal system in March, 1993. Visibility was so poor because of blowing sand that numerous roads in the area had to be closed. Sand-rich Lake Cahuilla shoreline sediments occur to the west and northwest of the dune fields and are also likely candidates for source sands (Figure 2). Alluvial fan deposits of the Chocolate and Cargo Muchacho Mountains occur mostly to the northeast of the dune fields (Figure 5) and would seem to be a less likely source based on drift potentials, but active washes carry abundant sand from these mountains and terminate at the northeastern side of the Algodones dunes. Therefore, the hypothesized source sediments sampled include: (1) modern alluvium collected from active channels of the Whitewater and San Gorgonio Rivers, which drain the San Bernardino Mountains; (2) modern sediments from active channels incised into alluvial fans that are derived from the Chocolate and Cargo Muchacho Mountains; (3) beach sediments of the shoreline of Lake Cahuilla that occur at about +12-14 m above sea level, as mapped by Waters (1983); and (4) Colorado River sediments, collected from the modern floodplain in the vicinity of Yuma, Arizona, and from deposits of Yuma Mesa, which is thought to be an old terrace of the Colorado River, possibly of last-interglacial age (Olmsted et al. 1973,

Dohrenwend and Smith 1991). Aeolian sediments from the Algodones dunes and East Mesa include samples from barchanoid-ridge, barchan, linear, and coppice dunes, as well as aeolian sand sheet deposits (Table 1).

MINERALOGY AND TRACE ELEMENT GEOCHEMISTRY

Mineralogy

We conducted simple mineralogical studies in our efforts to identify the source sediments of the Algodones and East Mesa aeolian sands. All source-sediment samples were sieved to include only the fine sand fraction, which is the modal particle size for aeolian sediments in the Algodones dune field (Sweet et al. 1988). Silts and clays were removed by wet-sieving after dispersion with Na-pyrophosphate.

Semiquantitative estimates of the relative amounts of quartz, K-feldspar, and plagioclase in the fine sand fractions of the Algodones dunes, East Mesa aeolian sediments, and hypothesized source sediments were made using X-ray diffractometry. Mineral peak heights were measured for the 20.8° (quartz), 27.4° (K-feldspar: microcline and sanidine), and 27.8° (plagioclase) peaks. These values were summed for each sample and relative proportions were calculated using the peak height of each mineral as a fraction of the sum. The data are presented in the form of ternary diagrams (Figures 7, 8, and 9), but are not intended to be interpreted as precise quantification of mineral content.

Van Andel (1964) and Olmsted et al. (1973) give data on the composition of modern and ancient Colorado River alluvium, respectively; their results are similar to each other and indicate that this alluvium is mineralogically quite mature. Colorado River sediments are characterized mostly by quartz (65%-75%), with lesser amounts of feldspar (10%-20%), and rock fragments (1%-8%). Our data support these earlier observations, and also indicate that Lake Cahuilla sediments have a composition that is not significantly different from Colorado River sediments, but is significantly more quartz-rich and lower in plagioclase than San Bernardino Mountains-derived alluvium (Figures 7a and 8b). Our observations differ from those of Van de Kamp (1973), who found that quartz abundances in Lake Cahuilla sediments are lower than in Colorado River sediments. Lake Cahuilla and Colorado River sediments are also generally higher in quartz than Chocolate Mountains-derived alluvium, although there is considerable overlap between these sediment groups (Figure 7b). Thus, Lake Cahuilla sediments are most likely derived from the Colorado River, as inferred by most previous workers (other than Van de Kamp, 1973), with possibly some contributions from the Chocolate Mountains, but little or no contribution from the San Bernardino Mountains. The higher quartz and lower plagioclase contents in Colorado River sediments compared to sediments derived from the local basin-bounding mountain ranges may be explained by the fact that Colorado River sediments have been derived from numerous rocks in the river's drainage basin, many of which have undergone several cycles of

Table 1
Elemental concentrations in Algodones and East Mesa aeolian sediments, Colorado River sands, Lake Cahuilla sands, Chocolate and Cargo Muchacho Mountain alluvial fan sands, and sands from rivers draining the San Bernardino Mountains

ID#	K (%)	Ca (%)	Ti	Rb	Sr	Y	Zr	Nb	Ba	La	Ce	Geomor phology**
			ppm									
Algodones aeolian sediments												
ALG-1	2.08	1.58	903	52	144	8	49	2	514	14	29	C
ALG-2	1.90	1.40	858	54	137	7	46	1	486	10	26	B
ALG-3	2.04	1.86	834	49	133	7	46	1	486	12	28	BR
ALG-4	1.74	1.57	798	47	129	7	56	2	447	21	28	B
ALG-5	2.07	1.89	864	48	128	7	47		451	13	16	B
ALG-6	1.90	2.36	1433	51	164	9	70	4	529	18	28	L
ALG-7	1.56	1.59	1099	48	143	9	61	1	457	7	24	S
ALG-8	1.96	1.59	959	48	131	7	52	2	465	10	26	C
ALG-9	1.74	1.47	883	48	40	7	49		482	14	28	S
ALG-10	1.67	1.48	987	49	141	8	56	2	482	17	20	C
ALG-11	1.66	2.23	1487	47	152	10	71	5	438	14	31	L
ALG-12	1.64	2.93	1965	46	163	13	90	7	436	21	33	L
ALG-13	2.08	3.24	1535	47	157	10	58	5	454	18	29	L
ALG-14	1.53	2.24	1673	47	154	11	93	6	437	18	32	L
ALG-15	1.57	1.88	1236	50	151	8	71	4	460	7	35	S
ALG-16	1.80	2.48	1296	48	151	9	71	2	457	17	36	S
ALG-17	1.69	1.93	1146	48	148	8	60	3	468	6	24	L
ALG-18	1.83	1.76	1076	51	148	7	58	2	486	8	21	L
ALG-19	1.85	1.97	1084	51	150	9	55	1	477		25	L
ALG-20	2.27	2.15	1140	48	136	8	50	1	471	16	35	BR
ALG-21	1.63	1.25	858	56	147	6	55	2	490	11	12	BR
ALG-22	2.10	2.76	1266	49	145	8	68	3	439	16	31	BR
East Mesa aeolian sediments												
EM-1	1.90	2.41	1322	50	167	10	76	2	567	7	29	S
EM-2	1.84	2.09	1388	50	163	9	81	3	580	7	23	C
EM-3	2.19	2.56	1654	55	162	11	117	2	548	16	30	C
EM-4	1.63	2.64	1597	49	178	12	86	5	531	16	30	S/C
EM-5	1.74	2.44	980	46	152	6	53	3	450	10	22	S/C
EM-6	1.56	1.70	1022	48	138	9	62	2	445	7	13	S/C
EM-7	1.88	2.31	1397	51	163	9	70	2	522	18	28	S/C
EM-8	2.03	2.30	1262	49	154	8	59	3	512	13	31	S/C
EM-9	1.99	2.27	1131	56	148	8	62	2	496	9	8	S
EM-10	1.66	2.62	1168	34	124	8	47	3	383	19	44	S
EM-11	1.83	1.75	887	49	134	7	54	3	477	4	14	C
EM-12	2.01	2.21	1193	50	138	8	68	3	472	14	24	C
EM-13	2.01	1.80	952	51	136	7	62	2	480	12	32	C
EM-14	1.70	1.97	963	48	132	8	53	2	446	5	16	B

Colorado River sediments												
CO-1	1.32	1.70	1635	36	131	10	113	5	370	15	30	YM
CO-2	1.88	1.80	1148	50	119	9	72	1	479	11	18	YM
CO-3	2.14	2.80	1978	52	168	9	89	5	470	15	48	YM
CO-4	1.87	1.65	1028	42	96	7	76	3	416	9	22	YM
CO-5	1.92	2.20	1064	49	106	11	87	2	437	10	24	YM
CO-6	1.92	1.97	996	50	332	8	67	2	676	23	35	YM
CO-7	1.66	1.39	968	46	108	9	65	2	470	7	13	YM
CO-8	2.18	1.96	994	51	113	7	76	2	485	7	33	YM
CO-9	2.37	1.51	2548	73	196	13	98	7	582	22	42	YM
CO-10	2.35	1.39	2940	73	196	15	102	8	578	16	38	YM
CO-11	2.29	1.31	3074	78	212	17	95	10	602	10	38	YM
CO-13	2.01	2.30	1784	50	138	9	98	2	468	17	33	FP
CO-14	1.86	2.52	1208	50	128	10	116	1	465	5	19	FP
CO-15	2.20	3.19	1546	58	152	10	101	2	530	13	22	FP
CO-16	2.11	3.69	1717	56	197	11	136	2	500	8	35	FP
CO-17	2.26	3.72	1661	59	150	12	110	3	548	19	26	FP
CO-18	2.13	3.90	1544	52	151	9	130	3	493	8	13	FP
CO-19	1.86	2.80	1759	50	147	12	184	3	467	14	22	FP
CO-20	1.79	2.86	2422	50	153	15	286	5	469	16	18	FP
CO-20A	2.14	3.31	1769	57	137	13	103	4	504	14	25	FP
CO-21	1.87	2.72	1651	52	158	9	167	4	480	11	28	FP
Lake Cahuilla sediments												
LC-1	1.46	1.87	842	48	160	5	52		462	10	15	
LC-2	1.75	3.58	1659	43	168	10	93	2	437	20	37	
LC-3	1.67	3.21	1552	52	181	12	89	5	480	18	32	
LC-4	2.05	3.10	1423	50	169	9	73	3	499	16	28	
LC-5	2.22	3.63	1360	54	171	9	60	4	533	14	34	
LC-6*	0.69	1.58	25229	21	105	85	1626	63	251	115	205	
LC-7	1.33	2.91	3579	39	150	17	142	10	400	26	47	
LC-8*	0.91	2.47	15539	26	122	50	580	46	319	84	148	
LC-9	1.58	2.24	1771	46	152	9	80	4	446	12	39	
LC-10	1.56	2.11	2709	40	140	14	87	7	424	21	47	
LC-11	2.07	2.97	1381	54	172	8	68	2	543	11	24	
LC-11A	2.44	2.87	1708	68	195	9	91	4	578	10	36	
LC-12	1.88	1.64	824	58	169	7	62	1	598	14	28	
LC-13	1.92	1.75	931	54	57	7	55	2	558	12	21	
LC-14	1.77	2.64	1488	41	149	10	74	4	434	20	37	
LC-15	1.65	2.07	1074	50	134	9	59	2	444	13	25	
LC-16	1.41	1.98	1654	40	145	11	83	6	374	11	27	
LC-17	1.96	2.47	1849	48	155	10	73	3	461	12	24	
Alluvial fan sediments from the Chocolate and Cargo Muchacho Mountains												
CM-1	1.78	2.42	1811	56	195	15	214	4	638	22	47	CA
CM-2	1.96	2.71	3101	56	208	38	589	15	834	63	110	CA
CM-3	2.34	2.70	2653	69	229	25	318	13	786	36	68	CA
CM-4	2.72	2.51	1698	84	244	18	161	8	893	12	30	CA
CM-5	2.86	2.23	2382	92	243	28	338	13	782	36	71	CA
CM-6	2.37	2.12	2189	76	198	21	294	8	634	23	43	CA
CM-7	2.68	2.52	2779	100	249	29	354	12	735	34	71	CA

CM-8	2.18	2.42	2559	82	201	21	312	10	580	20	42	CA
CM-9	2.33	2.73	3153	92	225	36	385	18	603	41	84	CA
CM-10	3.42	2.18	1860	121	299	19	195	9	843	8	30	CA
CM-11	2.69	2.06	2583	99	257	24	264	11	661	28	57	CA
CM-11A	2.36	2.04	2200	73	197	18	332	6	594	29	54	CH
CM-12	2.66	2.14	1846	93	272	16	109	6	740	21	35	CH
CM-14	2.38	2.68	2488	73	243	16	170	7	680	19	45	CH
CM-15	2.29	2.42	2294	74	297	14	126	8	779	5	39	CH
CM-16	2.40	2.99	2809	69	235	17	242	7	679	15	36	CH
CM-17	2.13	2.64	2858	64	195	18	299	7	586	18	42	CH
CM-18	2.48	2.81	2682	84	232	18	229	8	756	16	40	CH
CM-19	2.22	2.18	2966	73	192	20	391	7	626	24	52	CH
CM-20	2.72	2.19	2316	106	214	16	123	6	715	13	40	CH
CM-21	2.26	2.70	2608	69	213	16	235	6	639	23	37	CH
Sediments from rivers draining the San Bernardino Mountains												
SB-1	2.10	2.35	4722	70	333	40	321	9	767	64	120	SG
SB-2	2.08	2.39	4946	67	336	40	307	12	749	55	119	SG
SB-3	2.30	2.51	4030	72	390	36	197	13	856	71	142	SG
SB-4	2.23	2.31	3722	71	381	32	266	10	836	40	92	SG
SB-5	1.94	3.20	4770	67	519	19	201	11	718	21	53	SG
SB-6	2.33	2.44	3975	81	399	35	271	11	852	43	98	SG
SB-7	2.66	2.61	3136	109	440	20	143	6	966	22	60	SG
SB-8	2.40	2.50	3940	88	399	34	350	12	862	42	101	SG
SB-9	2.54	2.17	2880	89	409	18	129	7	925	22	58	SG
SB-10	2.45	2.27	3647	87	388	26	249	9	860	33	79	SG
SB-11	2.87	2.05	3428	104	387	22	180	9	959	29	61	W
SB-12	2.63	2.14	3324	91	391	28	271	10	960	46	92	W
SB-13	2.59	2.49	3991	96	384	39	357	11	868	43	95	W
SB-14	2.80	2.07	3523	102	402	30	288	10	964	48	103	W
SB-15	2.63	2.20	3567	96	413	35	265	13	913	66	112	W
SB-16	2.85	1.94	3466	102	402	34	277	11	930	73	141	W
SB-17	2.77	1.88	3025	100	404	29	236	11	986	55	102	W
SB-18	2.57	2.01	4000	94	383	42	433	14	892	90	166	W
SB-19	2.95	1.84	3185	112	389	30	289	11	1010	45	105	W

* Not included in element concentration plots because of unusually high concentrations of heavy minerals.

** Abbreviations for landforms: C, coppice dune; B, barchan dune; BR, barchanoid-ridge; L, linear dune; S, sand sheet; YM, Yuma Mesa; FP, modern floodplain of Colorado River; CA, modern fan channel sediments of Cargo Muchacho Mountains; CH, modern fan channel sediments of Chocolate Mountains; SG, modern channel sediments of San Gorgonio River; W, modern channel sediments of Whitewater River.

(a)

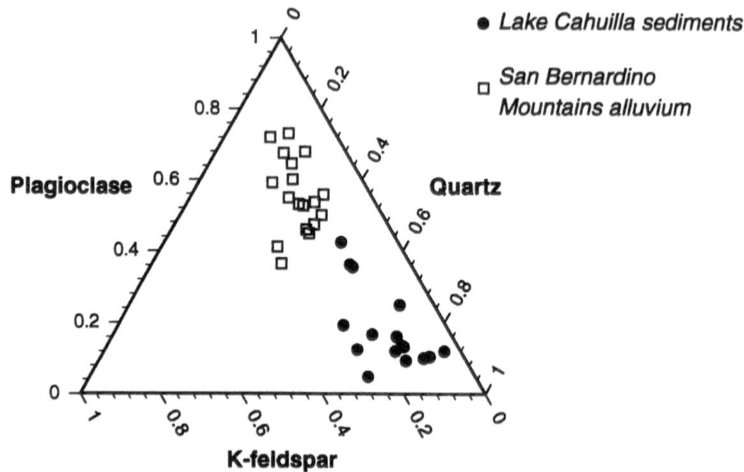

(b)

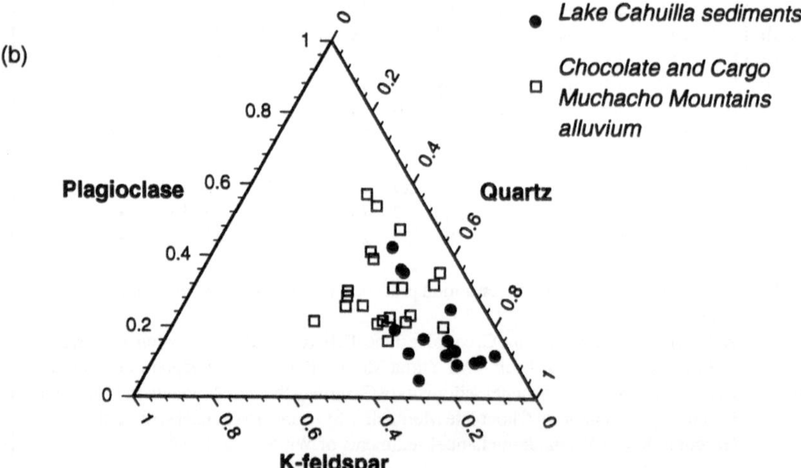

Figure 7. Ternary diagram comparing relative abundances of quartz, K-feldspar, and plagioclase in sediments from Lake Cahuilla, the San Bernardino Mountains, and the Chocolate and Cargo Muchacho Mountains.

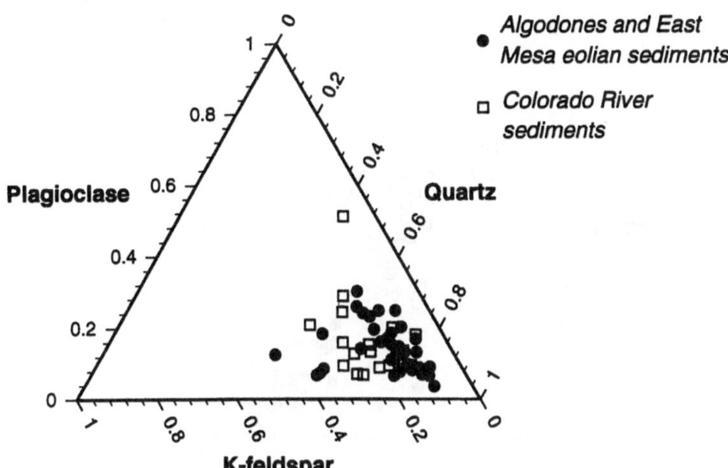

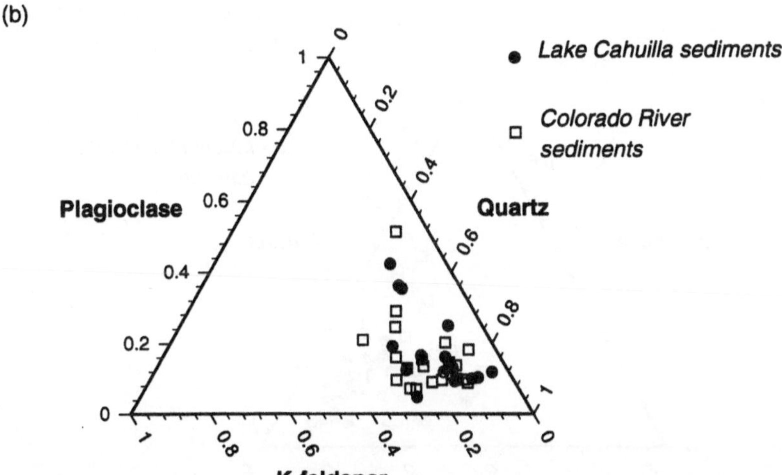

Figure 8. Ternary diagram comparing relative abundances of quartz, K-feldspar, and plagioclase in sediments from the Colorado River and Lake Cahuilla, and aeolian sediments from the Algodones dunes and East Mesa.

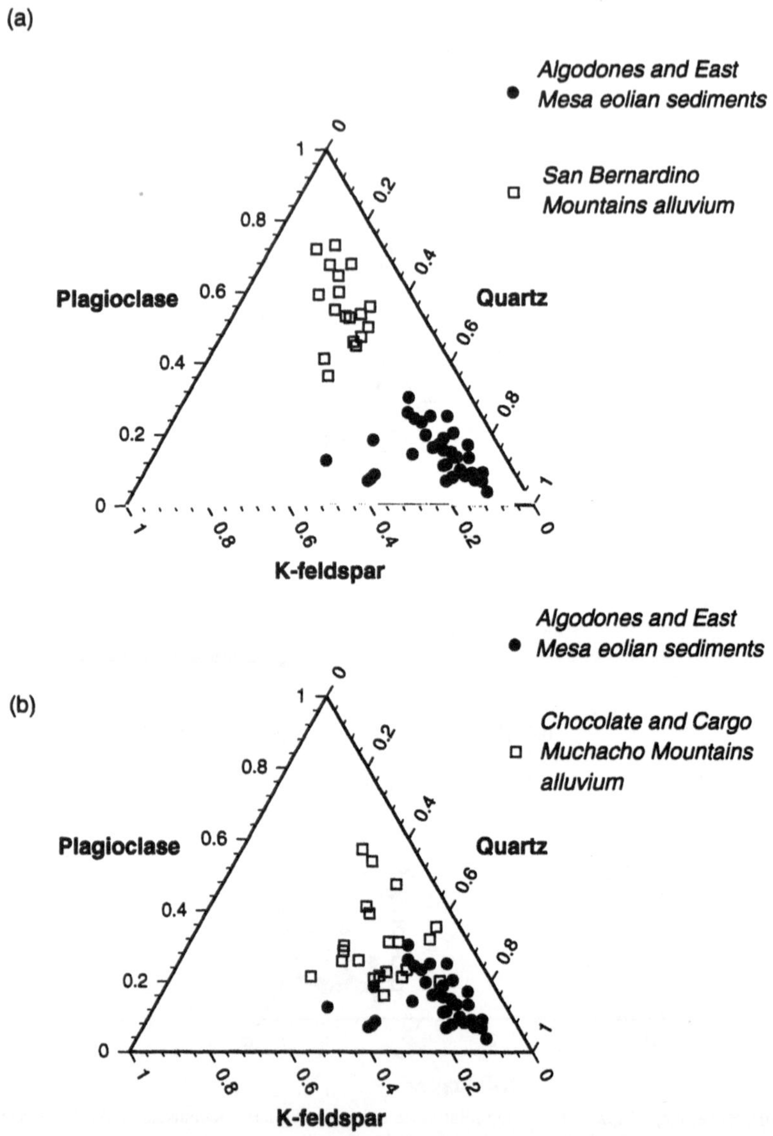

Figure 9. Ternary diagram comparing relative abundances of quartz, K-feldspar, and plagioclase in aeolian sediments from the Algodones dunes and East Mesa, and alluvial sediments derived from the San Bernardino Mountains and Chocolate and Cargo Muchacho Mountains.

Table 2
Mean values and standard deviations for USGS rock standard GSP-1 (n = 15 runs) and published values (Govindaraju 1989) for GSP-1.

Element	Mean value	Standard deviation	Standard deviation (%)	Published value
K (%)	4.86	0.04	0.9	4.57
Ca (%)	1.47	0.02	1.4	1.48
Ti (ppm)	3806	82	2.2	3897
Rb (ppm)	255	3	1.2	254
Sr (ppm)	237	3	1.3	234
Y (ppm)	34	3	8.8	26
Zr (ppm)	513	12	2.3	530
Nb (ppm)	28	1	3.6	28
Ba (ppm)	1320	10	0.8	1310
La (ppm)	181	9	5.0	184
Ce (ppm)	407	8	2.0	399

weathering, erosion, and sedimentation, and are thus mineralogically more mature.

Mineralogical data indicate that there is no significant difference between aeolian sediments of the Algodones dunes and East Mesa. Therefore, on subsequent ternary diagrams, we combine these two sediment groups because they appear to have had a common origin. Aeolian sediments of East Mesa and the Algodones dunes are enriched in quartz and depleted in plagioclase when compared to alluvium derived from the San Bernardino Mountains, which suggests that sediments of the Whitewater and San Gorgonio Rivers are not important sources of the aeolian sediments (Figure 9a). The same conclusion can be inferred from a comparison of aeolian sediments with Chocolate Mountains alluvium, although there is a slight amount of overlap between the sediment groups (Figure 9b). The greatest similarities are seen in a comparison of aeolian sediments of East Mesa and the Algodones dunes with a pooled sediment group that consists of Colorado River alluvium and Lake Cahuilla sediments (Figure 8a). From this comparison and those made above, we conclude that the best candidate for a source of the aeolian sediments is Lake Cahuilla sediments, which in turn appear to be derived mainly from the Colorado River (Figures 8a, 8b).

Geochemical Methods

We used trace element concentrations, measured by energy-dispersive X-ray fluorescence spectrometry, to "fingerprint" the potential source sediments and compare them to aeolian sands from the two dune fields. A similar approach, using ratios of trace element concentrations, was employed by Muhs et al.

(1990) in determining the source of clay-rich soils on Quaternary limestones on Caribbean and western Atlantic islands, and has been commonly used by petrologists studying the origins of volcanic and metavolcanic rocks (Pearce and Cann 1973, Brooks and Coles 1980). As with the mineralogical determinations, only the fine sand fraction was used for analyses. Concentrations of most trace elements can be measured with relatively high precision and accuracy with the exceptions of Y (low precision and uncertain accuracy) and Nb and La, which show relatively low precision (Table 2). However, concentrations of Y, Nb, and La are still useful for semiquantitative discrimination. The trace elements studied occur in a variety of mineral phases. In aeolian and alluvial sands derived from granitic rocks, the most common constituents are quartz, K-feldspar, plagioclase, rock fragments, and minor amounts of heavy minerals. Rb substitutes for K and is therefore found in K-feldspar and biotite. Sr substitutes for Ca, and is present in plagioclase, but can be found in some potassium minerals as well. La and Ce are light rare earths, and Y has behavior similar to the heavy rare earths. The rare earth elements are found in a wide variety of minerals, but on an overall basis, most of the rare earth abundances in these samples are probably accounted for by the feldspars. Zr is present almost exclusively in zircon. Ti is found mainly in ilmenite, anatase, rutile, titanomagnetite, and sphene. Nb is present in sphene and ilmenite, and to a lesser extent in biotite; Nb concentrations are often highly correlated with concentrations of Ti.

Trace Element Concentrations

Concentrations of Sr, Rb, Ti, Zr, Y, Nb, La, and Ce are not significantly different in aeolian sediments of East Mesa and the Algodones dunes, which suggests a common origin, consistent with the mineralogical data discussed above (Figure 10). In further discussions, therefore, we combine the East Mesa and Algodones dunes geochemical data, and consider them as one sediment population. Comparison of Lake Cahuilla beach sands with San Bernardino Mountains-derived alluvial sands shows that the former are significantly lower in concentrations of all trace elements, which is consistent with the mineralogical data discussed above. Relatively high concentrations of Sr, La, Ce, and Y in San Bernardino Mountains-derived alluvium are consistent with relatively high amounts of plagioclase in these sediments; high Rb also indicates relatively high amounts of K-feldspar. Concentrations of heavy minerals, as shown by the Ti, Zr, and Nb data, are also much higher in alluvium derived from the San Bernardino Mountains (Figure 11). Lake Cahuilla sediments do not appear to have a significant component of alluvial sands derived from the Chocolate Mountains and the Cargo Muchacho Mountains. Although the mineralogical data discussed above show that some component of alluvium derived from these mountains could be present in Lake Cahuilla sediments, data for Sr, Rb, Ti, and Zr show that Lake Cahuilla sediments all have much lower concentrations of these elements (Figure 12). Concentrations of La and Ce, and to a lesser degree, Y and Nb, show some overlap between alluvium

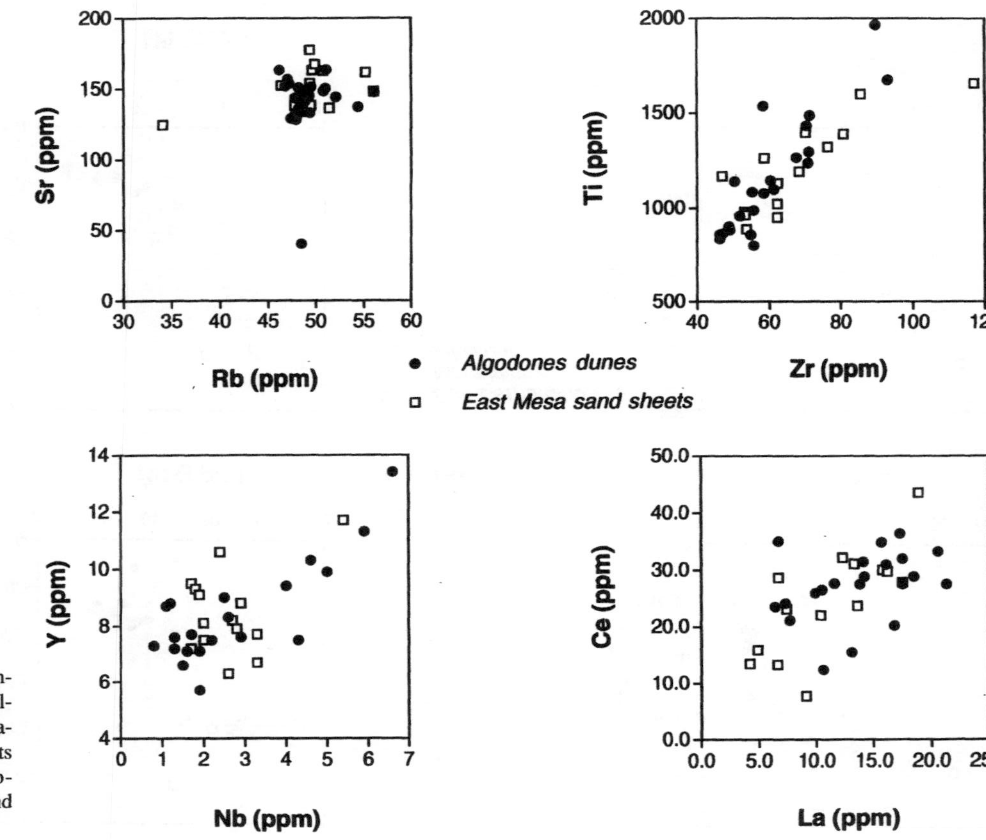

Figure 10. Comparison of trace element concentrations in sediments from the Algodones dunes and East Mesa.

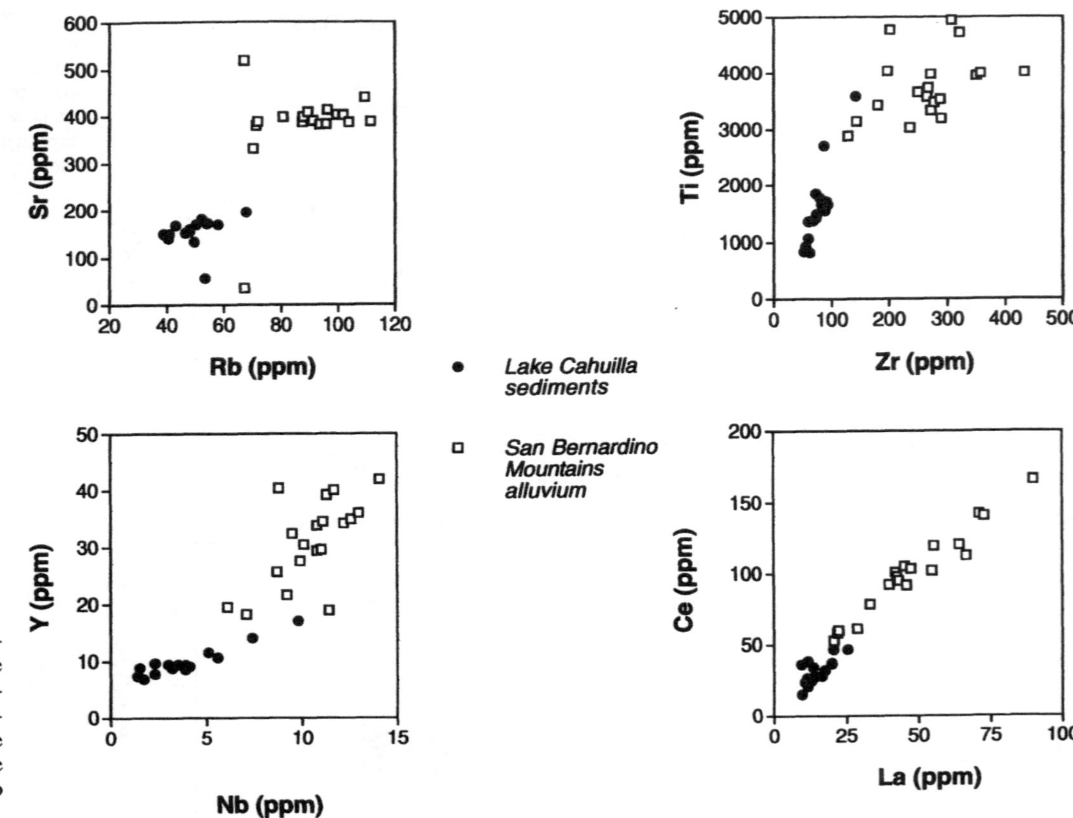

Figure 11. Comparison of trace element concentrations in sediments from Lake Cahuilla and the San Bernardino Mountains.

from the Chocolate and Cargo Muchacho Mountains and Lake Cahuilla sediments. Thus, both mineralogical and geochemical data suggest that alluvium derived from the Chocolate and Cargo Muchacho Mountains could be a minor, but not significant, component of Lake Cahuilla sediments. Overall, the best geochemical match is between Lake Cahuilla sediments and Colorado River sediments, which is consistent with the mineralogical data. Concentrations of Sr, Rb, Ti, Zr, Y, Nb, Ce, and La are not significantly different between Lake Cahuilla and Colorado River sediments, although it appears that the Ti/Zr value is somewhat higher in Lake Cahuilla sediments (Figure 13). Based on these data, and the geochemical comparisons between Lake Cahuilla sediments and alluvium derived from the basin-bounding mountain ranges, we conclude that the lake sediments were derived primarily from Colorado River sediments. Hence, in further geochemical discussions, we combine the Lake Cahuilla and Colorado River data and consider them as one sediment group.

The trace element concentrations of the combined East Mesa and Algodones dunes sediment group can be compared to the three other sediment groups of San Bernardino Mountains alluvium, Chocolate and Cargo Muchacho Mountains alluvium, and Lake Cahuilla/Colorado River sediments. The East Mesa and Algodones dunes all have significantly lower concentrations of all trace elements when compared to San Bernardino Mountains alluvium, consistent with the mineralogical data (Figure 14). There is a slight amount of overlap in the concentrations of Nb, Ce, and La between East Mesa and Algodones dunes and alluvium derived from the Chocolate and Cargo Muchacho Mountains, but concentrations of Rb, Sr, Ti, Zr, and Y are all significantly higher in the alluvial sediments (Figure 15). However, concentrations of all trace elements in East Mesa and Algodones dunes are not significantly different from those found in Lake Cahuilla/Colorado River sediments (Figure 16). Combined with the mineralogical data, these observations lead us to conclude that both East Mesa and Algodones dunes are derived from Lake Cahuilla beach sediments, and the beach sediments in turn are ultimately derived from Colorado River sediments. Geochemical and mineralogical data *permit* the interpretation that alluvial fan sediments from the Chocolate and Cargo Muchacho Mountains could have made minor contributions to both Lake Cahuilla sediments and East Mesa and Algodones dunes, but it is unlikely that the bulk of the aeolian sediments are derived from this source.

It can be inferred from Van de Kamp's (1973, p. 841) studies that the Algodones dunes were derived from Colorado River sediments in the southern part of the basin, based on his mineralogical data showing the similarities of these two sediment groups, and his observed differences with Lake Cahuilla sediments. In such an interpretation, sediments of the Algodones dunes would have to have been derived from the Colorado River floodplain during the summer, because this is the only time of the year that the monsoonal flow of air would generate winds from the south, based on the sand rose data presented earlier (Figures 2 and 3a). Although sand rose data show that drift potentials are relatively high for the Yuma area during this period, the geomorphic data

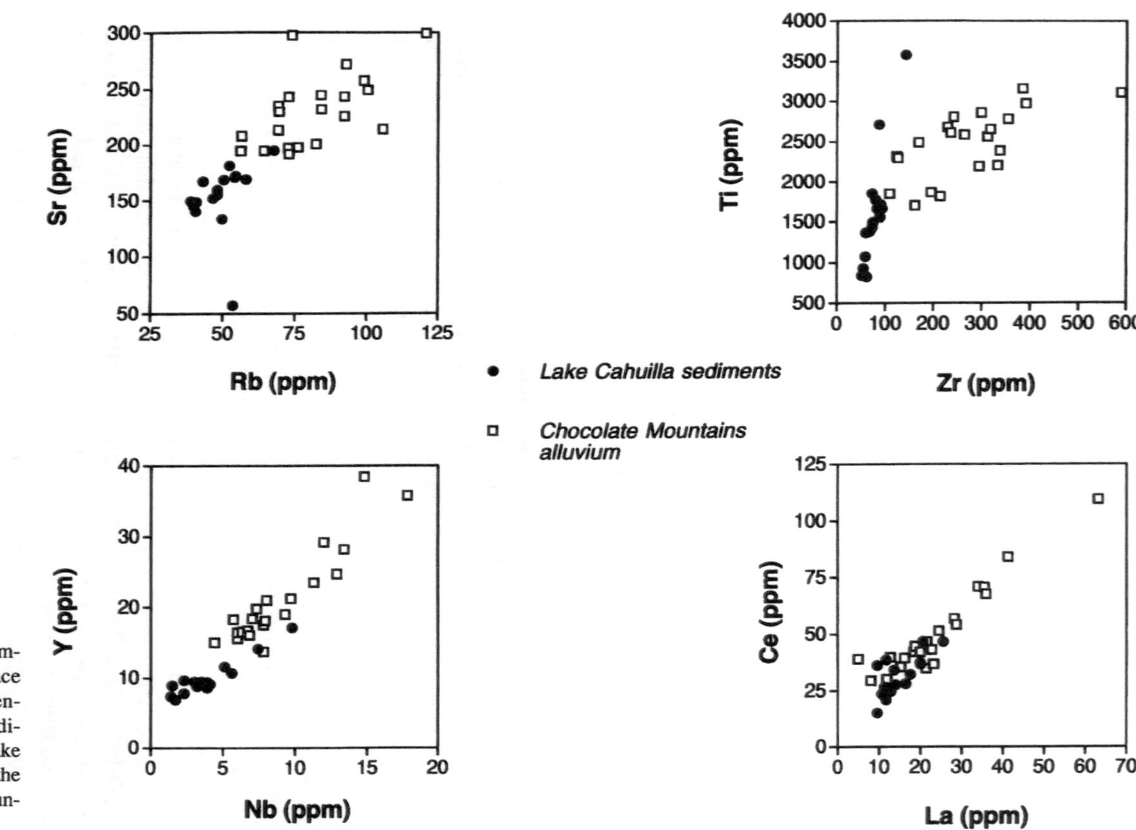

Figure 12. Comparison of trace element concentrations in sediments from Lake Cahuilla and the Chocolate Mountains.

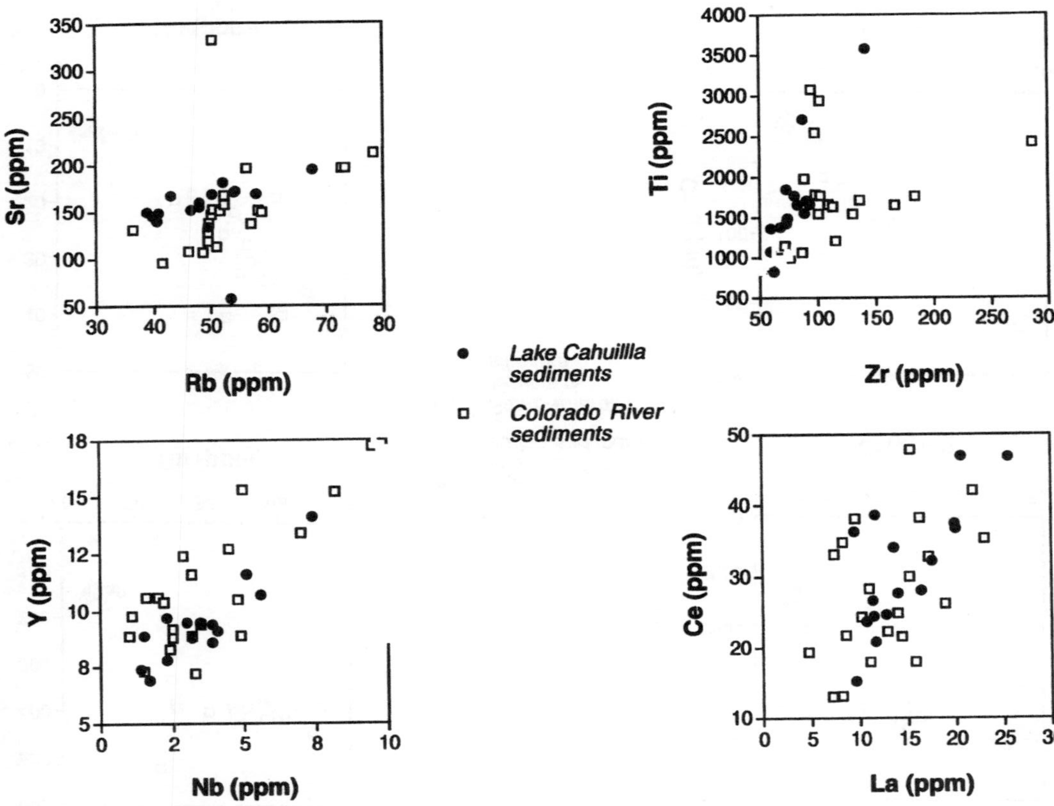

Figure 13. Comparison of trace element concentrations in sediments from Lake Cahuilla and the Colorado River.

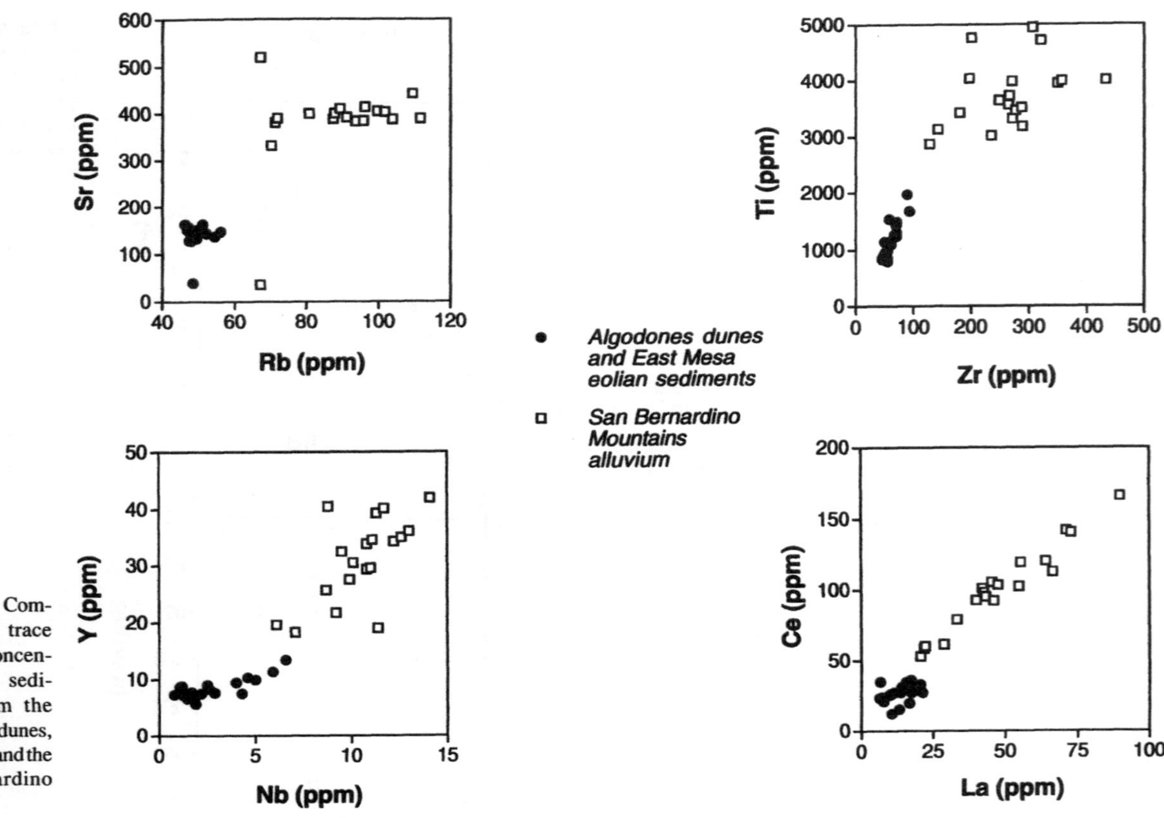

Figure 14. Comparison of trace element concentrations in sediments from the Algodones dunes, East Mesa, and the San Bernardino Mountains.

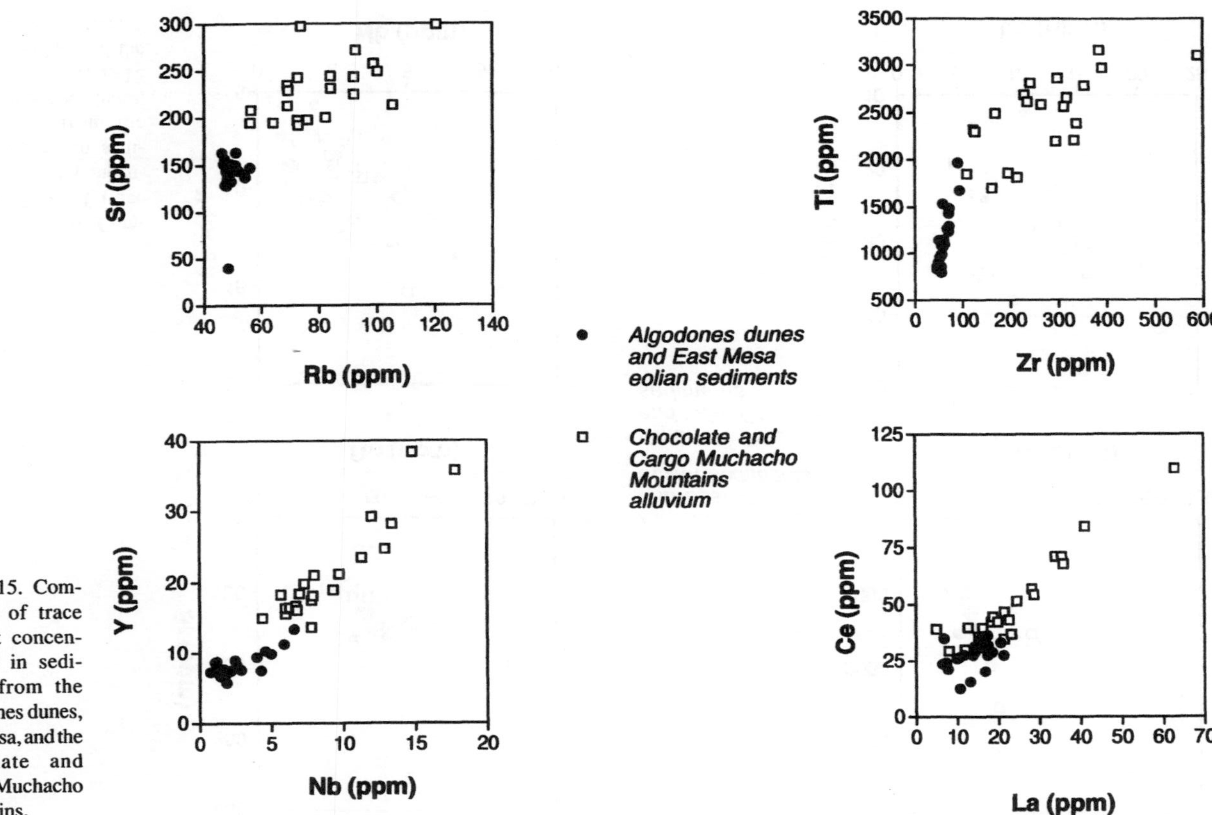

Figure 15. Comparison of trace element concentrations in sediments from the Algodones dunes, East Mesa, and the Chocolate and Cargo Muchacho Mountains.

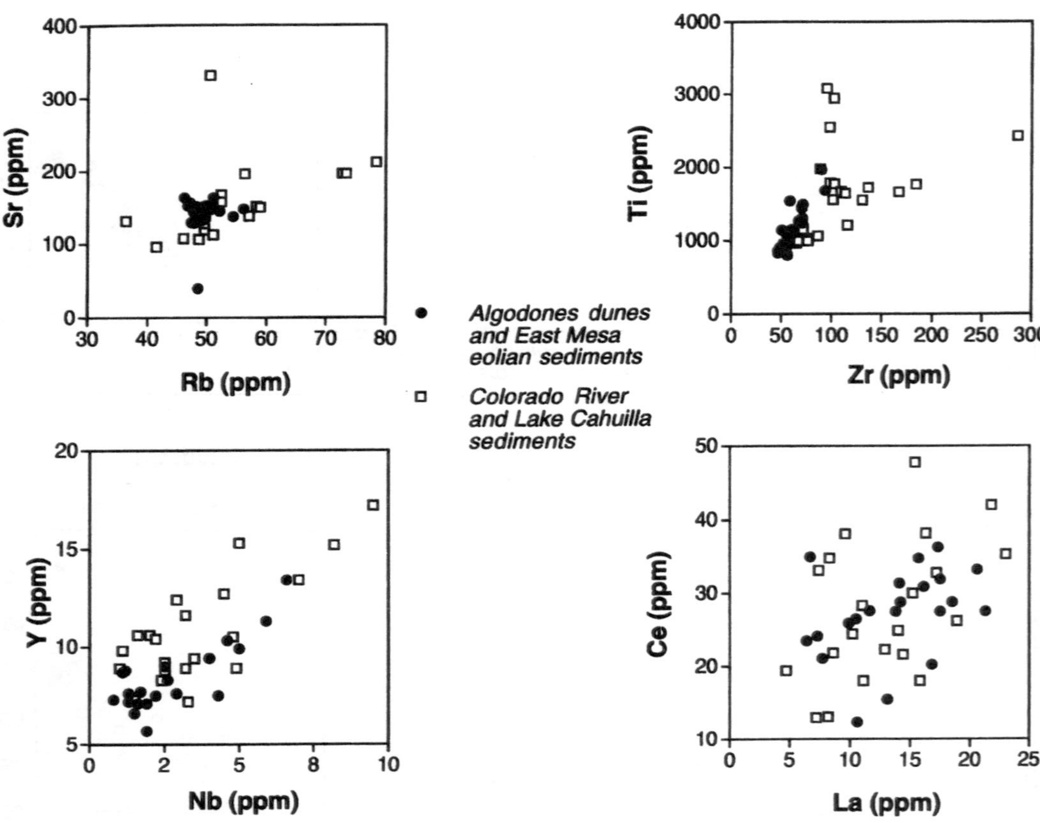

Figure 16. Comparison of trace element concentrations in sediments from the Algodones dunes, East Mesa, Lake Cahuilla, and the Colorado River.

presented by Havholm and Kocurek (1988) and in Figures 4 and 5 show that the dominant dune-forming winds have been from the northwest, both on East Mesa and in the Algodones dune field. It does not seem likely that all of the aeolian sand in the two areas was derived directly from the Colorado River during southerly, monsoonal flows of air in the summer and then simply reworked by northwesterly winds during the fall, winter, and spring. Given the geochemical and mineralogical data which show that Lake Cahuilla sediments are not significantly different from the aeolian sediments, and given that the late Holocene Lake Cahuilla shorelines are to the west and northwest of the dunes, we conclude that the immediate source of the aeolian sediments was the Lake Cahuilla shoreline deposits, and that northwesterly winds during the fall, winter, and spring were responsible for the movement of aeolian sand from the shoreline to the dune fields.

SUMMARY AND CONCLUSIONS

Climatic data indicate that winds in the Colorado Desert of southeastern California have low-to-intermediate drift potential for aeolian sand, but because of low precipitation and high evapotranspiration, a dune mobility index predicts that most aeolian sand in the area should be active, and this is supported by field observations. The major exceptions to predicted degree of activity are the aeolian deposits of East Mesa, which may have been recently stabilized by vegetation owing to human-caused rises in the ground water. This hypothesis requires testing by examination of older aerial photographs, however.

Orientations of the Algodones dunes indicate that the dominant winds of formation are from the northwest, which is consistent with resultant drift directions at Indio, to the northwest of the dune field, and Yuma, to the southeast of the dune field. Orientations of dunes and sand streaks on East Mesa also indicate winds of formation from the northwest, but with a more westerly component, which is consistent with resultant drift directions at El Centro, immediately to the west of East Mesa.

The various potential source sediments for the Algodones dunes vary in their degree of mineralogical maturity, and thus can be easily differentiated from one another. Colorado River sediments are high in quartz and relatively low in plagioclase because of the incorporation of grains from rocks that have undergone many cycles of weathering, erosion, and deposition. In contrast, sands from the local mountain ranges that surround the Salton Trough are mostly first-cycle sediments, and have high feldspar and heavy mineral contents, and relatively low quartz contents. Mineralogical data indicate that the Algodones dunes, East Mesa aeolian sediments, Lake Cahuilla sediments, and Colorado River sediments are indistinguishable from one another. In contrast, alluvium derived from the San Bernardino Mountains is significantly lower in quartz and higher in plagioclase than aeolian sediments from the Algodones dunes and East Mesa. Alluvium from the Chocolate Mountains is closer in composition to the aeolian sediments than is alluvium from the San Bernardino Mountains, but does not have the similarities to them that sediments from the Colorado River and Lake Cahuilla have.

Trace element concentrations (Rb, Sr, Ti, Zr, Y, Nb, La, and Ce) in the sediments studied support the conclusions drawn from the mineralogical data. With the exception of La and Ce concentrations from the Chocolate Mountains, concentrations of all trace elements from the Chocolate and San Bernardino Mountains are significantly higher than those in sediments from the Algodones and East Mesa dunes, and Lake Cahuilla and Colorado River sediments. In contrast, Algodones and East Mesa dunes, Lake Cahuilla sediments, and Colorado River sediments have trace element concentrations that are not significantly different from one another, indicating similar feldspar contents (based on Rb, Sr, La, Ce, and Y concentrations) and similar heavy mineral contents (based on Ti, Zr, and Nb concentrations). We conclude from the combined mineralogical and geochemical data that Lake Cahuilla sediments were the immediate source of sand for both the East Mesa and Algodones dunes. Lake Cahuilla sediments, in turn, were ultimately derived from the Colorado River, when its course shifted at times in the past and emptied into the Salton Trough rather than the Gulf of California.

Many additional questions about the origin of the Algodones dunes are raised by our results. It is not known if shoreline sand was deflated from beaches to the dunes while Lake Cahuilla was present or after the lake waters receded. In addition, it is not clear whether aeolian sediment inputs to the dune fields were associated with each lake stand; stratigraphic and radiocarbon data in Waters (1983) and Rockwell and Gurrola (1993) indicate that there were several late Holocene high stands. Given that the overall volume of sediment in the Algodones dunes is high (McCoy et al. 1967) and the present surface area of the late Holocene Lake Cahuilla shorelines is limited, we hypothesize that the present volume of sediment in the Algodones dunes is probably the result of multiple episodes of aeolian deflation from beach sources. If this is the case, there may be stratigraphic evidence of multiple episodes of aeolian sediment accumulation in the form of buried soils between dune or sand sheet deposits in the Algodones or East Mesa, similar to what Lancaster (1993, and this volume) has described from the Gran Desierto sand sea to the south. Future studies of the history of the Algodones dunes and East Mesa could explore the possibility of a stratigraphic record of multiple episodes of aeolian sand accumulation.

Our results indicate that trace element geochemistry is a promising technique for provenance studies of aeolian sand. The best results should be obtained in areas where hypothesized source sediments have very different geochemical signatures such as silicic vs. mafic sediments or mineralogically mature vs. mineralogically immature sediments. Many other dune fields and sand sheets whose source sediments are uncertain, controversial, or of multiple origins could be studied using trace element geochemistry, such as those in the Mojave Desert (Smith 1984), the Colorado Plateau (Price et al. 1988, Wells et al. 1990), the Great Plains (Ahlbrandt and Fryberger 1980, Muhs 1985), and Alaska (Lea and Waythomas 1990).

ACKNOWLEDGMENTS

This study was supported by the Global Change and Climate History Program of the U.S. Geological Survey. We thank Milan Pavich for encouraging us to investigate the Algodones dunes. Benn, Joyce, and Joanna Silverman provided logistical support and hospitality while field work was being conducted. We thank Gene Whitney, Carl Hedge, Chris Schenk, Brian Marshall, and Hugh Millard for helpful discussions. Nick Lancaster, Gary Kocurek, and Robert Zielinski read an earlier version of the paper and made helpful comments for its improvement.

REFERENCES

Ahlbrandt, T. S., and Fryberger, S. G. (1980) *Aeolian deposits in the Nebraska Sand Hills.* U.S. Geological Survey Professional Paper 1120-A, 24 p.

Beheiry, S. A. (1967) Sand forms in the Coachella Valley, southern California. *Annals of the Association of American Geographers*, v. 57, p. 25-48.

Blake, W. P. (1857) Geological Report. In *Reports of Explorations and Surveys to Ascertain the Most Practicable and Economical Route for a Railroad from the Mississippi River to the Pacific Ocean.* 33rd Congress, 2nd Session, Senate Executive Document No. 78.

Blount, G., and Lancaster, N. (1990) Development of the Gran Desierto sand sea, northwestern Mexico. *Geology*, v. 18, p. 724-728.

Brazel, A. J., and Nickling, W. G. (1986) The relationship of weather types to dust storm generation in Arizona (1965-1980). *Journal of Climatology*, v. 6, p. 255-275.

Breed, C. S. (1977) Terrestrial analogs of the Hellespontus dunes, Mars. *Icarus*, v. 30, p. 326-340.

Breed, C. S., Fryberger, S. G., Andrews, S., McCauley, C., Lennartz, F., Gebel, D., and Horstman, K. (1979) *Regional studies of sand seas using Landsat (ERTS) imagery.* U.S. Geological Survey Professional Paper 1052K, p. 305-397.

Brooks, E. R., and Coles, D. G. (1980) Use of immobile trace elements to determine original tectonic setting of eruption of metabasalts, northern Sierra Nevada, California. *Geological Society of America Bulletin*, v. 91, p. 665-671.

Brown, J. S. (1923) *The Salton Sea Region, California.* U.S. Geological Survey Water-Supply Paper 497.

Bull, W. B. (1991) *Geomorphic Responses to Climatic Change.* Oxford University Press, New York.

Cutts, J. A., and Smith, R.S.U. (1973) Aeolian deposits and dunes on Mars. *Journal of Geophysical Research*, v. 78, p. 4139-4154.

Dibblee, T. W., Jr. (1964) *Geologic map of the San Gorgonio Mountain quadrangle San Bernardino and Riverside Counties, California.* U.S. Geological Survey Miscellaneous Geologic Investigations Map I-431.

Dohrenwend, J. C., and Smith, R.S.U. (1991) Quaternary geology and tectonics of the Salton Trough. In R.B. Morrison (ed.) *Quaternary Nonglacial Geology: Conterminous U.S.* Geological Society of America, Boulder, Colorado, The Geology of North America, v. K-2, p. 334-337.

Ezcurra, E., and Rodrigues, V. (1986) Rainfall patterns in the Gran Desierto, Sonora, Mexico. *Journal of Arid Environments*, v. 10, p. 13-28.

Fryberger, S. G., and Ahlbrandt, T. S. (1979) Mechanisms for the formation of aeolian sand seas: *Zeitschrift fur Geomorphologie*, v. 23, p. 440-460.

Fryberger, S. G., and Dean, G. (1979) *Dune forms and wind regime*: U.S. Geological Survey Professional Paper 1052-F, p. 137-169.

Govindaraju, K. (1989) 1989 compilation of working values and sample description for 272 geostandards. *Geostandards Newsletter*, v. 13, p. 1-113.

Havholm, K. G., and Kocurek, G. (1988) A preliminary study of the dynamics of a modern draa, Algodones, southeastern California, USA. *Sedimentology*, v. 35, p. 649-669.

Haxel, G. B., Budahn, J. R., Fries, T. L., King, B.-S.W., White, L. D., and Aruscavage, P. J. (1987) Geochemistry of the Orocopia schist, southeastern California: summary. *Arizona Geological Digest*, v. 18, p. 49-64.

Jennings, C. W. (1967) *Geologic Map of California Salton Sea sheet*. California Division of Mines and Geology, Sacramento, California.

Jorgensen, D. W. (1992) Use of soils to differentiate dune age and to document spatial variation in aeolian activity, northeast Colorado, U.S.A. *Journal of Arid Environments*, v. 23, p. 19-34.

Kocurek, G., and Nielson, J. (1986) Conditions favorable for the formation of warm-climate aeolian sand sheets. *Sedimentology*, v. 33, p. 795-816.

Lancaster, N. (1988) Development of linear dunes in the southwestern Kalahari, southern Africa. *Journal of Arid Environments*, v. 14, p. 233-244.

Lancaster, N. (1990) Palaeoclimatic evidence from sand seas. *Palaeogeography, Palaeoclimatology, Palaeoecology*, v. 76, p. 279-290.

Lancaster, N. (1993) Origins and sedimentary features of supersurfaces in the northwestern Gran Desierto sand sea. In K. Pye and N. Lancaster (eds.) *Aeolian Sediments Ancient and Modern*. International Association of Sedimentologists Special Publication No. 16, p. 71-83.

Lancaster, N., Greeley, R., and Christensen, P. R. (1987) Dunes of the Gran Desierto sand-sea, Sonora, Mexico. *Earth Surface Processes and Landforms*, v. 12, p. 277-288.

Lea, P. D., and Waythomas, C. F. (1990) Late-Pleistocene aeolian sand sheets in Alaska. *Quaternary Research*, v. 34, p. 269-281.

Lee, J. A. (1991) The role of desert shrub size and spacing on wind profile parameters. *Physical Geography*, v. 12, p. 72-89.

Loeltz, O. J., Irelan, B., Robison, J. H., and Olmsted, F. H. (1975) *Geohydrologic reconnaissance of the Imperial Valley, California*. U.S. Geological Survey Professional Paper 486-K.

Long, J. T., and Sharp, R. P. (1964) Barchan-dune movement in Imperial Valley, California. *Geological Society of America Bulletin*, v. 75, p. 149-156.

MacKinnon, D. J., Elder, D. F., Helm, P. J., Tuesink, M. F., and Nist, C. A. (1990) A method of evaluating effects of antecedent precipitation on duststorms and its application to Yuma, Arizona, 1981-1988. *Climatic Change*, v. 17, p. 331-360.

Madole, R. F. (1992) Recurring deposition of aeolian sand during the late Quaternary in northeastern Colorado. *Geological Society of America Abstracts with Programs*, v. 24, p. A314.

McCoy, F. W., Jr., Nokleberg, W. J., and Norris, R. M. (1967) Speculations on the origin of the Algodones dunes, California. Geological Society of America Bulletin, v. 78, p. 1039-1044.

McKee, E. D. (1979) *Introduction to a Study of Global Sand Seas*. U.S. Geological Survey Professional Paper 1052-A, p. 1-19.

Merriam, R. (1969) Source of sand dunes of southeastern California and northwestern Sonora, Mexico. *Geological Society of America Bulletin*, v. 80, p. 531-534.

Merriam, R., and Bandy, O. L. (1965) Source of upper Cenozoic sediments in Colorado delta region. *Journal of Sedimentary Petrology*, v. 35, p. 911-916.

Morton, D. M., Cox, B. F., and Matti, J. C. (1980) *Geologic map of the San Gorgonio Wilderness, San Bernardino County, California*. U.S. Geological Survey Miscellaneous Field Studies Map MF-1161A.

Muffler, L.J.P., and Doe, B. R. (1968) Composition and mean age of detritus of the Colorado River delta in the Salton Trough, southeastern California. *Journal of Sedimentary Petrology*, v. 38, p. 384-399.

Muhs, D. R. (1985) Age and paleoclimatic significance of Holocene sand dunes in northeastern

Colorado. *Annals of the Association of American Geographers*, v. 75, p. 566-582.
Muhs, D. R., and Maat, P. B. (1993) The potential response of aeolian sands to greenhouse warming and precipitation reduction on the Great Plains of the U.S.A. *Journal of Arid Environments*, v. 25, p. 351-361.
Muhs, D. R., Bush, C. A., Stewart, K. C., Rowland, T. R., and Crittenden, R. C. (1990) Geochemical evidence of Saharan dust parent material for soils developed on Quaternary limestones of Caribbean and western Atlantic islands. *Quaternary Research*, v. 33, p. 157-177.
Nielson, J., and Kocurek, G. (1986) Climbing zibars of the Algodones. *Sedimentary Geology*, v. 48, p. 1-15.
Norris, R. M. (1966) Barchan dunes of Imperial Valley, California. *Journal of Geology*, v. 74, p. 292-306.
Norris, R. M., and Norris, K. S. (1961) Algodones dunes of southeastern California. *Geological Society of America Bulletin*, v. 72, p. 605-620.
Olmsted, F. H., Loeltz, O. J., and Irelan, B. (1973) *Geohydrology of the Yuma area, Arizona and California*. U.S. Geological Survey Professional Paper 486-H.
Pearce, J. A., and Cann, J. R. (1973) Tectonic setting of basic volcanic rocks determined using trace element analyses. *Earth and Planetary Science Letters*, v. 19, p. 290-300.
Price, A. B., Nettleton, W. D., Bowman, G. A., and Clay, V. L. (1988) Selected properties, distribution, source, and age of aeolian deposits and soils of southwest Colorado. *Soil Science Society of America Journal*, v. 52, p. 450-455.
Pye, K., and Tsoar, H. (1990) *Aeolian Sand and Sand Dunes*: Unwin Hyman, London.
Rempel, P. J. (1936) The crescentic dunes of the Salton Sea and their relation to the vegetation. *Ecology*, v. 17, p. 347-358.
Rockwell, T., and Gurrola, L. (1993) Dating of earthquakes and determination of the slip rate for the Superstition Mountain strand of the San Jacinto fault, southern California. U.S. Geological Survey Technical Report 14-08-0001-G1669.
Rubin, D. M., and Hunter, R. E. (1987) Bedform alignment in directionally varying flows. *Science*, v. 237, p. 276-278.
Schmidt, R. H., Jr. (1989) The arid zones of Mexico: climatic extremes and conceptualization of the Sonoran Desert. *Journal of Arid Environments*, v. 16, p. 241-256.
Sharp, R. P. (1979) Intradune flats of the Algodones chain, Imperial Valley, California. *Geological Society of America Bulletin*, v. 90, p. 908-916.
Shelton, J. S., Papson, R. P., and Womer, M. (1978) Aerial guide to geological features of southern California. In R. Greeley, M. B. Womer, R. P. Papson, and P. D. Spudis (eds.) *Aeolian Features of Southern California: A Comparative Planetary Geology Guidebook*. Office of Planetary Geology, National Aeronautics and Space Administration, U.S. Government Printing Office, Washington, D.C., p. 216-249.
Smith, R.S.U. (1982) Sand dunes in the North American deserts. In G.L. Bender (ed.) *Reference Handbook on the Deserts of North America*. Greenwood Press, Westport, Connecticut, p. 481-524.
Smith, R.S.U. (1984) Eolian geomorphology of the Devils Playground, Kelso dunes and Silurian Valley, California. In J.C. Dohrenwend (ed.) *Surficial Geology of the Eastern Mojave Desert, California*. Geological Society of America, 1984 Annual Meeting, Field Trip 14 Guidebook, p. 162-174.
Strand, R. G. (1962) *Geologic Map of California San Diego-El Centro sheet*. California Division of Mines and Geology, Sacramento, California.
Sweet, M. L., Nielson, J., Havholm, K., and Farrelley, J. (1988) Algodones dune field of southeastern California: case history of a migrating modern dune field. *Sedimentology*, v. 35, p. 939-952.
Sweet, M. L., Kocurek, G., and Havholm, K. (1991) A field guide to the Algodones dunes of southeastern California. In M. J. Walawender and B. B. Hanan (eds.) *Geological Excursions in Southern California and Mexico* (Guidebook for 1991 annual meeting of the Geological Society of America). Department of Geological Sciences, San Diego State University, San Diego, California, p. 171-185.
Sykes, G. (1937) *The Colorado Delta*. American Geographical Society Special Publication No. 19.

Thomas, D.S.G., and Shaw, P. A. (1991) "Relict" desert dune systems: interpretations and problems. *Journal of Arid Environments*, v. 20, p. 1-14.

Thornthwaite, C. W., and Mather, J. R. (1957). Instructions and tables for computing potential evapotranspiration and the water balance. *Publications in Climatology* (Laboratory of Climatology, Centerton, NJ), v. 10, p. 185-311.

Van de Kamp, P. C. (1973) Holocene continental sedimentation in the Salton Basin, California: a reconnaissance. *Geological Society of America Bulletin*, v. 84, p. 827-848.

Waters, M. R. (1983) Late Holocene lacustrine chronology and archaeology of ancient Lake Cahuilla, California. *Quaternary Research*, v. 19, p. 373-387.

Wells, S. G., McFadden, L. D., and Schultz, J. D. (1990) Eolian landscape evolution and soil formation in the Chaco dune field, southern Colorado Plateau, New Mexico. *Geomorphology*, v. 3, p. 517-546.

Zimmerman, R. P. (1981) *Soil Survey of Imperial County, California, Imperial Valley Area.* U.S. Government Printing Office, Washington, D.C.

4 POTENTIAL TRANSPORT OF WINDBLOWN SAND: INFLUENCE OF SURFACE ROUGHNESS AND ASSESSMENT WITH RADAR DATA

Ronald Greeley,[1] Dan G. Blumberg,[1] Anthony R. Dobrovolskis,[2] Lisa R. Gaddis,[3] James D. Iversen,[4] Nicholas Lancaster,[5] Keld R. Rasmussen,[6] R. Stephen Saunders,[7] Stephen D. Wall,[7] and Bruce R. White[8]

[1]Department of Geology, Arizona State University; [2]NASA-Ames Research Center; [3]Astrogeology Branch, U.S. Geological Survey; [4]Department of Aerospace Engineering, Iowa State University; [5]Quaternary Sciences Center, Desert Research Institute; [6]Institute of Geology, Aarhus University; [7]Jet Propulsion Laboratory; [8]Department of Mechanical Engineering, University of California at Davis

ABSTRACT

The transport of windblown sand is controlled by many factors, including wind regime and sediment supply. Surface roughness at the sub-meter scale is also important because it influences both the threshold conditions for particle entrainment and the flux of sand once it is set into motion. In general, increases in surface roughness result in higher threshold speeds for particle movement and decreases in sand fluxes. Aerodynamic roughness (z_0) is the aeolian parameter related to surface roughness and is defined as the height above some mean level at which average wind speed is zero. Values of z_0 are derived from wind measurements through the boundary layer, but few z_0 values have been obtained over natural surfaces because of the expense and limitations of making such measurements. Rather, remote sensing using radar systems has the potential for addressing this problem. In this investigation, we derived z_0 values for a wide variety of surfaces in the southwestern United States and obtained radar data for these sites in P-band (wavelength = 68 cm), L-band (wavelength = 24 cm) and C-band (wavelength = 5.6 cm). We show that there are good correlations among z_0, the RMS height of the surface, and the radar backscatter coefficient, σ^0, with the best correlation for L-band HV polarized radar data. This study shows the potential for mapping large regions with radar in order to derive aerodynamic roughness values, which in turn can be used in predictive models of sand transport.

INTRODUCTION

Aeolian processes play an important part in modifying the Earth's surface in many areas. The degree of aeolian activity is governed by complex factors, including wind regime, particle supply, moisture, and vegetative cover. Although surface roughness on the order of only a few meters is recognized as an important parameter in aeolian processes, few studies have been made of this parameter. Moreover, determinations of surface roughness relevant for aeolian processes are very limited. In this study, we show that radar imaging systems

Desert Aeolian Processes. Edited by Vatche P. Tchakerian. Published in 1995 by Chapman & Hall, London.
ISBN 978-94-010-6519-1

can provide a rapid means for obtaining relevant roughness data for large areas and have the potential for use in predictive models of sand transport.

Intuitively, one would expect a rough surface, such as a lava flow, to inhibit aeolian processes and to act as a trap for saltating sand. On the other hand, widely spaced roughness elements, such as isolated boulders, could generate local turbulence and enhance aeolian activity. Thus, both the size and spacing of roughness elements are important considerations. The parameter that links ~sub-meter surface roughness to aeolian processes was described by Bagnold (1941) as the height above the surface at which the average effective wind speed is zero. This parameter, termed aerodynamic roughness (Greeley et al. 1991) and abbreviated z_0, was found to be ~1/30 of the grain size for a uniform sandy surface (Bagnold 1941), and on the order of 1/10 the average height of the surface relief for a rocky, heterogeneous surface (Greeley and Iversen 1987). These are only crude approximations and are inadequate to provide quantitative assessments for aeolian processes.

Aerodynamic roughness is determined from wind velocity profiles measured through the atmospheric boundary layer. Typically, such profiles are obtained from multiple anemometers placed on meteorological masts of a few to >50 m height. Values of z_0 derived from these measurements are valid only for the homogeneous surface upwind of the mast. Because of these limitations, relatively few field measurements of z_0 have been made. Yet, in order to assess the potential for aeolian processes, it is desirable to know z_0 as a function of terrain type and to be able to map z_0 values regionally. Preliminary studies suggest that radar imaging systems could be used to map z_0 (Greeley et al. 1988). The relevant parameter is the radar backscatter coefficient, σ^0, which is partly a function of sub-meter surface roughness. Consequently, one could expect a correlation between radar backscatter and aerodynamic roughness because both are functions of the same surface roughness, at least to first order. Such a correlation was demonstrated in initial studies (Greeley et al. 1988, 1991, Blumberg and Greeley 1993).

In this chapter, we assess the effect of aerodynamic roughness on both the ability of wind to set particles into motion (i.e., threshold) and the flux of windblown sand. We also give results for field sites (Figure 1) where measurements were made for aerodynamic roughness and microtopography, and compare the results with radar data (Figure 2). We then discuss the potential application of radar data to the study of aeolian processes.

EFFECT OF ROUGHNESS ON AEOLIAN PROCESSES

Effect on Threshold

Surface roughness influences the vertical structure of the wind profile close to the surface which, in turn, determines the value of the friction speed for a given wind speed above the boundary-layer. This is demonstrated as

$$u(z) = u_*/k \ln[(z-d_0)/z_0] \qquad (1)$$

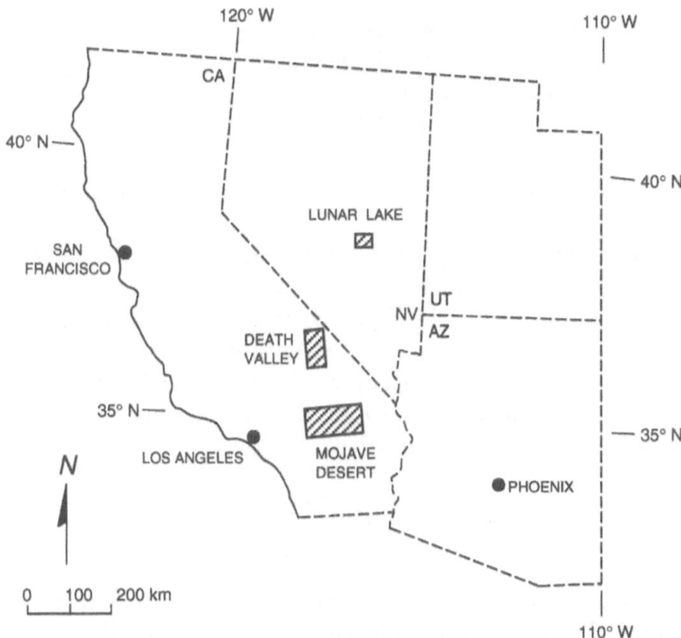

Figure 1. Location map of field sites.

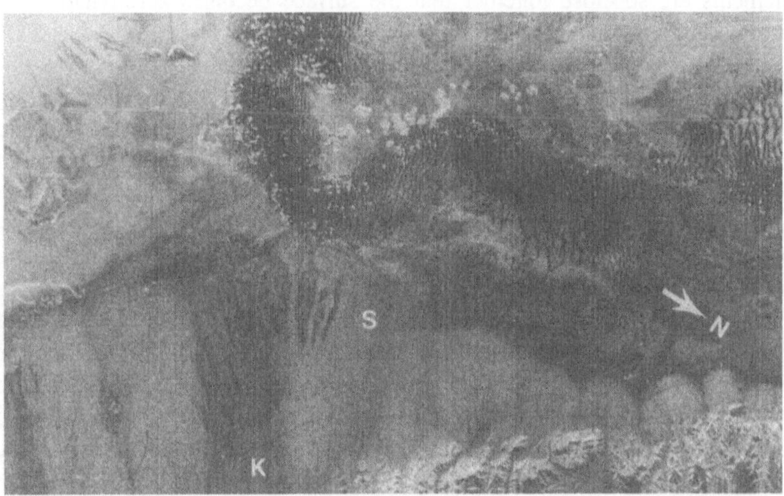

Figure 2. Radar image of the Stovepipe Wells area, Death Valley, California, showing the location of the Stovepipe Wells (S) and Kit Fox Fan (K) sites. The dark zone in the upper part of the image is the Stovepipe Wells dune field, with bright speckles being reflections from star dunes. Calibration corner reflectors are visible as bright spots along the road on the left side of the image. Data are from the total power of C-band, L-band, and P-band in HH polarization. Area shown is 7 km by 11 km (photograph IPF-891).

in which u is wind speed at height z above the surface, u_* is the friction velocity, k is the von Karman constant, d_0 is the zero plane displacement (a measure of the effective "base level" for rough surfaces), and z_0 is aerodynamic roughness length. From Equation 1 it can be shown that for the same undisturbed wind speed (i.e., outside the boundary-layer) over two surfaces (designated 1 and 2), for $z_{01} > z_{02}$, it follows that u_{*1} also is greater than u_{*2}, or the surface drag coefficient, $C_{D1} > C_{D2}$.

An assessment of the drag on individual roughness elements was made by Marshall (1971). The ratio of intervening surface friction speed (u_{*g}) to overall friction speed (u_*) is shown in Figure 3 for arrays of hemispherical and cylindrical roughness elements as a function of an equivalent roughness density ($C_D\lambda$) (Iversen et al. 1991), in which λ is the windward silhouette or frontal area of the roughness elements per unit floor area. An empirical equation made to fit the data of Marshall (1971) is

$$\frac{u_{*g}}{u_*} = \frac{1 - 16.67(C_D\lambda)}{1 + 1.37\ [1 - e^{-47.8C_D\lambda}]} \tag{2}$$

The data were for "sparse" arrays of roughness elements, i.e. the values of λ are less than 0.1. The aerodynamic roughness height z_0 for uniform arrays of similar roughness elements is approximately equal to the element height H times λ for sparse arrays, i.e., $z_0 = H\lambda$ (Raupach et al. 1991). The roughness height reaches a maximum for $\lambda \simeq 0.12$ and then decreases as the number of roughness elements increases beyond that value; in effect, the roughness elements are so close together that the surface becomes aerodynamically smooth.

Because surface roughness influences the wind velocity profile, it influences threshold speeds for particle entrainment (Chepil and Woodruff 1963). Multiple roughness elements increase u_* for a given atmospheric pressure gradient. As their frequency increases, the roughness elements take up more of the average shear stress, and the shear stress decreases on the intervening surface area among the elements. As the turbulent mixing increases, the time-averaged mean wind speed decreases near the surface. Consequently, as the number of roughness elements increases if sands are located only on the intervening surface areas, the effective (average) value of saltation threshold friction speed u_{*t} increases and mass transport decreases. Data showing the effect of roughness on threshold (Figure 4) were compiled by Lyles and Allison (1975), Gillette and Stockton (1989), Iversen et al. (1991), and from our wind tunnel experiments. These results contradict Marshall (1971) in that the intervening shear stress reaches zero at the rather small value of $\lambda = 0.06$. Consequently, we derived an empirical equation of the same form to fit the data more closely (Figure 4):

$$\frac{u_{*t}}{u_{*t0}} = \frac{u_*}{u_{*g}} = \frac{1 + 1.8(1 - e^{-10C_D\lambda})}{1 - 3C_D\lambda} \tag{3}$$

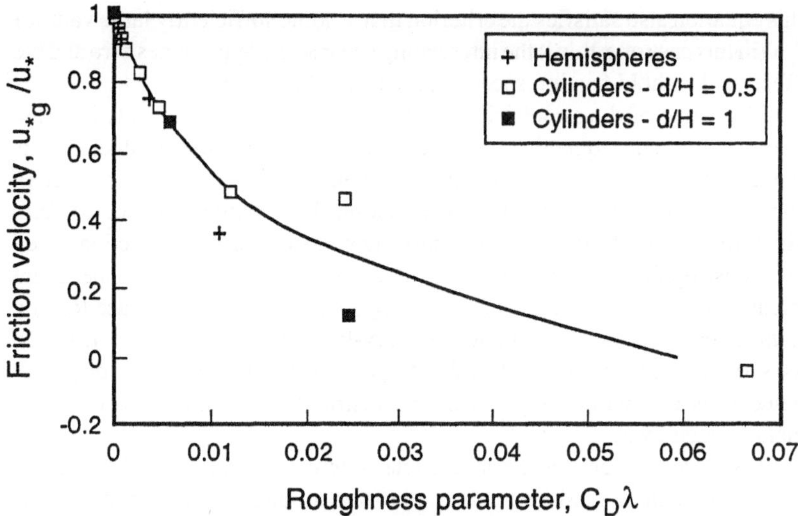

Figure 3. Data from Marshall (1971) for the ratio of intervening surface friction speed u_{*g} to overall friction speed u_* for uniform roughness element arrays as a function of drag coefficient, C_D, times λ, the windward frontal area of the roughness element per unit surface area of the floor (or surface). The solid line is Equation 2.

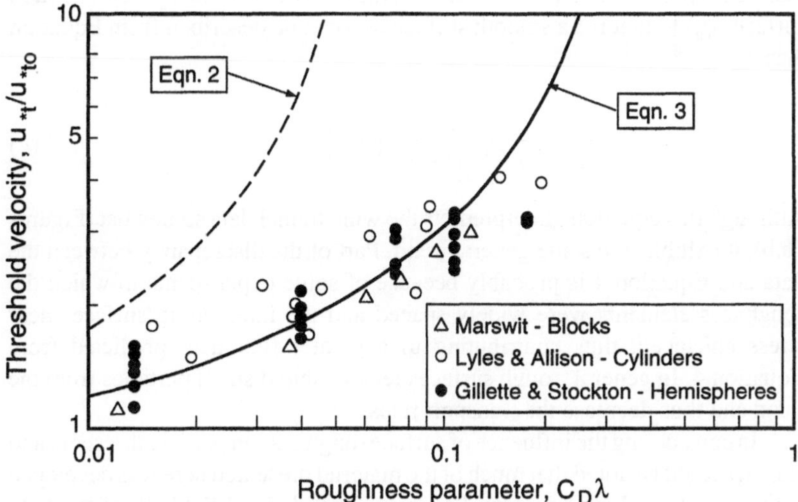

Figure 4. Ratio of threshold friction speed (u_{*t}/u_{*to}) in the presence of roughness elements as a function of element shape and array density (λ); data for rectangular blocks, circular cylinders (Gillette and Stockton 1989), and hemispheres (Lyles and Allison 1975). The solid line is Equation 3 and the inverse of Equation 2.

This equation also satisfies the criterion that at some sufficiently large value of $C_D\lambda$ (in this case one-third) the intervening shear stress approaches zero and the effective threshold friction speed approaches infinity.

The value of drag coefficient C_D for most roughness elements is of the order of one; consequently, in subsequent calculations the product $C_D\lambda$ is replaced by λ (the ratio of roughness element frontal area to surface area). Threshold curves corresponding to Equation 3 are plotted in Figures 5a,b, which illustrate the effect of roughness elements on threshold friction speed and geostrophic wind speed. The values of geostrophic wind speed are calculated from a plot of geostrophic drag coefficient versus surface Rossby number for a neutral atmospheric boundary-layer (Csanady's figure 4, 1972). As shown in Figure 5, the threshold speed in the presence of roughness elements is a function of the size, and distribution of roughness elements characterized by λ.

It should also be noted that a single roughness element, such as a rock placed on a bed of sand, can enhance erosion around the element and cause deposition over the rest of the bed close to the element (Iversen et al. 1990, 1991).

Effect on Flux

In addition to threshold, surface roughness influences the flux of sand transport. In order to assess the influence of roughness on flux, wind tunnel experiments were run under Earth (1 bar) and equivalent Mars (11 mbar) atmospheric pressures for test-bed floors of different roughnesses. Figures 6a,b show results for sand particles with an average diameter 145 μm. Bagnold (1941) suggested that flux is proportional to u_*^3. From this, the ratio of mass transport for a rough surface (q_R) to that for a smooth surface (q_S) can be described from Equation 3 as

$$\frac{q_R}{q_S} = \left(\frac{1 - 3(C_D\lambda)}{1 + 1.8(1 - e^{-10C_D\lambda})} \right) \qquad (4)$$

Although this equation underpredicts the wind tunnel data somewhat (Figures 6a,b), it exhibits the same general trend. Part of the discrepancy between the data and Equation 4 is probably because of some experiments in which the roughness elements were widely spaced and the intervening surface shear stress enhanced, thus contributing to a greater flux than predicted from Equation 4. In general, rough surfaces tend to shield small particles from the wind and thus decrease the transport rates.

In considering the influence of surface roughness on particle threshold and flux, it should be noted that much of the material presented here was developed from wind tunnel experiments. Rigorous, well-defined field experiments to assess flux as a function of surface roughness have not been conducted as yet, but are planned for the future.

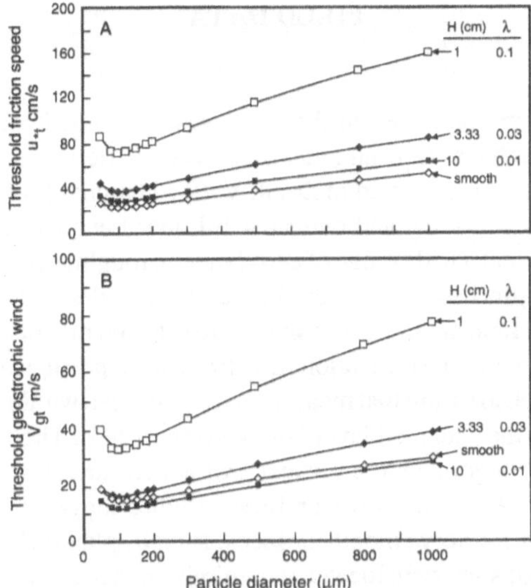

Figure 5. (a) Threshold friction speed, u_{*t}, for smooth (sand-only) surface and rough surfaces versus particle diameter, D_p. The threshold speed decreases as element height increases and the array density (λ) decreases. The smooth (sand-only) case is derived from Equation 1 of Iversen et al. (1976). (b) Threshold geostrophic wind speed, V_{gt}; same conditions as Figure 5a. Note that the smooth (sand-only) case has a lower threshold friction speed but a higher threshold geostrophic wind speed than for the H = 10 cm case. The rougher the surface the lower the necessary geostrophic wind speed to create a given surface shear stress.

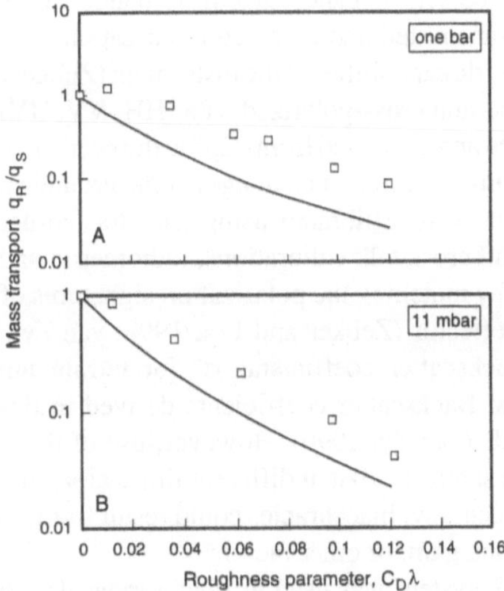

Figure 6. Mass transport ratio (rough, q_R, to smooth, q_S, surfaces) as a function of roughness parameter, $C_D\lambda$. The solid line is prediction from Equation 4. Data are from wind tunnel experiments conducted at Earth atmospheric pressure (1 bar) and Martian atmospheric pressure (11 mbar).

FIELD DATA

Field Sites

Field studies were conducted in the southwestern United States to acquire aerodynamic roughness (z_0) and radar data over a variety of terrains. In 1990 and 1991, 10 sites were studied in Death Valley National Monument and the Mojave Desert, California, and two sites at Lunar Lake, Nevada (Figure 1). These sites represent a wide range of aerodynamic roughnesses and are subject to active aeolian processes. Appendix 1 gives detailed descriptions of the sites.

Within each field area, specific sites were selected in order to obtain valid wind data. In general, the derivation of z_0 from wind profiles requires a large homogenous fetch to assure that measurements are taken within an equilibrium boundary-layer (the "adapted layer" of Wieringa 1993). The fetch should be about 100 times longer than the vertical profile measured for wind velocities (e.g., Peterson et al. 1978). However, recent studies suggest that this distance may be too short for very smooth surfaces, such as playas (Wieringa 1993). Consequently, our sites were located to minimize or avoid the effects of large-scale (>tens of meters) topography on the wind and to maximize homogeneous fetch for the direction of prevailing sand-transporting winds.

Radar Data

Radar data were acquired by the NASA/JPL Airborne Synthetic Aperture Radar (AIRSAR) polarimeter system (Evans et al. 1986). The data have a spatial resolution of ~12 m and were obtained at three wavelengths simultaneously (P-band: 68 cm; L-band: 24 cm; C-band: 5.6 cm). Although any combination of transmitted and received polarization states may be derived using the polarimetric capabilities of the instrument (Zebker et al. 1987, van Zyl et al. 1987), direct- and cross-polarized data (HH, VV, HV) were used in this analysis. Incidence angles ranged from ~25° in the near-range (top of an image) to 55° in the far-range (bottom of an image) of the ground swath over the sites.

AIRSAR data were calibrated using trihedral corner reflectors. After standard phase and cross-talk calibrations, radiometric and absolute calibrations were made by matching the polarization signatures of the observed and an ideal corner reflector (Zebker and Lou 1990, van Zyl 1990), permitting calculation of backscatter coefficients σ^0 for terrain units shown on the calibrated images. Backscatter coefficients derived in this manner have an accuracy of ±3 dB (van Zyl 1990). However, use of this technique for data acquired over the same area but at different times along different flight paths are of unknown accuracy. Inaccuracies could result from variations in aircraft altitude and antenna gain, or other factors.

The AIRSAR system was used to obtain radar data for the Pisgah and Amboy lava fields in June 1988 during the Mojave Field Experiment (Wall et al. 1988) and for Death Valley and Lunar Lake playa in September 1989 during the Geologic Remote Sensing Field Experiment (Arvidson and Evans 1990). Backscatter coefficients were calculated from calibrated AIRSAR images for these sites. For example, Figure 2 is an AIRSAR image of the Stovepipe Wells site in Death Valley, taken 14 September 1989. In addition to the study sites, a dune field is visible as the dark region in the upper half of the image; bright speckles are star dunes that return an intense signal to the radar antenna because of the dune geometry. Also visible are the bright radar returns from the corner reflectors used to calibrate the data.

Wind Profiles

Near-surface wind profiles were obtained using field-portable anemometer masts 9.8 m high. Cup anemometers were placed at 0.75, 1.25, 2.07, 3.44, 5.72, and 9.5 m above the surface to give a logarithmic spacing (Figure 7). Pairs of temperature sensors were placed in shielded and naturally ventilated housings at heights of 1.3, 5.8, and 9.6 m. Wind directions were measured at heights of 9.7 and 1.5 m. Data were segmented and sampled every 20 seconds, averaged for 20-minute segments, and recorded for periods ranging from four to six weeks. A data subset was extracted for wind speeds >4 m/sec (i.e., well above the anemometer threshold) and sorted by direction and atmospheric-stability characteristics (based on the bulk Richardson number, Rb). Wind data were reduced in order to extract aerodynamic roughness (ζ_0) values (Table 1) and friction velocities (u_*) using the method described in Appendix 2.

Surface Characterization

The general geology of each site was described and the characteristics of particles exposed on the surface noted. The proportions of clay and silt, sand, gravel, cobbles, and boulders in a 1-m-square area were estimated at four or five representative locations at each site, all within a 24- by 24-m-square centered on the anemometer mast. The proportion of exposed bedrock was also estimated. The amount and type of vegetation was noted, although sites were selected to be relatively vegetation-free.

Submeter-scale topography was measured using a profiling template and a laser-photo device. The template consisted of a 2-m-long horizontal bar through which 200 1-m-long by 0.5-cm-diameter vertical aluminum rods protruded, giving a horizontal resolution of 1 cm (Wall et al. 1991). After the bar was leveled, the vertical rods slid into contact with the surface and their positions recorded photographically. The height of the bar above the ground was measured at each end, so that adjacent template measurements could be linked. In this manner, two profiles, each 24 m long, were obtained parallel and perpendicular to the prevailing wind directions. In the laboratory, photographs of the rod tops were digitized so that standard deviations in the changes in elevation (RMS heights) could be derived. Although at some sites the rods protruded into soft ground, we consider the template accuracy for vertical measurements to be ±5 mm. Data were not used at sites where the rods penetrated >5 mm.

The laser-profiling device was used to determine sub-centimeter scale roughness. This device consisted of a triangle with 1.2-m-long sides, raised

Table 1
Study sites, mean z_o values, and RMS heights

Site+	Direction from which wind originated	z_o (m)	RMS height (m) Geometric NW-SE	NE-SW	Mean	Laser
Playa						
Lunar Lake 1(1)	Southwest	0.000077	-	-	-	0.0006*
Lunar Lake 2 (2)	Southwest	0.000126	-	-	-	0.0012*
Alluvium						
Kit Fox T2 (3)	Northeast	0.000850	-	-	-	-
Kit FoxTl (4)	Northeast	0.001696	0.0615	0.0329	0.0450	0.0097*
Stovepipe Wells, main (S)	Northwest	0.000760	0.0715	0.01853	0.0364	0.0031
StovepipeWells, Nl (6)	Northwest	0.002080	0.0532	0.0131	0.0264	0.0033
Stovepipe Wells, Sl (7)	Northwest	0.002880	0.0353	0.0292	0.0321	0.0041
Stovepipe Wells, S2 (8)	Northwest	0.003100	0.0455	0.024	0.0328	0.0028
Amboy 1(9)	Northwest	0.001730	-	-	-	-
Golden Canyon (10)	Northeast	0.001100	0.0772	0.0394	0.0552	0.0150*
Golden Canyon (11)	Southeast	0.002450	-	-	-	-
Pahoehoe lava						
Pisgah 4 (12)	Northwest	0.001315	0.0708	0.1910	0.1163	-
Pahoehoe lava with mantle						
Amboy 2 (13)	Northwest	0.003610	0.1380	0.0430	0.0770	-
Pisgah 3 (14)	Northwest	0.014830	0.0708	0.0547	0.0622	-
Pisgah 5 (15)	Northwest	0.073508	0.1670	0.0250	0.0646	-
Amboy 4 (16)	Northwest	0.028630	0.0260	0.0870	0.0476	-

+ Numbers in parentheses correspond to Figure 8.

* These laser profiles were averaged over three sides of an equilateral triangle, other laser profiles are geometric means of a transect parallel and normal to the prevailing wind direction.

above the ground ~50 cm. A small laser mounted vertically on a traversing stage was moved by an electric motor along the rails with a traverse time of ~45 seconds. A camera mounted on the vertex opposite the laser stage recorded the laser trace on the ground using a time exposure of ~45 seconds. An electronic flash illuminated a scale bar and the surface. The result was a photograph of the ground showing a red profile line from the laser trace. This procedure was repeated for each side of the triangle, and the images were then digitized to produce a topographic profile with a 2- to 3-mm horizontal and vertical resolution.

The digital terrain data produced from the template and laser device yielded statistics on the microtopography after the overall surface slope was

Figure 7. Typical micrometeorology system used to collect wind and atmospheric data. Towers were 9.8 m high, equipped with six anemometers (spaced logarithmically), three shielded temperature sensors, and two wind vanes. This tower was deployed at Stovepipe Wells, Death Valley (photograph 3395-D37).

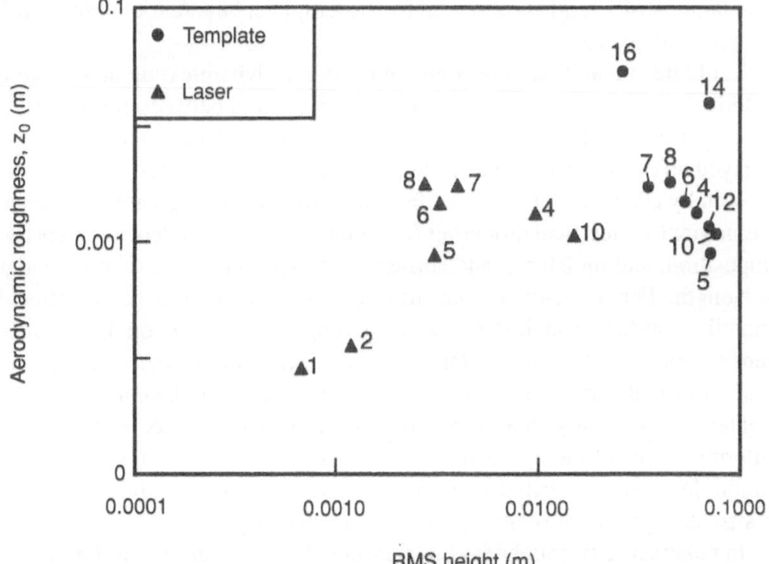

Figure 8. Correlation of aerodynamic roughness, z_0, with RMS height for field sites (numbers are keyed to Table 1).

removed. In general, the RMS heights of the laser data are about an order of magnitude less than those of the template data, suggesting that the laser measured sand and gravel particle roughness. In addition to RMS heights (Table 1), slope distributions, correlation lengths, and power spectra were determined. Thus far in our study, only RMS heights have been used for comparisons with aerodynamic roughness and radar backscatter coefficients.

ANALYSIS AND DISCUSSION

In this section we compare field values of aerodynamic roughness (z_0), surface roughness as described by the RMS height, and radar backscatter coefficients (σ^0) for various wavelengths and polarizations. We then discuss the results in terms of aeolian processes.

Preliminary studies by Lancaster et al. (1991) showed a modest correlation between z_0 and RMS height. Figure 8 provides additional field measurements from the laser profiling device and the template and shows the same general trend. Except for four sites (see Table 1), the RMS height data were taken upwind from the meteorological masts, parallel to the wind direction. Consequently, the RMS heights represent the surface roughnesses responsible for generating the z_0 values. Note that six sites are represented by both laser device and template data, which produce different values of RMS height. These are alluvial fan sites which have gravels and cobbles on the surface and, in some cases, are dissected by small gullies. These roughness elements are responsible for the primary development of the aerodynamic roughness and the generation of the larger RMS heights measured by the template, but are too large for the laser profiling device.

Calibrated radar data were compared with aerodynamic roughness (Table 2). Radar backscatter represents a complex interaction between properties of the surface and radar system parameters. These surface effects are well displayed on radar images of the field sites in this study (Figures 9 and 10). As reviewed by Ulaby et al. (1982), surface properties with the greatest influence are topography, electrical properties (including the effects of density, porosity, composition, and moisture), and surface roughness on the scale of the radar wavelength. For a relatively flat, dry surface with limited compositional variability, surface roughness has the strongest influence on backscatter intensity. Radar system parameters include wavelength, incidence angle, and polarization of the transmitted signal. For cross-polarized radar data, backscatter intensity is largely determined by the amount of surface reflection or scattering at the wavelength scale. Cross-polarized radar data show sensitivity to multiple or volume scattering from a surface, and thus give an indication of the size and spacing of scatterers at a given wavelength.

In radar surveys, roughness is a measure of the irregularity of the surface compared with the radar wavelength. Although vertical relief (typically characterized by RMS height or RMS slope) only approximates complex natural

Table 2
Correlation of z_0 and σ^0 (Blumberg and Greeley 1993)

Band	Correlation Coefficient R	Significance (f)*
P-HH	0.80	0.0003
P-HV	0.74	0.0014
P-VV	0.78	0.0007
L-HH	0.80	0.0003
L-HV	0.81	0.0003
L-VV	0.78	0.0007
C-HH	0.79	0.0005
C-HV	0.80	0.0004
C-VV	0.74	0.0016

*The f-test significance is an indication of the probability that the regression results are random; hence, a low f-test significance means a low probability that the results of the regression are random (f-test significance values range from Ø to 1). Thus, in all of the above cases the f-test shows that the regression is significant and is not random.

geometries, it is commonly used to assess surface roughness (e.g., Schaber et al. 1980, Gaddis et al. 1990, Gaddis 1992, Arvidson et al. 1993). Figure 11 shows RMS heights for our sites versus σ^0 values for C, L, and P bands in HH, VV, and HV polarizations. There is a clear correlation for all radar configurations, consistent with results of Evans et al. (1986), Wang et al. (1986), and van Zyl et al. (1988). Our best correlation corresponds to L-band HV (correlation coefficient = 0.81).

Figure 12 shows ζ_0 versus σ^0 values for different radar wavelengths and polarizations. Where multiple incidence angles are available, data between 40° to 45° range were selected, based on previous analyses (Greeley et al. 1988). The general increase of aerodynamic roughness with increasing backscatter is observed in all cases. Good correlations are observed for all three wavelengths, with the highest for L-band HV polarization (correlation coefficient R = 0.81). The lowest correlation (R = 0.74) is for C-band VV and P-band HV. We attribute the relatively high correlation for L-band to the sensitivity of this wavelength to surface roughness ~24 cm. The longer wavelength P-band data could have penetrated into the subsurface a few tens of centimeters and been reflected by subsurface rocks. If so, there would be a contribution to the backscatter coefficient from subsurface "roughness" that would be "invisible" to the wind and lead to poor correlation. Because of the small range of correlation coefficients between the direct and cross-polarized backscatter at each wavelength, evaluations of polarization as a parameter were not made.

Equations 3 and 4 are preliminary steps toward the formulation of expressions that can use field data to estimate u_{*t} and q_1. Unfortunately, the

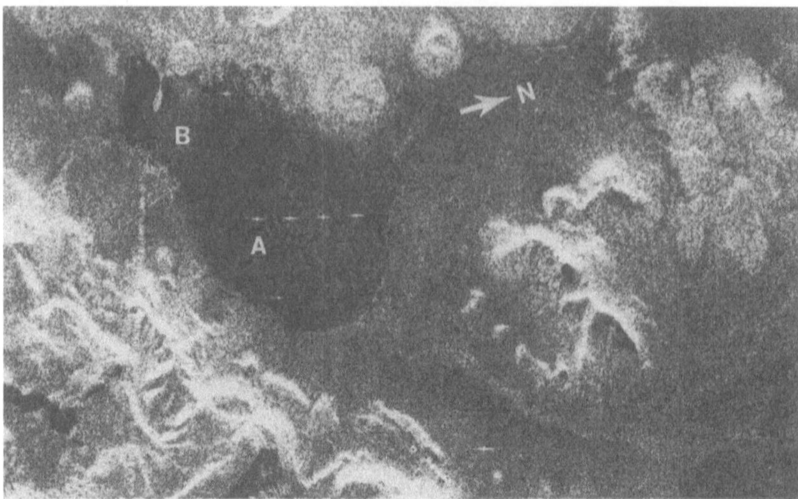

Figure 9. AIRSAR image acquired in C band VV of the Lunar Lake playa, Nevada, showing the smooth playa surface (A), the cobble-covered playa (B), and the bright signatures of the corner reflectors (horizontal trace of bright spots). The eastern third of the playa was wet when the data were acquired and accounts for the slightly brighter return in comparison to the rest of the playa surface. Area shown is 7 km by 11 km (photograph IPF-870).

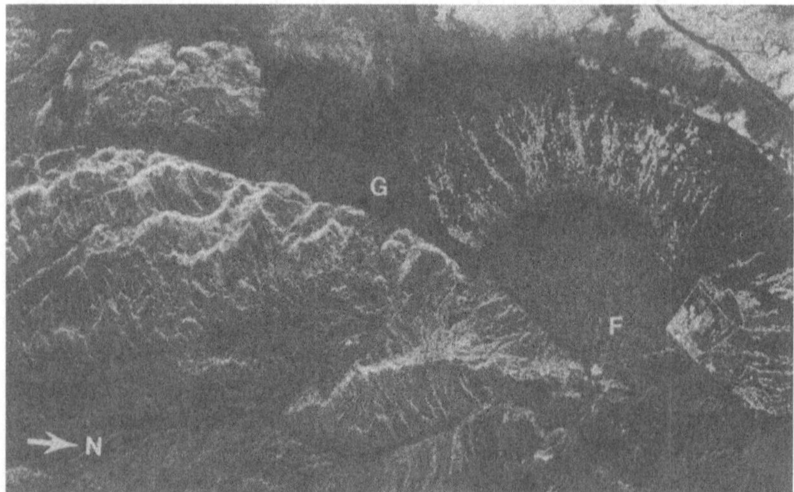

Figure 10. AIRSAR image acquired in L-band HV of the Golden Canyon site (G). The fan associated with Furnace Creek Wash is designated "F"; the dark band at the distal end (west) of the fan corresponds to sand deposits; the bright radar returns beyond the fan correspond to vegetation. The moderate-to-bright radar returns at the top (west) of the image are from salt deposits; the dark stripe in the upper right corner is Salt Creek. Furnace Creek Ranch is to the right of "F." Area shown is 7 by 11 km (photograph IPF-861).

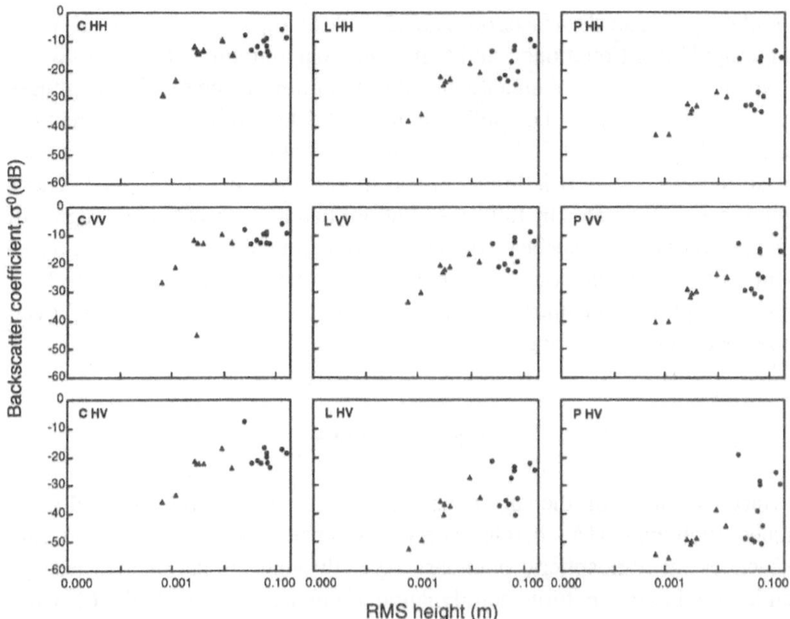

Figure 11. RMS heights versus radar backscatter coefficient (σ^0) for different wave lengths and polarizations.

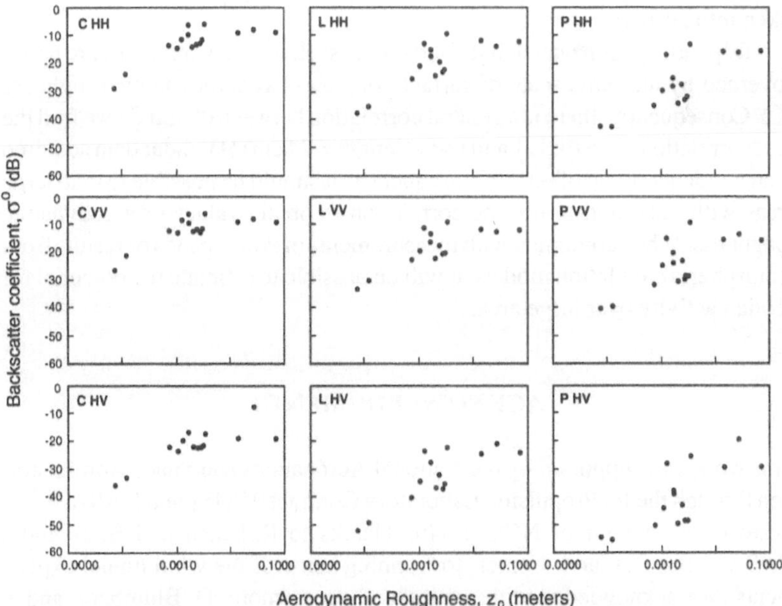

Figure 12. Aerodynamic roughness (z_0) versus radar backscatter coefficient (σ^0) for different wave lengths and polarizations, for all sites listed in Table 1.

complexity of natural surfaces does not provide a simple estimation of $C_D\lambda$ that can be applied to Equations 3 and 4. It is our goal to improve Equations 3 and 4 (which were developed under controlled conditions using regular roughness arrays) so that they will be applicable to field data. Until then, we direct the reader to Figures 4 and 5, which demonstrate that for progressively rougher surfaces, sand transport is hindered and more sediments are trapped. These figures are corroborated by field observations that lava flows tend to trap sand, whereas intermediate surfaces such as smooth alluvial fans with desert pavement will accumulate some sand but will then be blown clean during high winds, and playas and other smooth surfaces will be transport "runways," even under relatively weak winds.

CONCLUSIONS

Surface roughness on the sub-meter scale can be described by several techniques, including RMS height. Surface roughness on the sub-meter scale influences the atmospheric boundary layer through the parameter, aerodynamic roughness. In turn, aerodynamic roughness affects both the wind threshold for sand entrainment in saltation and the flux of material after it is set into motion. In general, "rough" surfaces tend to increase threshold and to decrease flux. However, isolated roughness elements, such as a boulder, can lower threshold locally and enhance flux by creating a turbulent zone in its wake. Thus, both the size and the areal density of roughness elements must be taken into account.

Depending upon wavelength, radar backscatter coefficients (σ^0) are partly governed by the same scale of surface roughness as aerodynamic roughness (ζ_0). Consequently, there is a general correlation between σ^0 and ζ_0; we find the best correlation to be for L-band (wavelength = 24 cm) HV radar data acquired at an incidence angle of ~35°. Consequently, it should be possible to map large areas with radar and to use the correlation to obtain values of aerodynamic roughness. When combined with measurements of wind speeds or results from atmospheric circulation models, it will be possible to estimate the potential for aeolian activity over large areas.

ACKNOWLEDGMENTS

This work was supported by the National Aeronautics and Space Administration through the Jet Propulsion Laboratory Contract 958451 and NASA-Ames Research Center Grant NCC 2-346. Thanks to R. Leach and B. Boundy, NASA-Ames Research Center, for running some of the wind tunnel experiments. We acknowledge with gratitude G. Beardmore, D. Blumberg, and J. Lancaster for helping with the field work. We also thank S. Blixt, M. Geringer, and B. Erwin for word processing, D. Ball for photographic support, and S.

Selkirk for drafting. Jim Zimbelman and an anonymous reviewer provided very useful suggestions for improving the manuscript.

APPENDIX 1: FIELD SITE DESCRIPTIONS

Field sites in this study were selected in the southwestern United States (Figure 1) to provide a wide variety of surface roughness where aeolian processes occur.

Death Valley

Sites in Death Valley are on alluvial fans of different ages and surface roughnesses, and on a silt-clay playa at the south end of the valley (Figure 13). The surface at the playa site, Confidence Mill, is composed of a soft to locally hard, clay-rich crust. Although generally flat, the site also includes south-north trending rills that are 5- to 10-cm-deep. The Stovepipe Wells site is on the distal part of an alluvial fan and consists of a flat to gently undulating sand and gravel surface. Small washes, 10- to 20-cm-deep and 1- to 2-m-wide, cross the surface, which at the time of study was unvegetated. Active sand transport was evidenced by 10- to 20-cm-high sand drifts and wind ripples in the washes. Some rocks on the surface show evidence of wind abrasion. Kit Fox Fan site is on the active mid- to proximal-part of the same fan as the Stovepipe Wells site. The surface consists of gravel to cobble-sized rocks in a sandy matrix, and ephemeral plants spaced >1 m apart. Multiple small channels 5 to 10 cm deep and 1 to 2 m wide cross the site from east to west. During field studies, active sand transport was observed at the site. The Golden Canyon site is on an active alluvial fan and former mudflow (Figure 14). The fan has a well-developed bar and channel topography with 20 to 50 cm relief. Channels are partly filled with sand and silt. Bars are composed of gravel and small boulders of various lithologies set in a sandy silt matrix. The site is free of vegetation.

Pisgah Lava Field

The Pisgah lava field is about 65 km east of Barstow, California (Figure 15), and covers ~83 km^2. It includes a 60-m-high cinder cone and basaltic lava flows with pahoehoe, transitional, and aa surface textures (Figure 16). Lava was erupted in three phases (Wise 1966): Phase I (pahoehoe), Phase II (aa and pahoehoe), and Phase III (pahoehoe). Five sites were selected to provide a variety of roughness values. Assigned site numbers are from previous studies but site 2 was not used in the current investigation. Study site 1 is on Phase II aa lava flows and features raised ridges and platforms separated by depressions; the maximum relief is 8 m. The platforms are covered with gravel-sized lava fragments and scoria that in places form a desert pavement surface. The depressions are generally filled with larger, more massive blocks of lava which resulted from fracturing of cooled parts of the flow during emplacement. Although rare, some silt and sand occurs in the matrix of the desert pavement. Vegetation includes isolated patches of grasses. Site 3 is on the western part of

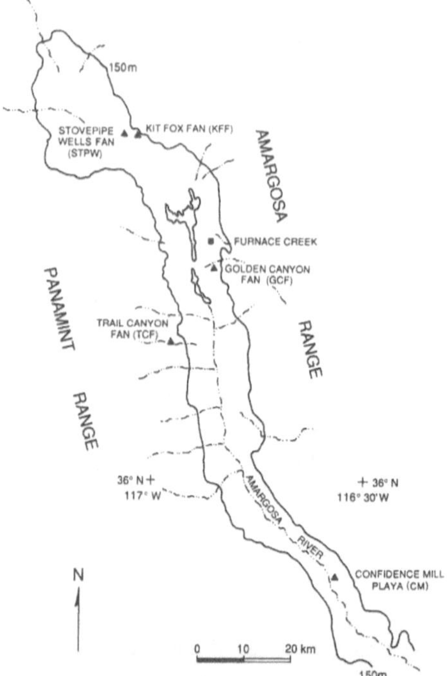

Figure 13. Sketch map of Death Valley, California, showing field sites.

Figure 14. Golden Canyon site on an alluvial fan, Death Valley, California, viewed toward the west. Rocks in foreground are 10 to 30 cm across (photograph 3372-D20A).

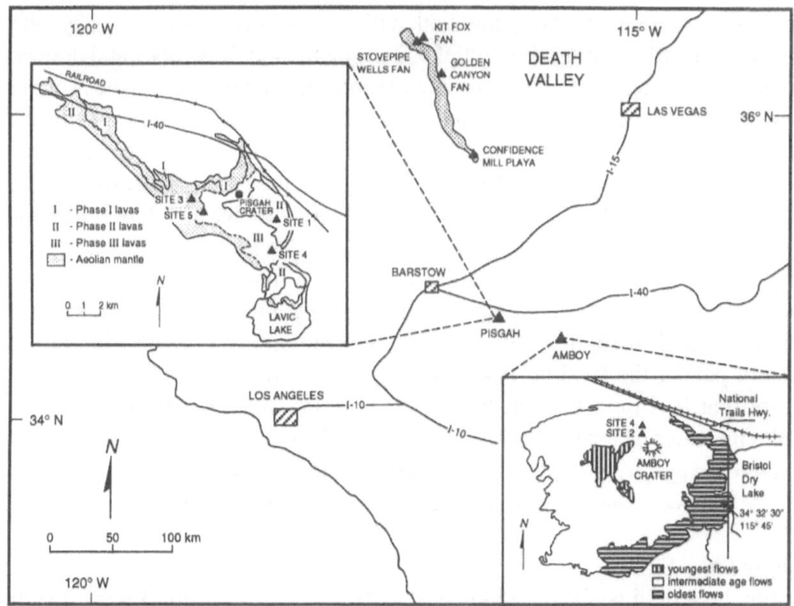

Figure 15. Sketch map of Pisgah and Amboy lava fields, California, showing field sites; also shown is Death Valley.

Figure 16. Pisgah lava field, California. Site 5 viewed toward the northeast. Small bushes in the foreground are about 30 cm high (photograph 3461-D11).

Phase III lavas and is partly mantled with windblown sediments. The site consists of gently sloping, low-relief (~2 m) pahoehoe flow lobes and domes, separated by large gravel-covered silt and sand deposits that cover ~60% of the surface. About 10% of the surface includes sand patches with little or no gravel. During field studies, vegetation was well developed, with bursage, creosote bush, and grasses growing in the sand. Site 4, near the eastern edge of the lava field, is also on Phase III pahoehoe flows. The flow in this area has 3- to 5-m-high pahoehoe lobes and tumuli 3 to 5 m high. The tumuli are separated by rubble-covered sand and silt patches that form a mantle as thick as 10 cm. Sparse vegetation consists of bursage and creosote bushes which had silt and sand deposited in shallow depressions and crevices. Study site 5 is on the western part of the field on Phase III pahoehoe flows. It is similar to site 3, except that it is on a thicker deposit of sand and silt. The surface includes moderate-relief (~3 m) pahoehoe flow lobes and tumuli separated by gravel-covered silt and sand deposits that cover ~75% of the surface. Vegetation is well developed, consisting of bursage, creosote bush, and grasses.

Amboy Lava Field

Amboy lava field is about 125 km east of Barstow, California (Figure 15). It covers 70 km^2 and includes a prominent cinder cone and flows of vesicular pahoehoe lava that were erupted on a playa surface. The flows form a hummocky topography (2 to 5 m relief) with depressions partly filled by windblown sand. Site numbers were assigned from a previous study (Greeley and Iversen 1987), but sites 1 and 3 were not used in this investigation. Site 2 is ~300 m north of the cinder cone in a shallow depression within a platform unit of pahoehoe lava that is partly mantled by sand. Site 4 is ~250 m northeast of site 2 on a pahoehoe platform that is surfaced with desert pavement and mantled by aeolian deposits. Vegetation was sparse at both sites at the time of field work.

Lunar Lake

Lunar Lake is a playa in the Quaternary-age Lunar Crater volcanic field, Nye County, Nevada, about 130 km east of Tonopah (Figure 17). The playa covers 14 km^2 and has a silty-clay surface. Patches of gravel occur on the margins of the playa, particularly in the southwestern part. Site 1 is near the center of the playa where the surface included small desiccation cracks at the time of the field work. Site 2 is near the southwestern edge of the playa where basaltic gravels partly mantle the playa surface (Figure 18).

APPENDIX 2: REDUCTION OF WIND DATA

Wind profiles over a horizontally homogeneous and thermally neutral surface can be described by

$$u(z) = u_*/k \ln [(z-d_0)/z_0], \tag{1}$$

in which u is the wind speed at height z, u_* is the friction velocity, k is the von

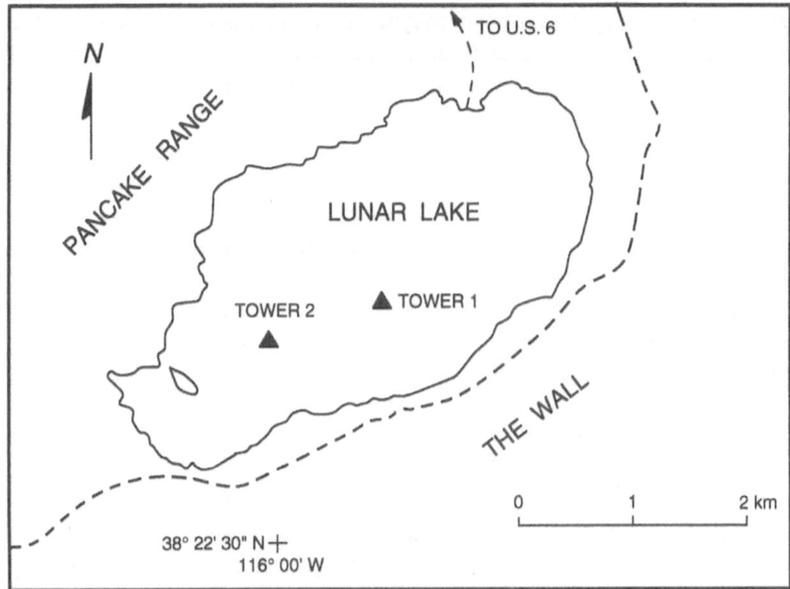

Figure 17. Sketch map of Lunar Lake, Nevada, showing field sites.

Figure 18. Lunar dry lake, Nevada, viewed toward the east. Foreground is the gravel-mantled part of the playa where site 2 was deployed. Site 1 was located on the bright part of the playa in the distance (photograph 3489-D4A).

Karman constant (0.4), z_0 is the aerodynamic roughness length, and d_0 is the zero plane displacement. Because most of our field sites lacked vegetation and had small or moderately sized roughness elements, d_0 was not included in the reduction.

Most of the wind data were obtained under conditions of strong solar radiation. Consequently, influences of buoyancy on the wind profile must be considered. Non-neutral conditions were determined by the bulk Richardson number (Rb). Rb is an approximation of the flux Richardson number (Ri), which is the ratio of buoyant turbulence force-to-energy gained by turbulence acting on the mean velocity gradient (Stull 1988). The stability correction can be expressed in terms of the buoyancy length scale $\zeta = z/L$, in which L is the Monin-Obukhov length (Monin and Yaglom 1971), so that the wind profile in non-neutral conditions was expressed as

$$u(z) = u_*/k\,[\ln (z/z_0) - \psi], \tag{2}$$

in which ψ is the stability function. The derived characteristics of the atmosphere are functions of its stability. Values of z_0 and u_* were determined from field measurements using the following technique. Measurements of velocity and temperature at various heights provided the definition of the bulk Richardson number (Rb), which is calculated as

$$Rb = gz^2/T\,((\partial T/\partial z + g/C_p)/(u^2)), \tag{3}$$

in which g is acceleration due to gravity, C_p is specific heat constant at constant pressure, and T is temperature.

Although temperatures were measured at only three heights, the linear relation between the natural logarithm of the height, ln(z), and the temperature, T(z), define the temperature at all heights. Thus, if we obtain

$$T(z) = A \ln(z) + B, \tag{4}$$

the slope of this curve is

$$\partial T/\partial z = A/z, \tag{5}$$

The iterative nature of this technique requires an initial estimate of z_0. Unfortunately, the stability function, ψ, and the stability length, L, are known accurately only for values of the flux Richardson number, not the bulk Richardson number. The conversion to the flux Richardson number follows the method of Golder (1972):

Step 1: Calculate Rb using Equation 3; the ratio $\partial T/\partial z$ was approximated from the log-linear profile.

Step 2: Estimate $z_0 = \exp(u)$, where u is the y-intercept of the plot of u versus ln z (z being the anemometer height).

Step 3: Calculate Ri (initial estimate is Ri = Rb); if Ri < 0 (air is unstable) then

$$x = (1\text{-}16\ \mathrm{Ri})^{1/4} \tag{6}$$

$$\Psi = 2\ln\left(\frac{1+x}{2}\right) + \ln\left(\frac{1+x^2}{2}\right) - 2\tan^{-1}x + \frac{\pi}{2} \tag{7}$$

$$\phi = (1\text{-}16\ \mathrm{Ri})^{-1/4}, \tag{8}$$

in which ϕ is the dimensionless wind shear in the surface layer (Stull 1988).

$$\mathrm{Ri} = \mathrm{Rb}\left[\frac{\ln\frac{z}{z_0} - \Psi\{\mathrm{Ri}\}}{\phi\{\mathrm{Ri}\}}\right]^2 \tag{9}$$

These equations were iterated until Ri converges; if Ri > 0 (air is stable), then

$$\phi = \sqrt{\frac{\mathrm{Rb}}{\mathrm{Ri}}}\left|\ln\left(\frac{z}{z_0}\right) - \Psi\{\mathrm{Ri}\}\right| \tag{10}$$

$$\psi = (-\beta\ \mathrm{Ri})/(1 - \beta\ \mathrm{Ri}), \tag{11}$$

in which the empirical coefficient $\beta = 7$.

$$\mathrm{Ri} = \mathrm{Rb}\left[\frac{\ln\frac{z}{z_0} - \Psi\{\mathrm{Ri}\}}{\phi\{\mathrm{Ri}\}}\right]^2 \tag{12}$$

These equations were iterated until Ri converges.

Step 4: Calculate the stability length, L; if Ri < 0 (air is unstable), then

$$L = z/\mathrm{Ri}. \tag{13}$$

If Ri > 0 (air is stable) then

$$L = z\ (1 - 5\ \mathrm{Ri})/\mathrm{Ri}. \tag{14}$$

Step 5: Calculate $\psi\ \{z/L\}$; if $L < 0$ (air is unstable), then

$$x = (1\text{-}16\ \mathrm{Ri})^{1/4} \tag{15}$$

$$\Psi = 2\ln\left(\frac{1+x}{2}\right) + \ln\left(\frac{1+x^2}{2}\right) - 2\tan^{-1}x + \frac{\pi}{2}. \tag{16}$$

if L > 0 (air is stable) then

$$\psi = (-5\ z)/(L). \tag{17}$$

Step 6: Calculate $u*$

$$u_* = \frac{ku}{\ln\frac{z}{z_0} - \Psi}, \tag{18}$$

in which u is wind speed at one specific height.

Step 7: Plot [ln z-ψ] versus u; least-squares fit where z_0 is the y-intercept value.

Step 8: Select a new value of z_0, which is the average between the initially estimated z_0 value (step 2) and the least-square determined z_0 value (step 7). Repeat steps 1 through 7 with the updated value of z_0 as the new initial estimate in step 2. Repeat this process until z_0 converges to acceptable limits or a specified uncertainty, i.e., (z_0 initial)/(z_0 final) less than a numerical value such as 0.0001.

REFERENCES

Arvidson, R. E., and Evans, D. L. (1990) Geologic remote sensing field experiment (abstract). *Lunar and Planetary Science XXI*, p. 26-27.

Arvidson, R. E., M. K. Shepard, E. A. Guiness, S. B. Petroy, J. J. Plaut, D. L. Evans, T. G. Farr, R. Greeley, N. Lancaster, L. R. Gaddis. (1993) Characterization of lava-flow degradation in the Pisgah and Cima volcanic fields, California, using Landsat Thematic Mapper and AIRSAR data. *Geological Society of America Bulletin*, v. 105, p. 175-188.

Bagnold, R. A. (1941) *The Physics of Blown Sand and Desert Dunes*. Methuen, London.

Blumberg, D. G., and Greeley, R. (1993) Field studies of aerodynamic roughness. *Journal of Arid Environments*, v. 25, p. 39-48.

Chepil, W. S., and Woodruff, N. P. (1963) The physics of wind erosion and its control. *Advanced Agronomy*, v. 15, p. 211-302.

Csanady, G. T. (1972) Geostrophic drag, heat and mass transfer coefficients for the diabatic Ekman layer. *Journal of Atmospheric Science*, v. 29, p. 488-496.

Evans, D. L., Farr, T. G., Ford, J. P., Thompson, T. W., and Werner, C. L. (1986) Multipolarization radar images for geologic mapping and vegetation discrimination. *IEEE Transactions of Geoscience Remote Sensing*, v. 24, p. 774-789.

Gaddis, L. R. (1992) Lava-flow characterization at Pisgah volcanic field, California, with multiparameter imaging radar. *Geological Society of America Bulletin*, v. 104, p. 695-703.

Gaddis, L. R., Mouginis-Mark, P. J., and Hayashi, J. N. (1990) Lava flow surface textures: SIR-B radar image texture, field observations, and terrain measurements. *Photogrammetric Engineering and Remote Sensing*, v. 56, p. 211-224.

Gillette, D. A., and Stockton, P. H. (1989) The effect of non-erodible particles on wind erosion of erodible surfaces. *Journal of Geophysical Research*, v. 94, p. 885-893.

Golder, D. G. (1972) Relations among stability parameters in the surface layer. *Boundary-Layer Meteorology*, v. 3, p. 47-58.

Greeley, R., and Iversen, J. D. (1987) Measurements of wind friction speeds over lava surfaces and assessment of sediment transport. *Geophysical Research Letters*, v. 14, p. 925-928.

Greeley, R., Lancaster, N., Sullivan, R. J., Saunders, R. S., Theilig, E., Wall, S., Dobrovolskis, A., White, B. R., and Iversen, J. D. (1988) A relationship between radar backscatter and aerodynamic roughness: Preliminary results. *Geophysical Research Letters*, v. 15, p. 565-568.

Greeley, R., Gaddis, L., Lancaster, N., Dobrovolskis, A., Iversen, J., Rasmussen, K., Saunders, S., van Zyl, J., Wall, S., Zebker, H., and White, B. (1991) Assessment of aerodynamic roughness via airborne radar observations. *Acta Mechanica*, Suppl. 2, p. 77-88.

Iversen, J. D., Pollack, J. B., Greeley, R., and White, B. R. (1976) Saltation threshold on Mars: the effect of interparticle force, surface roughness, and low atmospheric density. *Icarus*, v. 29, p. 381-393.

Iversen, J. D., Wang, W. P., Rasmussen, K. R., Mikkelsen, H. E., Hasiuk, J. F., and Leach, R. N. (1990) The effect of a roughness element on local saltation transport. *Journal of Wind Engineering and Industrial Aerodynamics*, v. 36, p. 509-516.

Iversen, J. D., Wang, W. P., Rasmussen, K. R., Mikkelsen, H. E., and Leach, R. N. (1991) Roughness element effect on local and universal saltation transport. *Acta Mechanica*, Suppl. 2, p. 65-75.

Lancaster, N., Greeley, R., and K. R. Rasmussen, K. R. (1991) Interaction between unvegetated desert surfaces and the atmospheric boundary layer: a preliminary assessment. *Acta Mechanica*, Suppl. 2, p. 89-102.

Lyles, L., and Allison, B. E. (1975) Wind erosion: Uniformly spacing non-erodible elements eliminates effects of wind direction variability. *Journal of Soil and Water Conservation*, v. 30, p. 225-226.

Marshall, J. K. (1971) Drag measurements in roughness arrays of varying density and distribution. *Agricultural Meteorology*, v. 8, p. 269-292.

Monin, A. S., and Yaglom, A. M. (1971) *Statistical Fluid Mechanics*, Vol. 1: *Mechanics of Turbulence*. The M.I.T. Press, Cambridge, Massachusetts.

Peterson, E. W., Busch, N. E., Jensen, N. O., Hojstrup, J., Kristensen, L., and Petersen, E. W. (1978) The effect of local terrain irregularities on the mean wind and turbulence characteristics near the ground. In *Symposium on Boundary-Layer Physics Applied to Air Pollution, Norrkoplng, WMO-No. 510*, p. 45-50.

Raupach, M. R., Antonia, R. A., and Rajagopalan, S. (1991) Rough-wall turbulent boundary-layers. *Applied Mechanic Reviews*, v. 44, p. 1-25.

Schaber, G. G., Berlin, G. L., and Pike, R. J. (1980) Terrain analysis procedures for modeling radar backscatter. *Radar Geology: An Assessment*, Jet Propulsion Laboratory Report 80-61, p. 168-181.

Stull, R. L. (1988) *An Introduction to Boundary Layer Meteorology*. Kluwer, Boston, p. 347-404.

Ulaby, F. T., Moore, R. K., and Fung, A. K. (1982) *Microwave Remote Sensing Active and Passive: Radar Remote Sensing and Surface Scattering and Emission Theory*, v. 2. Addison-Wesley, Reading, Massachusetts.

van Zyl, J. J. (1990) Calibration of polarimetric radar images using only image parameters and trihedral corner reflector responses. *Journal of Geophysical Research*, v. 28, p. 337-348.

van Zyl, J. J., Zebker, H. A., and Elachi, C. (1987) Imaging radar polarization signatures: Theory and observation. *Radio Science*, v. 22, p. 529-543.

van Zyl, J. J., Dubois, P., Zebker, H. A., and Farr, T. G. (1988) Inference of geologic surface parameters from polarimetric radar observations and model inversion. *Proceedings of the IGARSS '88 Symposium*, European Space Agency SP-284, p. 51-52.

Wall, S. D., van Zyl, J. J., Arvidson, R. E., Theilig, E., and Saunders, R. S. (1988) The Mojave field experiment: Precursor to the planetary test site (abstract). *Bulletin of the American Astronomical Society*, v. 20, p. 809.

Wall, S. D., Farr, T. G., Muller, J. P., Lewis, P., and Leberl, F. W. (1991) Measurement of surface microtopography. *Photogrammetric Engineering and Remote Sensing*, v. 57, p. 1075-1078.

Wang, J. R., Engman, E. T., Shiue, J. C., Rusek, M., and Steinmeier, C. (1986) The SIR-B observations of microwave backscatter dependence on soil moisture, surface roughness, and vegetation covers. *IEEE Transactions of Geoscience and Remote Sensing, GE-24*, p. 510-516.

White, B. R. (1979) Soil transport by winds on Mars. *Journal of Geophysical Research*, v. 84, p. 4643-4651.

Wieringa, J. (1993) Representative roughness parameters for homogeneous terrain. *Boundary Layer Meteorology*, v. 63, p. 323-363.

Wise, W. S. (1966) Geologic map of the Pisgah and Sunshine Cone lava fields. *NASA Technical Letter 11*.

Zebker, H. A., and Lou, Y. (1990) Phase calibration of imaging radar Stokes matrices. *IEEE Transactions of Geoscience and Remote Sensing*, v. 28, p. 246-252.

Zebker, H. A., van Zyl, J. J., and Held, D. N. (1987) Imaging radar polarimetry from wave synthesis. *Journal of Geophysical Research*, v. 92, p. 683-701.

5 SAND TRANSPORT PATHS IN THE MOJAVE DESERT, SOUTHWESTERN UNITED STATES

James R. Zimbelman,[1] Steven H. Williams,[2] and Vatche. P. Tchakerian[3]
[1]*Center for Earth and Planetary Studies, Smithsonian Institute*
[2]*Department of Space Studies, University of North Dakota*
[3]*Department of Geography, Texas A&M University*

ABSTRACT

Remote sensing and field evidence are used to describe sand deposits found in associated pathways of emplacement in the eastern Mojave Desert. Two separate pathways are identified here: one extending eastward from the Bristol Playa through the Cadiz and Danby Playas and Rice Valley to the Colorado River, and a second parallel path extending eastward from Dale Playa through the Palen and Ford Playas to the Mule Mountains near the Colorado River. The preferential location of sand ramps on the west slopes of mountains along each path suggests that the eastward moving, wind-driven sand was not confined by topographic divides between separate drainage basins around the individual playas and valleys. Sediment analysis of selected samples shows that there are discreet associations of sand characteristics along the sand pathways, with an inferred similarity between the stabilized (vegetated) sands in Rice Valley, west of the Colorado River, and stabilized sand dunes on Cactus Plain and La Posa Plain in Arizona, east of the Colorado River. Sand transport along the paths appears to have been episodic, based on multiple paleosols present in several dissected sand ramps. Future testing of the sand transport path hypothesis will require additional sediment analyses, spectral studies of remote sensing data, and obtaining dates for selected soil horizons along the sand paths.

INTRODUCTION

Wind has long been recognized as a powerful agent for sediment transport in arid environments. Sand transport in the hyper-arid Sahara Desert in northern Africa can be traced for thousands of kilometers, providing physical evidence of the wind patterns prevalent throughout the region (Wilson 1971, El-Baz et al. 1979, El-Baz and Maxwell 1982). However, significant aeolian transport is not restricted to hyper-arid deserts. Semi-arid regions also can preserve evidence of substantial deposits of aeolian sand, but many of these deposits may be stabilized at present by a variety of desert flora adapted to the intermittent rainfall.

The advent of airborne and satellite-based remote sensing data allow both the surface materials and their associated flora to be examined in a regional context. In particular, spacecraft images have been used to identify aeolian deposits throughout the Earth (Breed and Grow 1979), as well as on Mars (Sagan et al. 1972, Greeley and Iversen 1985) and Venus (Greeley et al. 1992).

Desert Aeolian Processes. Edited by Vatche P. Tchakerian. Published in 1995 by Chapman & Hall, London.
ISBN 978-94-010-6519-1

Conclusions derived from remote sensing data must be corroborated by "ground truth" investigations at key localities. The present study combines preliminary field observations with satellite remote sensing data to document aeolian deposits along hypothesized sand transport pathways in the eastern Mojave Desert of California. While a considerable amount of field work remains to be carried out, our intent here is to describe the primary features which suggest that an association exists between various sand deposits. Integrated pathways of sand transport would imply that aeolian processes have regional significance well beyond the confines of individual drainage basins. The time scale of this aeolian activity is not well constrained at present, but exposures described here suggest that the dissected sand ramps in the eastern Mojave Desert contain climatic information which predates the Holocene activity evidenced by the present isolated accumulations of active dunes.

BACKGROUND

The Mojave Desert is located in southern California at the southern end of the Basin and Range physiographic province. It is an important field geology study area because it contains numerous, accessible, well-exposed examples of a variety of geologic features (Dohrenwend 1987). The Garlock and San Andreas faults define sharp boundaries to the western margin of the Mojave Desert, while the eastern boundary with the arid region surrounding the Colorado River is more gradational. The sand transport paths described in this study lie in the eastern part of the Mojave Desert, possibly including sand deposits east of the Colorado River (Figure 1). A synopsis of the geology of the study region can be found in Jahns (1954) and in Bassett and Kupfer (1964).

Aeolian activity has formed sand sheets at several locations in the Mojave Desert. Sand ramps over 100 m thick occur in places where topography has impeded local sand migration (H.T.U. Smith 1967, R.S.U. Smith 1982). These sand ramp deposits include soil layers and other features that contain paleoclimatic information (Tchakerian 1991). The deposition of each layer presumably followed the desiccation of pluvial lakes lying upwind, with soil formation occurring between pulses of aeolian activity (Smith 1982, McFadden et al. 1987, Wells et al. 1987, Chadwick and Davis 1990, Tchakerian 1991). Some sand ramps are so large that they surmount the windward side of the topographic obstacle responsible for their formation. This study presents a hypothesis of regional aeolian transport that provides a unifying framework in which to interpret the results obtained from widely distributed sand ramps in the Mojave region.

A synoptic view of the Mojave Desert is best obtained from remote sensing data. Several recent remote sensing and field studies have focused on aeolian processes in the Mojave region (e.g., Blount et al. 1990, Paisley et al. 1991, Lancaster et al. 1992, Laity 1987, 1992, Zimbelman and Williams, in preparation). These efforts revealed that active sand can be distinguished from sand

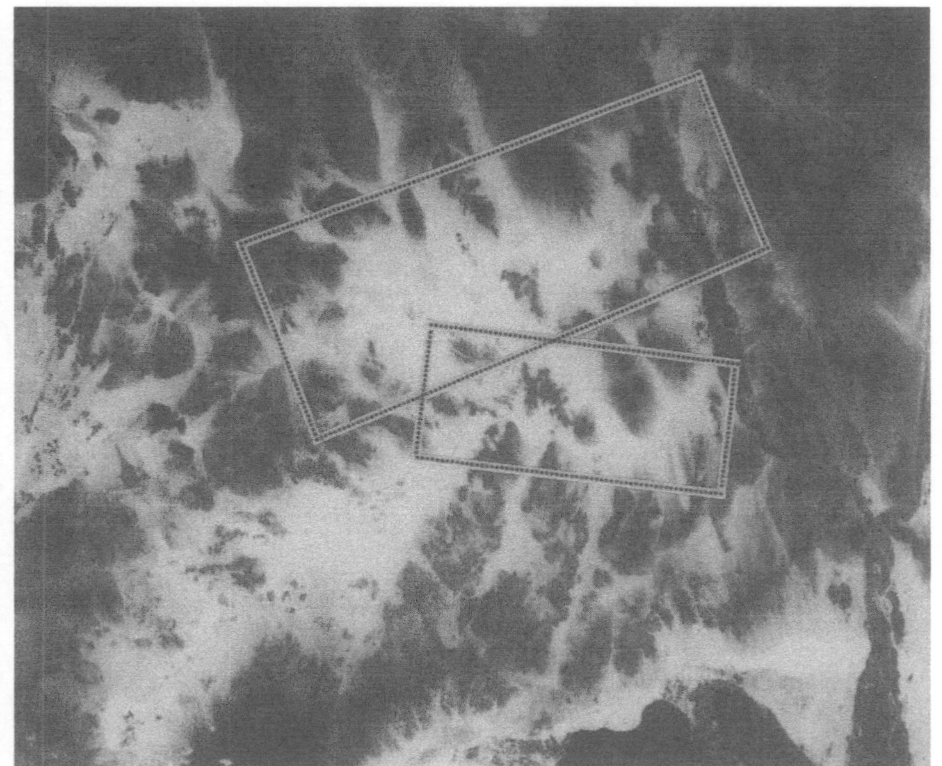

Figure 1. (a) Oblique view of the eastern Mojave Desert, taken with the Linhof camera on board the Space Shuttle. This view shows the Mojave Desert area from the outwash plains of the Mojave River (near Barstow, California) at left, to the agricultural fields along the Colorado River at right. The line of sight is nearly coincident with the paths of sand transport described in the text. Dotted lines show the locations of Figures 2 (top) and 7 (bottom). See Figure 1b for selected feature names near the pathways. Portion of frame 51B-146-111, obtained during Shuttle flight STS 51B, between April 26 and May 6, 1985.

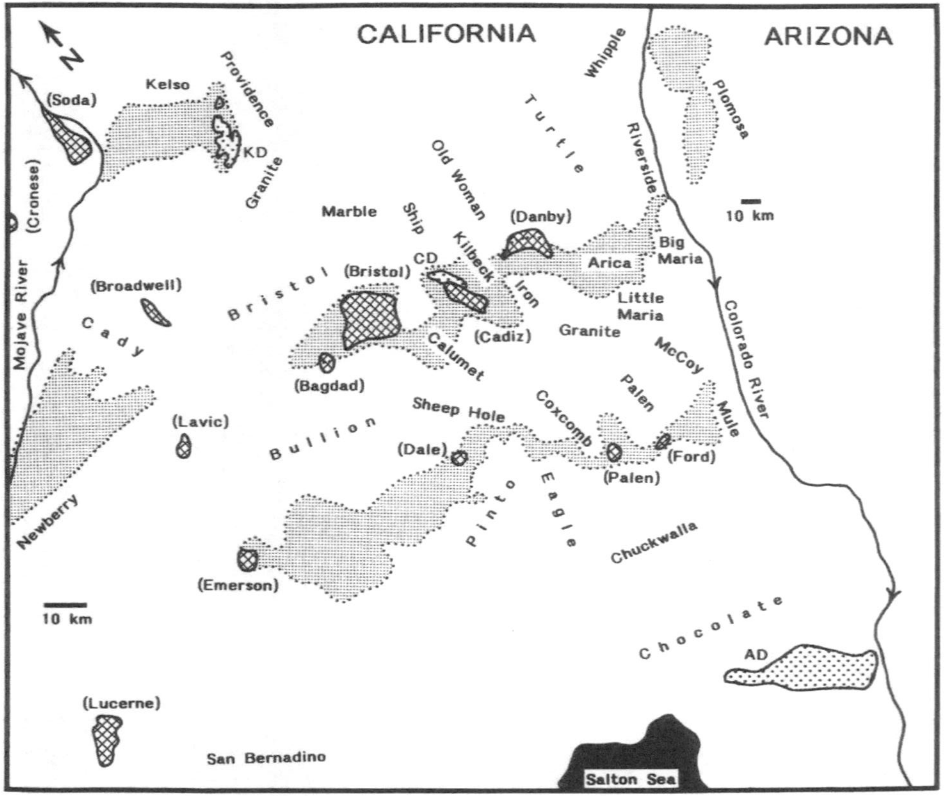

Figure 1. (b) Simplified sketch map of the area shown in Figure 1a. Selected playas (gridded pattern, names in parentheses), mountains (names listed at appropriate location), and rivers (arrows show direction of flow) are labeled for reference. Sand deposits are shown in dotted patterns; the open pattern represents active (relatively unvegetated) dunes (KD = Kelso Dunes, CD = Cadiz Dunes, AD = Algodones Dunes) and the dense pattern represents inactive (stabilized by vegetation) deposits. Note that foreshortening due to the oblique viewing geometry causes horizontal scale variations across the area.

stabilized by vegetation through subtle but consistent differences in reflectance properties between the Landsat Thematic Mapper spectral bands. Similar spectral differences exist in the Landsat data used in the present study, although the differing plant populations appear to play a significant role in the reflectance properties of aeolian deposits in the Mojave region (Zimbelman and Williams, in preparation). Consequently, seasonal variations may prove to be critical to the spectral response of certain Mojave sand deposits. These relationships are still under active investigation, so the results presented here will be based primarily on morphology as observed in a single spectral band.

SAND TRANSPORT PATHS

Three principal locations of aeolian deposits in the eastern Mojave Desert are described here: the Bristol Trough (which includes the Bristol, Cadiz, and Danby playas), Clark's Pass (which includes the Dale, Palen, and Ford playas), and the Cactus and La Posa Plains in Arizona (Figure 1). Both the active and stabilized (vegetated to the point of nonmobility) sand deposits observed at these locations are hypothesized here to be part of regional sand transportation paths which cross the Mojave Desert southeast to the Colorado River, and possibly beyond the river. These locations are all south and east of the Kelso Dunes (Figure 1), the most prominent and intensively studied dune field in the Mojave Desert (Sharp 1966, 1978, Paisley et al. 1991, Lancaster et al. 1992, Lancaster 1993). The sand in the Kelso Dunes originated in broad outwash plains associated with the Mojave River, was transported to the southeast by the prevailing winds, and collected at the base of the >1800-m Providence and Granite Mountains, which formed an insurmountable barrier to the windblown sediments (Sharp 1978). The sands associated with the Bristol Trough and Clark's Pass also are oriented along the prevailing northwest-to-southeast wind direction (Greeley and Iversen 1985), but these sand deposits have traversed several distinct drainage basins on their way to the Colorado River. Sand from the Bristol Trough may have even contributed to a third major sand deposit, a field of stabilized dunes on the Cactus Plain and La Posa Plain east of the Colorado River. Each of the three sand localities is described in greater detail in the following sections.

Bristol Trough

The most prominent association of sand deposits begins at the Bristol Playa near the head of a broad topographic low called the Bristol Trough (Thompson 1929). Sand occurs continuously over a distance of almost 150 km, eventually terminating at the Colorado River (Figure 2). The sand deposits concentrate around three large playas (Bristol, Cadiz, and Danby), and consist of both active dunes and vegetation-stabilized sand sheets and linear dunes (Figure 2). The relation of the sand deposits to the mountains they traverse indicates that the sand movement was toward the east-southeast, with prominent sand ramps present on the west side of several mountain ranges along the pathway.

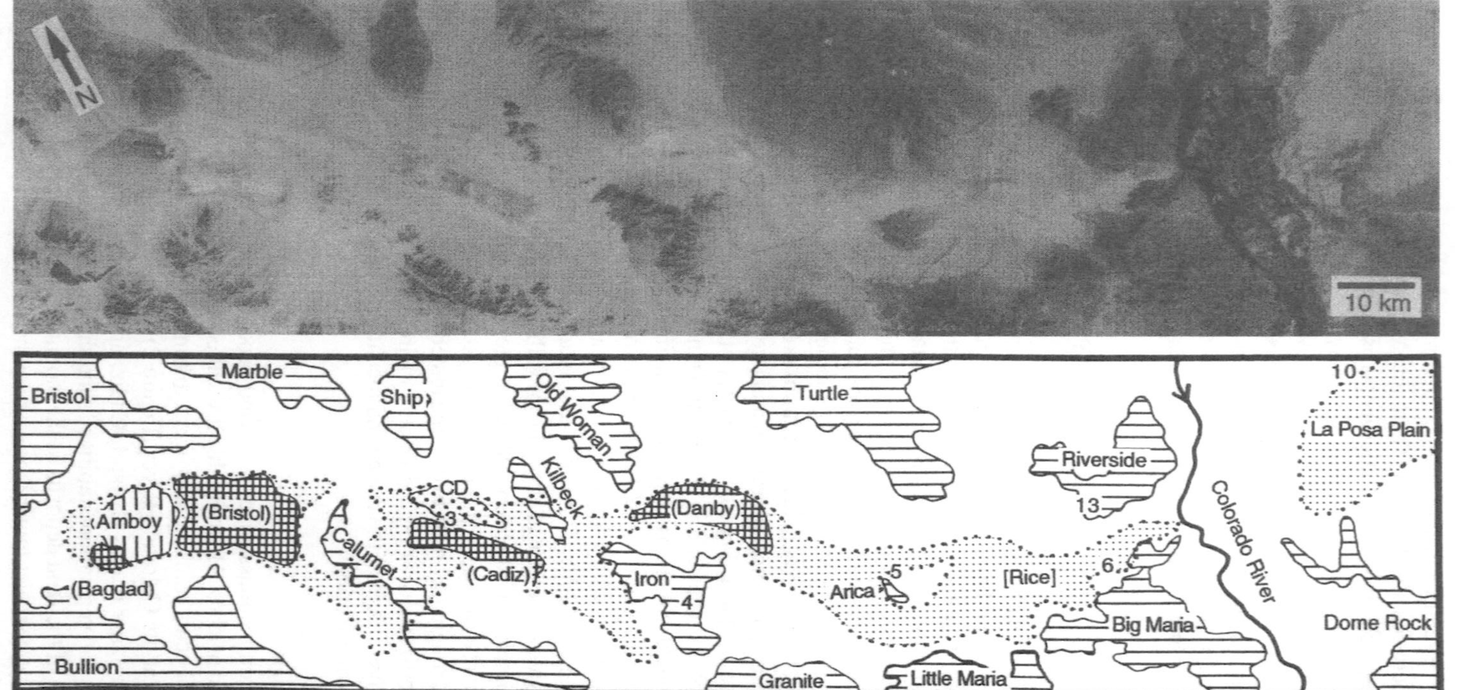

Figure 2. Bristol Trough sand path, seen on a portion of a Large Format Camera photograph, frame 2063, taken during the STS 41-G Space Shuttle flight between October 5 and 13, 1984. The sketch map below the photograph labels mountains (horizontal lined pattern), playas (gridded pattern), and sand deposits (dotted pattern). The active Cadiz Dunes (CD) are shown in a large dotted pattern. Rice Valley is identified by the name within square brackets. The Amboy lava flow (vertical lined pattern) is a Holocene basaltic eruption that covered the western portion of Bristol Playa. Numbers indicate the approximate centers of the orbital views shown in the corresponding figures.

Sand accumulations first become discernible west of the Bristol Playa, in the broad valley between the Bristol and Bullion Mountains (Figure 2). The Holocene basalt flow associated with the Amboy cinder cone covered the western portion of the Bristol Playa, leaving the small Bagdad Playa west of the Amboy flow as a remnant of the ancestral Bristol Playa (Bassett and Kupfer 1964). Sand derived from Bristol Mountains alluvium traverses the Amboy lava flow from WNW to ESE (Greeley and Iversen 1985), consistent with the annual wind flow in the region during the Holocene (Laity 1992). A prominent low-albedo wind streak is present downwind from the Amboy cinder cone (Figure 2). Sand transport across the flow is obstructed by the cinder cone, and enhanced turbulent wind scour in the lee of the cone aids in inhibiting sand migration into the wind streak (Greeley and Iversen 1978, 1985). There is no evidence, either in remote sensing data or on the ground, that sand from the Mojave River has traversed the Cady Mountains to enter the Bristol Playa basin from the west (Figure 1); the Bristol area is interpreted here to represent the beginning of the aeolian sand deposits that extend east to the Colorado River (Figure 2).

Sand is abundant southeast of the Bristol Playa, where it has built large ramps against the western slopes of the Calumet Mountains (Figure 2). The sand ramps provide shallow slopes for saltating sand to climb the western flanks of the mountains, as well as shallow slopes along which the sand moves

Figure 3. (a) Portion of a Landsat Thematic Mapper image showing the northern end of Cadiz Playa and the Cadiz Dunes north of the playa. The largest dunes (open arrow) have 30 m of relief. Transverse dunes are present along the northern margin of the playa; the dark arrow shows the location and orientation of Figure 3b. Landsat TM band 5, obtained on September 26, 1986.

Table 1
Mean values of grain size, sorting, and percent silt and clay for selected samples from the Mojave Desert, California

Aeolian unit	Mean (ϕ)	Standard deviation	Skewness	Kurtosis	% silt and clay
Dale Lake					
Unit 1	2.24	0.83	0.10	1.17	2.90
Unit 2	1.91	0.91	0.14	1.15	3.10
Unit 3	2.15	0.85	0.08	1.10	3.25
Unit 4	1.70	1.21	0.05	1.05	1.25
Unit 5	2.23	0.78	-.07	0.93	2.40
Calumet Mtns	2.91	0.97	0.22	1.29	4.83
Cadiz Dunes	2.35	0.29	-.03	0.89	
Iron Mtns	2.45	0.88	0.15	1.10	7.20
Rice Valley	2.95	0.75	0.20	1.25	6.25
Cactus Plain	2.87	0.87	0.25	1.35	5.98

down the eastern flanks of the mountains. Neither climbing nor falling dunes are observed on the Calumets, but Landsat spectral data indicates that stabilized sand dominates a 10-km-long reach of the central portion of the mountains, where sand ramps were evident on the ground. Alluvial fan deposits around the northern end of the Calumets lack any prominent sand accumulations, leading to the interpretation that most sand from the Bristol area crossed the central Calumets instead of going around the northern alluvial fans.

East of the Calumet Mountains, sand deposits are concentrated around the Cadiz Playa, which is in the broad valley between the Calumet Mountains and the Kilbeck Hills (Figure 2). The sand deposits attain a considerable thickness on the northern margin of the Cadiz Playa; individual dunes display 30 m of relief and are clearly resolved in Landsat Thematic Mapper data (Figure 3a). Ground investigation showed that the dunes north of Cadiz Playa are the only substantive area of active dunes observed along the Bristol Trough path. The active sand gradually thins to the east, where transverse dunes with 1-2 m of relief become the dominant aeolian landform (Figure 3b).

Stabilized sand sheets extend eastward from the Cadiz Valley across the southern end of the Kilbeck Hills into the Danby Valley (Figure 2). Extensive sand ramps are present around the southern Kilbeck Hills, and are particularly well developed on the western slopes of the Iron Mountains (Figure 4a). Where ephemeral streams dissect the Iron Mountain sand ramps, tens of meters of sand are exposed within the channels, both in the western sand ramps (some of which have active dunes on the channel crest, Figure 4b), and in stabilized sand ramps on the northern flanks (Figure 4c).

Figure 3. (b) Oblique stereo pair of transverse dunes north of the Cadiz Playa, looking southwest. Stereo view shows exaggerated topography; the dunes have 1 to 2 m of vertical relief and an average spacing of 40 m. Photographs taken on September 26, 1986, from the top of a small hill north of the playa.

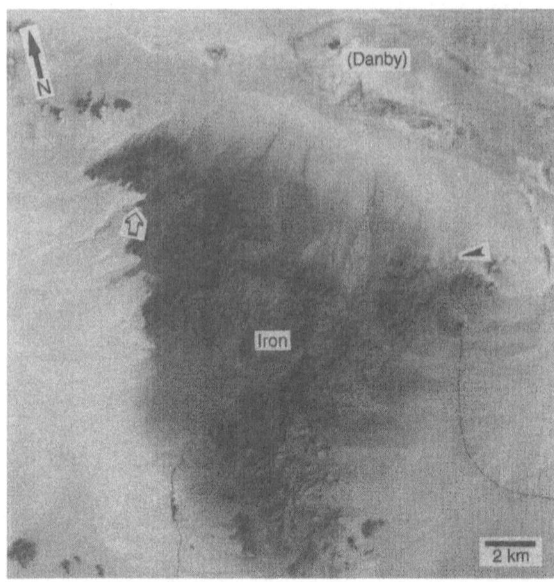

Figure 4. (a) Portion of a Landsat Thematic Mapper image showing the Iron Mountains south of Danby Playa. Prominent sand ramps are present on the western slope of the mountains; open arrow shows the location and orientation of the photograph in Figure 4b. Less active (more vegetated) sand ramp on the northern slope was sampled in 1991; the dark arrow shows location and orientation of the photograph in Figure 4c. Landsat TM band 5, obtained on September 26, 1986.

Figure 4. (b) View of entrenched sand ramp on the western slope of the Iron Mountains, taken from the channel floor. The 25-m-high channel wall has considerable vegetation cover at this locality, but the channel crests consist of active dune patches. Photograph was taken on October 9, 1993.

Figure 4. (c) Entrenched sand ramp on the northern slope of the Iron Mountains, with a paleosol complex (open arrow) capping the deposit. Note the outstretched arms of a 1.6 m field assistant (dark arrow) in the ephemeral wash, which exposes the sediments of the sand ramp. The top of the sand ramp is mantled with taluvium (talus and alluvium). Photograph was taken in 1992.

South of Danby Playa, the sand deposits spread to cover much of the Rice Valley with linear dunes and sand sheets, both of which are stabilized by desert vegetation (Figure 2). Sand surrounds the 500-m-high Arica Mountains (Figure 5a); a prominent sand ramp on the western slope almost reaches the top of the highest peaks, while the eastern slope is essentially sand-free (Figure 5b). Fields of stabilized linear dunes cover the southern side of the Rice Valley (Figure 6a). Sand ramps terminate at stream margins within the Big Maria Mountains, exposing up to 10 m of accumulated sand (Figure 6b). A narrow strip of sand exits the eastern end of Rice Valley, extending east to the Colorado River. Basic sedimentological characteristics of four sand deposits sampled along the pathway are listed in Table 1.

Clark's Pass

A second association of sand deposits roughly parallels the Bristol Trough path, but along a more southerly route (Figures 1 and 7). Sand ramps east of the Dale Playa at the eastern end of the Twentynine Palms Valley allowed sand to exit the valley through Clark's Pass, a narrow gap between the Sheep Hole and Pinto Mountains (Figure 8a).

The orientation of the Sheep Hole and Pinto Mountains acted like a funnel to concentrate migrating sand rather than trapping it completely, as at the Kelso Dunes. The sand ramps developed between the mountains allow wind-blown sand to climb more than 250 m from the level of Dale Playa to Clark's Pass.

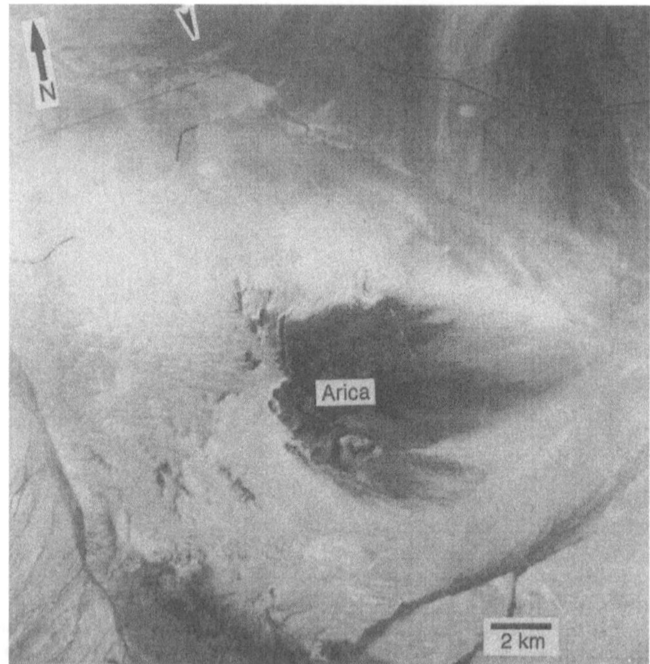

Figure 5. (a) Portion of a Landsat Thematic Mapper image showing the Arica Mountains at the west end of Rice Valley. The dark arrow at the top shows the orientation of the photograph in Figure 5b, taken from a position just off the northern edge of the image. Landsat TM band 5, obtained on September 26, 1986.

Figure 5. (b) Profile view of the Arica Mountains, looking south from a road that follows the railroad tracks north of Danby Playa. A prominent sand ramp is present on the west slope (right) while the east slope (left) is relatively sand-free, in the lee of the 500-m-high mountains. Photograph was taken in May 1991.

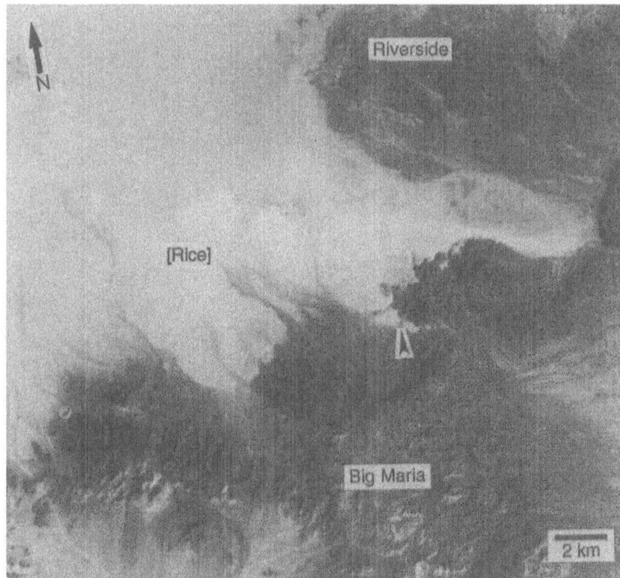

Figure 6. (a) Portion of a Landsat Thematic Mapper image showing stabilized dunes in Rice Valley. The sand deposits occur against the northern slopes of the Big Maria Mountains, and are cut by emphemeral channels from those mountains. The dark arrow shows the location and orientation of Figure 6b. Landsat TM band 5, obtained on September 26, 1986.

Figure 6. (b) Upper portion of the sand ramp on the northern slope of the Big Maria Mountians. The top of the section exposed by an ephemeral stream is stabilized by vegetation and taluvium, and capped by a prominent paleosol. Photograph was taken in 1991.

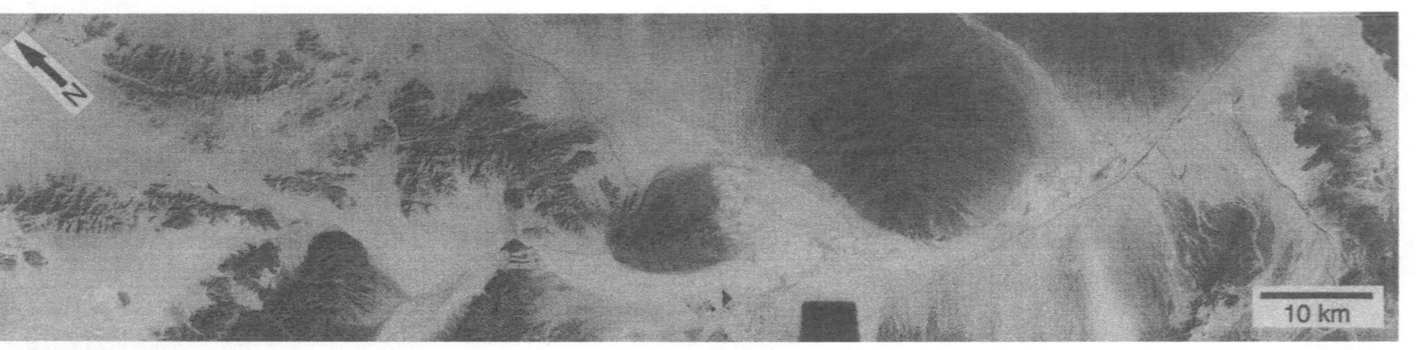

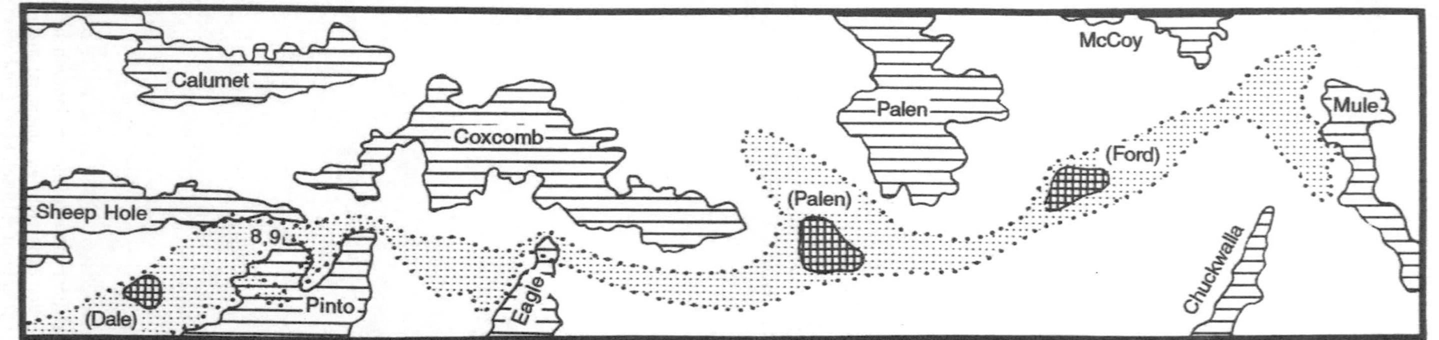

Figure 7. Clark's Pass sand path, seen on a portion of a Large Format Camera photograph, frame 2063, taken during the STS 41-G Space Shuttle flight between October 5 and 13, 1984. The sketch map below the photograph labels mountains (lined pattern), playas (gridded pattern), and sand deposits (dotted pattern).

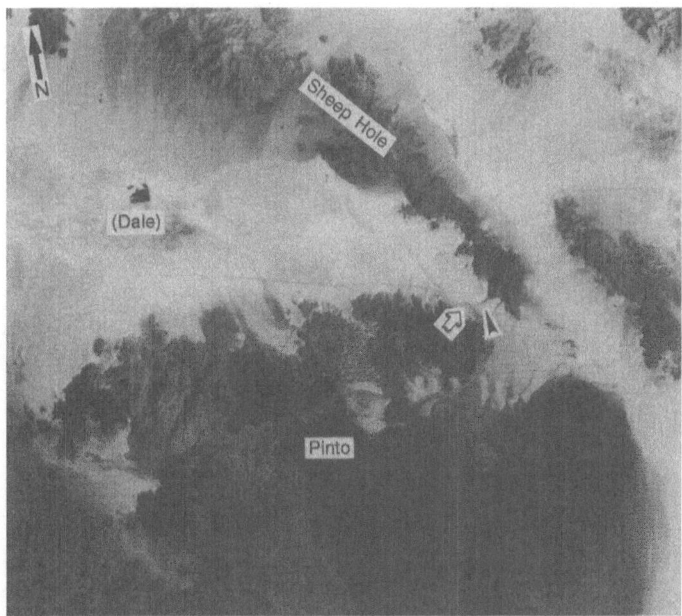

Figure 8. (a) Portion of a Landsat Thematic Mapper image showing the Clark's Pass area. Sand is ramped against the Sheep Hole (top) and Pinto (bottom) Mountains, providing an exit from the Twentynine Palms Valley, past Dale Playa, through Clark's Pass. The dark arrow shows the location and orientation of Figure 8b, and the open arrow shows the location and orientation of Figure 9a. Landsat TM band 5, obtained on September 26, 1986.

Some sand has been trapped in small valleys along the northern margin of the Pinto Mountains (Figure 8a), but the volume of sand deposits within the mountains appears to be much smaller than that of the sand ramps leading up to Clark's Pass. Ephemeral streams from the Sheep Hole Mountains cut into the sand ramps at several locations (Figure 8b), providing a cross-section through tens of meters of sand and exposing several soil horizons.

Several major stratigraphic units are identified within the Dale sand ramp (Tchakerian 1991) by the combination of geomorphic and soil-stratigraphic relations (Figure 9, Table 1). The units are predominantly aeolian in origin, with some intermixed fluvial deposits. Unit 1 contains fine to medium ($Mz = 2.24\ \phi$) moderately sorted ($\sigma = 0.83$) aeolian sands with grus and a silt/clay content of 2.9%. It is capped by a reddish yellow (5YR/6/6) paleosol with discontinuous carbonate nodules. Unit 2 comprises fine to medium ($Mz = 1.91\phi$) poorly sorted ($\sigma = 0.91$) aeolian sands and has a silt and clay content of 3.1%. It contains numerous fluvial cut and fill lenses. The unit is capped by a prominent reddish yellow (7.5YR/6/6) paleosol with calcrete disseminated throughout the matrix, and carbonate enriched root pseudomorphs. Unit 3 consists of mostly yellowish red (5YR/5/8) medium to coarse ($Mz = 2.15\phi$) moderately sorted ($\sigma = 0.85$) aeolian sands with large percentages of grus. It is capped by a poorly developed discontinuous paleosol with some calcareous

Figure 8. (b) Sand ramp on the western side of the Sheep Hole Mountains near Clark's Pass, at the east end of the Twentynine Palms Valley (Shelton et al. 1978, figure 9-21). Note the active dune along the crest of the channel. Oblique aireal photograph taken by R. Greeley.

rhizoliths. Unit 4 contains primarily fluvially redistributed dune sands, cut-and-fill structures, and coarse gravel alluvial channels. The sediments are mostly coarse sands (Mz = 1.70ϕ), and are poorly sorted (σ = 1.21).

The section is topped by weakly consolidated sand that forms the surface of the sand ramp. Unit 5 contains brownish yellow (10YR6/6) fine to medium (Mz = 2.23ϕ), moderately well sorted (σ = 0.78) aeolian sands, with a silt/clay content of 2.4%. The uppermost section of Unit 5 is obscured by loose aeolian sands. However, about 500 m to the west of this section, further aeolian depositional units have been identified which are stratigraphically equivalent to or younger than Unit 5. They consist of brownish yellow, fine to medium, moderately well sorted aeolian sands similar in composition to Unit 5. Additional units in the area are similar, with respect to grain size, sorting, percent silt/clay, and quartz grain surface micromorphologies (SEM analysis), to Unit 5, suggesting emplacement by a single aeolian episode with multiple depositional pulses (Tchakerian 1991).

After exiting through Clark's Pass, the sand traversed the northern end of the Eagle Mountains and passed the Palen and Ford Playas (Figure 7). The sand deposits around the Palen and Ford Playas are primarily in the form of broad sand sheets, with very limited development of isolated, stabilized dunes. There is no evidence at present that sand reached the Colorado River after passing the Mule Mountains (which have prominent sand ramps on their western slopes), but the lack of visible sand likely results from the extensive agricultural activity along the Colorado River in the vicinity of Blythe, California.

Cactus Plain–La Posa Plain

A third accumulation of sand deposits possibly may be related to the proposed sand transport pathways through the Mojave Desert. Large fields of stabilized linear dunes are present on the eastern bank of the Colorado River near the town of Parker, Arizona (Figure 10). These dunes are directly opposite the termination point of the Bristol Trough path at the Colorado River (Figure 1). There is no obvious source for the stabilized dunes on the Cactus Plain and the La Posa Plain (Figure 10); the adjacent mountains display typical alluvial fan development with no apparent accumulation of sand-sized materials to supply the sand to the extensive dune fields. The Colorado River could be a source for the sand, except that the Cactus Plain–La Posa Plain area is the only sand accumulation next to the river but not next to a large lake or playa (such as the Algodones Dunes near the Salton Sea; Figure 1).

The Arizona linear dunes generally have from 2 to 4 m of relief and are oriented approximately transverse to the prevailing wind direction evident within the Bristol Trough (Greeley and Iversen 1985, Laity 1992). No prominent sand ramps are evident around the dunes; the sand accumulation progressively thins leading up to the adjacent mountains. However, the silt/clay content of the Cactus Plain dunes is nearly twice as large as that of the Dale units exposed within the Clark's Pass path, but is similar to the silt/clay content of the Iron Mountain and Rice Valley sands from the eastern portion of the Bristol Trough path (Table 1). The increased silt/clay content does not appear to be pedogenic in origin; the loose sand covers the dunes but they are no longer mobile because of the desert vegetative cover. The adjacent locations and the overall similarities between the Rice Valley and Cactus Plain sands raise the possibility that the Arizona sands may be genetically related to the "apparent" termination of the Bristol Trough path at the Colorado River; this intriguing possibility is discussed in the following section.

DISCUSSION

The alignment of the Mojave sand paths is because of a combination of topography and prevailing winds. Our observations led to the hypothesis that the paths represent the aeolian part of a combined aeolian/alluvial/fluvial drainage system, as discussed below. Also described are the possibility of net sand migration across the Colorado River, and some paleoclimatic implications of the sand transport path hypothesis.

Figure 9. (a) The upper part of the Dale Lake sand ramp, with an ephemeral fluvial wash in the foreground (see also Figure 8). The Sheep Hole Mountains are in the background. The dark arrow points to the paleosols shown in Figures 9b and 9c, exposed by the streamcut in the sand ramp. Photograph taken in 1987.

Figure 9. (b) A close up view of the middle section of the sand column exposed in the wash of the ephemeral stream described in Figure 9a. The section exposed here is about 10 m thick. Photograph taken in 1987.

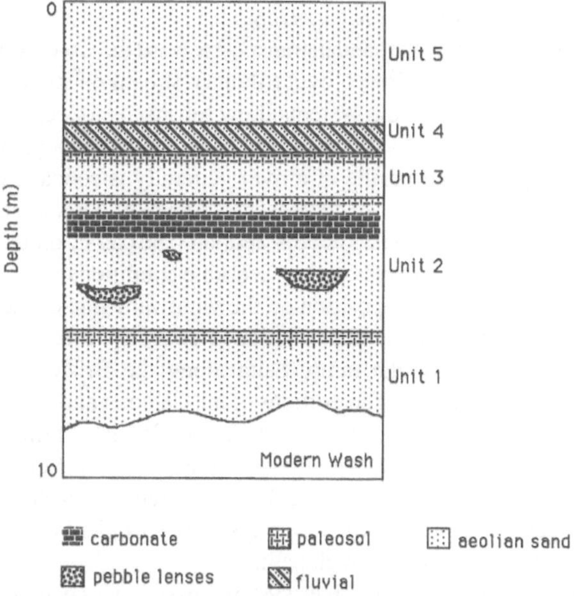

Figure 9. (c) A detailed geomorphic and soil-stratigraphic cross-section of the sand ramp exposure shown in Figure 9b.

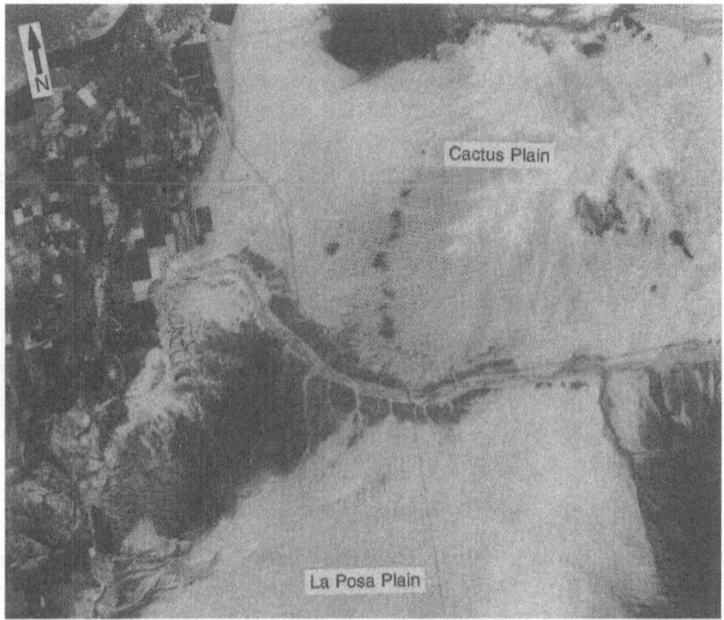

Figure 10. Portion of a Landsat Thematic Mapper image showing the stabilized dunes near Parker, Arizona. Agricultural use of the Colorado River floodplain is evident at left. The dunes are concentrated on the Cactus Plain (top) and the La Posa Plain (bottom), with no apparent source area evident in the sourrounding mountains. Landsat TM band 5, obtained on June 9, 1984.

Sand Transport Pathways As "Rivers Of Sand"

The pathways followed by the sand in transport toward the Colorado River can be compared to the path followed by a tributary stream on its way toward a higher order primary river. Both systems show sensitivity to local topography, but windblown sand is not forced always to flow down the local topographic gradient. The longitudinal profile of a river generally is concave upward, in accord with a steady downstream decrease in slope (e.g., Leopold et al. 1964, p. 248-255). In contrast to fluvial systems, aeolian sand can surmount or bypass significant topographic obstacles under favorable conditions of wind orientation and sand supply. Sand deposition on the windward side of the mountain ranges built sand ramps that facilitate continued access by saltating sand up the gentle windward slope of the ramp. Sand along the Bristol Trough path surmounts relief of up to 100 m on portions of the Calumet and Iron Mountains (Figure 11a), while the Clark's Pass path traverses 250 m of relief to provide an outlet for the sand from the Dale Playa (Figure 11b). Sufficient sand was available from the Dale Basin to build the large sand ramps that characterize the Clark's Pass path. In contrast, smaller ramps were sufficient to surmount the topography along the Bristol Trough path.

The Mojave and Colorado River systems may have been connected in earlier epochs (Blackwelder 1933, 1954, Miller 1946). Blackwelder (1954) postulated a Mojave/Colorado connection via the Bristol Trough that coincides exactly with the observed sand path (Figure 12). The hypothesized drainage connection was then disrupted by a combination of climate change, tectonic processes, and the eruption of the Pisgah volcanics. The association of the Mojave Desert sand deposits with (sometimes large) playas contributed to an assumption that the sand was locally derived, solely from the nearest paleolake. Our preliminary remote sensing analysis and field observations suggest that the present-day playas may be intermediate concentration points (at local topographic lows) for a more through-going movement of windblown sand. The tectonic trough enclosing the Bristol, Cadiz, and Danby Playas provides a preexisting trend along which the wind-blown sediments now encounter only minimum topographic obstacles, which were surmounted or bypassed through prolonged aeolian activity. In this sense, the sand transport paths might be considered "rivers of sand" that have reclaimed and actually shortened a possible drainage path from an earlier epoch. Considerable field work remains to be done to test the validity of this hypothesis, as well as much more extensive sediment analyses.

Possible Trans-Colorado River Sand Transport

Two intriguing questions are raised by the possibility of sand transport paths in the Mojave region: what is the ultimate fate of sand in transit along each path, and what is the source of the sands on the Cactus and La Posa Plains? Much of the sand entering the Colorado River is transported downstream, with much of it perhaps contributing to the Gran Desierto dune field in Mexico (Merriam 1969, Lancaster et al. 1987, Blount and Lancaster 1990). However, we

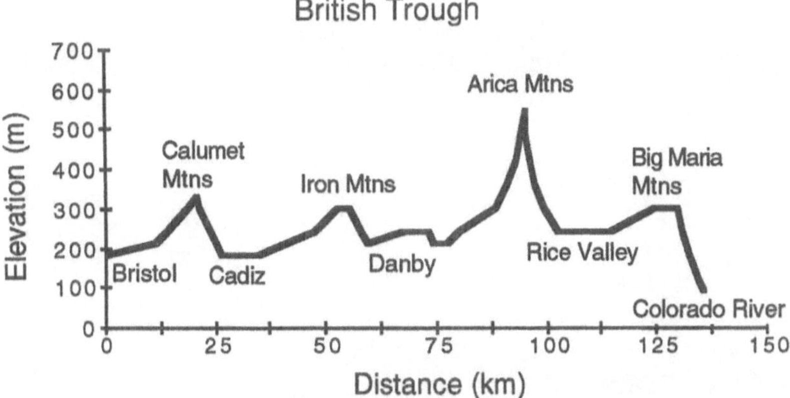

Figure 11. (a) Topographic profile along the Bristol Trough path. Vertical exaggeration is approximately 87X. The profile follows the approximate centerline of the sand path, including both active and inactive sand deposits. Mountain ranges encountered along the path are labeled above the profile; these topographic obstacles are crossed by the sand through emplacement of thick sand ramps. Playa names and Rice Valley are labeled below the profile, which ends at the Colorado River near Quien Sabe Point. Topographic data are from 1:250,000 Needles (USGS 1969a) and Salton Sea (USGS 1969b) map sheets.

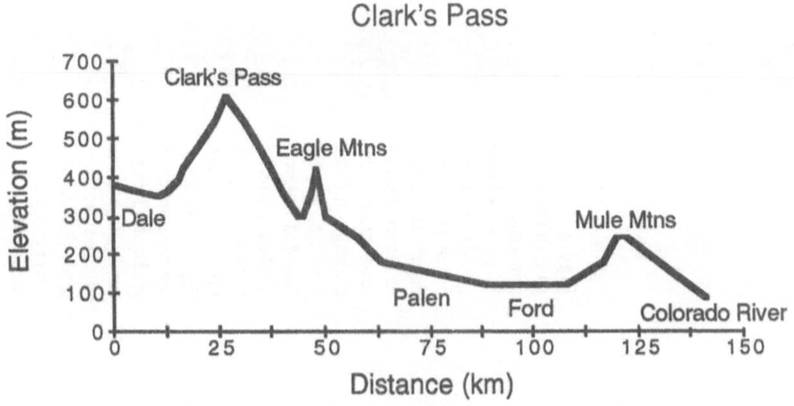

Figure 11. (b) Topographic profile along the Clark's Pass path. Vertical exaggeration is approximately 87X. The profile follows the approximate centerline of the sand path, including both active and inactive sand deposits. Mountain ranges encountered along the path are labeled above the profile; these topographic obstacles are crossed by the sand through emplacement of thick sand ramps. Playa names are labeled below the profile, which ends at the Colorado River near Blythe, California. Topographic data are from 1:250,000 Needles (USGS 1969a) and Salton Sea (USGS 1969b) map sheets.

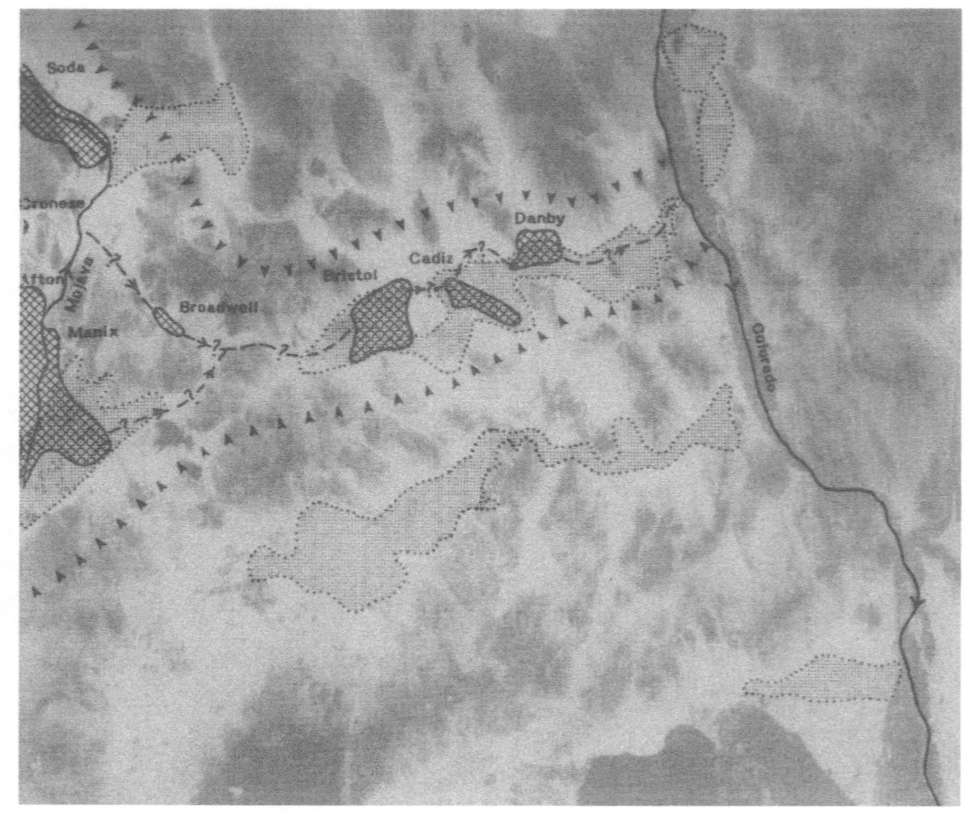

Figure 12. The paleodrainage reconstruction of Blackwelder (1954) is superposed on the oblique photograph in Figure 1a. Arrows outline the drainage into the paleolakes (gridded pattern). Dotted patterns show sand deposit locations from Figure 1b. Note the close match of the Bristol Trough sand path and the inferred drainage from the Mojave River to the Colorado River.

speculate that some of the sand transported down the pathways may have crossed the Colorado River and ended up on the Cactus and La Posa Plains (Williams et al. 1991, Tchakerian et al. 1992). The lack of other sand accumulations east of the Colorado River argues against the river itself as the primary sand source and argues for a mechanism which could bring mobile sand to this particular location. Batches of aeolian sand appear to enter the river floodplain at the western end of the sand transport paths (Figure 13). Eventually, the river may have opened a new meander channel west of aeolian sand deposits on the floodplain, which were then remobilized by the wind and exported onto the Cactus and La Posa dunefields. It is difficult to assess how effective such a process may have been prior to regulation of flow along the Colorado River. However, this mechanism could account for the presence of a large quantity of sand on the plains east of the Colorado River and is coincident with the termination of sand transport paths through the Mojave Desert.

Paleoclimatic Implications

The presence of well-developed paleosols and multiple aeolian depositional units within the sand ramps along the sand transport paths indicates that aeolian activity in the Mojave Desert has been widespread and episodic. A description of the units exposed in the Dale sand ramp was given earlier, and additional exposures of multiple soil-horizons were observed during a recent reconnaissance survey of the more remote portions of the Bristol Trough sand pathway. Such exposures within the sand pathways may be related to more extensively studied paleosols in the western Mojave Desert.

In the Silver Lake basin (part of Pleistocene Lake Mojave), an aeolian depositional period (Qe2) that took place between 12 and 8.7 ka, has been recognized by Wells et al. (1987) and Brown (1989). An older aeolian depositional episode prior to 22 ka has also been identified by Brown (1989) in sediment cores from the Silver Lake basin. The sedimentary record from Lake Mojave indicates that lake levels were low to intermediate between 13.5 and 9 ka, with final dessication around 8.7 ka (Brown 1989). It seems likely that the sand ramps observed along the sand transport pathways also witnessed increased levels of aeolian sediment input during low stands of the desert paleolakes, as sediments became available for transport from dried lake basins and their surrounding piedmont areas.

Sediment supply from desert lake basins was drastically curtailed after 9 ka, as a result of the changing environmental conditions which caused most lakes either to dry up or to reach very low water levels (Benson et al. 1990). The sand ramps probably underwent a period of stabilization through vegetation development and soil formation because of the reduction in sediment supply. They were subsequently mantled by rock debris from the adjoining mountains and later entrenched by ephemeral streams. In the middle Holocene, from about 7 to 5 ka, the Mojave Desert experienced a drier than present climatic regime, a period referred to as the climatic optimum or the Altithermal, first recognized

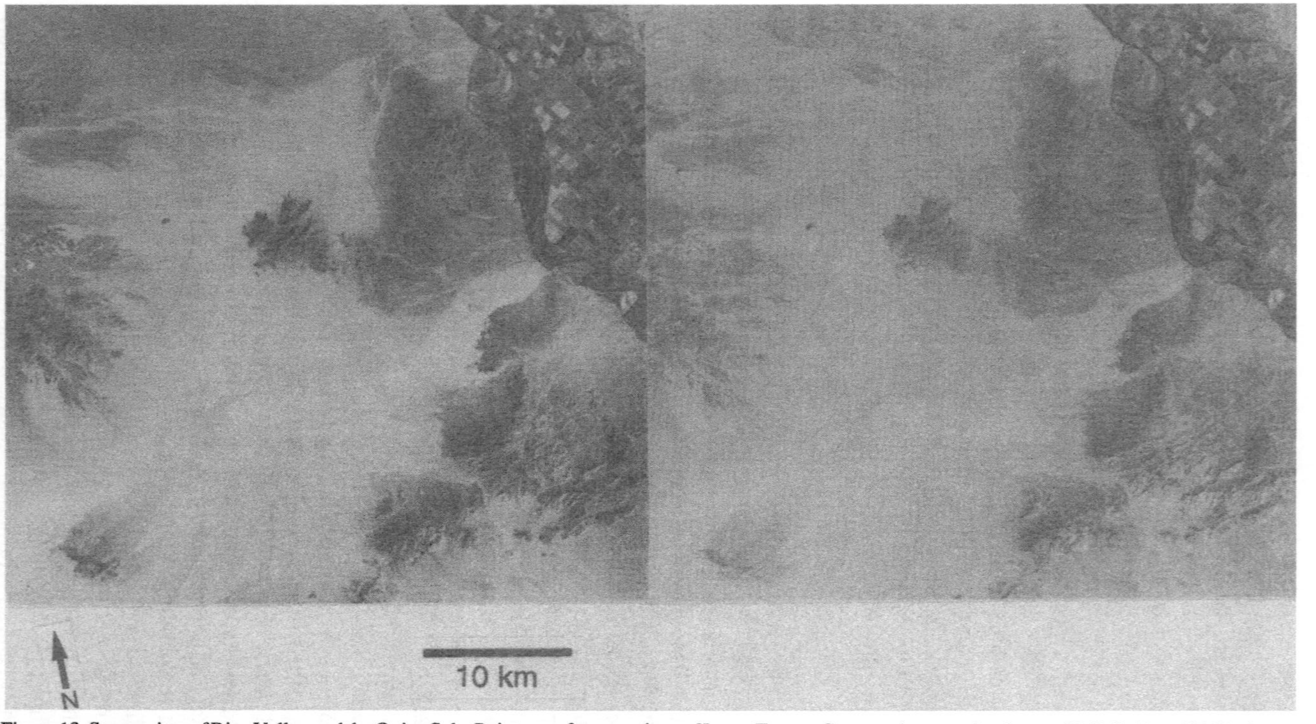

Figure 13. Stereo view of Rice Valley and the Quien Sabe Point area, from portions of Large Format Camera photographs, frames 2062 (left) and 2064 (right), taken during the STS 41-G Space Shuttle flight between October 5 and 13, 1984. The vertical relief is highly exaggerated in this stereo pair, but this view emphasizes the relation between the sand deposits, the mountains, and the Colorado River.

by Antevs (1962). According to Spaulding (1991), Middle Holocene macrofossil (packrat middens) records from the southern Mojave Desert indicate a more arid period than the present between 6800 and 5060 yr B.P. It is thus highly probable that the Middle Holocene period witnessed little aeolian activity as desert lakes were already dessicated by the time of the Altithermal, and most sand ramps fully stabilized.

Accelerator Mass Spectroscopy (AMS) ^{14}C and cation-ratio dating of varnished ventifacts on stabilized debris mantling sand ramps in the Cronese Basin in the Mojave Desert (see left margin of Figure 1b) indicate that aeolian activity ceased or was at a minimum, and that debris deposits were already stabilized, between 5.5 and 5 ka (Dorn et al. 1989). Hence (given the absence of numerical ages directly from dune deposits), most of the sand ramps were probably stable with rock talus and vegetation before the onset of more xeric conditions during the Altithermal, and aeolian activity was at a minimum or mostly restricted to those few desert basin areas that had active sediment input, such as the Mojave River Wash supplying sediments for the Kelso Dunes. Using luminescence dating measures, Lancaster et al. (1991) report a lack of ages older than 5000 yr B.P. from the main Kelso Dune fields, and suggest that the majority of the sediments have been extensively reworked prior to the mid-Holocene.

The entrenched sand ramps within the Bristol Trough and Clark's Pass sand pathways represent a valuable resource for studying paleoclimatic information preserved within the paleosols. The Dale sand ramp is presently the only locality within the sand pathways that has been thoroughly studied for sedimentological characteristics, but our field studies have identified other localities within the Bristol Trough path where entrenched sand ramps expose paleosol sequences. Comparison of the paleosol sequences, both along a given pathway and between adjacent pathways, should provide a test for the emplacement scenarios proposed here. Luminescence dating of key paleosol horizons is perhaps the most critical information required to quantify the climatic information recorded within the sand ramps.

FUTURE WORK

We have presented here the preliminary descriptions and interpretations of the sand deposits present in the eastern Mojave Desert. A considerable amount of field work remains to be carried out, particularly in terms of describing and documenting the sediment characteristics and internal stratigraphy within the thick sand ramps evident at several locations. The sand transport pathway hypothesis can be tested through additional analyses of samples already collected, particularly looking for mineralogical information which could indicate whether or not the sand at the proposed "upstream" end of the pathways could have supplied the sands observed at the termination of the pathways. Additional sedimentological studies may also help to test whether or not the sands along the pathways are consistent with transport away from the inferred source of each pathway. Obtaining samples specifically collected for luminescence dating measurements (Lancaster et al. 1991) from geographically separated soil horizons is essential to the development of a regional stratigraphic history of the eastern Mojave Desert.

The remote sensing data have been used in the present work for basic geomorphic and geographic descriptions, but spectral variations between different bands of Thematic Mapper data should be useful in refining the distribution of active and stabilized sand deposits (Blount et al. 1990). We also hope that the spectral information will be useful for estimating sand thickness throughout the region based on vegetation that is sensitive to particular sand thicknesses (Zimbelman and Williams, in preparation).

CONCLUDING REMARKS

We have presented both remote sensing and field evidence for the emplacement of aeolian sand pathways in the eastern Mojave Desert. Two pathways are described in detail: one extending eastward from the Bristol Playa past the Cadiz and Danby Playas through Rice Valley to the Colorado River, and a second path extending eastward from Clark's Pass past Palen and Ford Playas to the Mule Mountains by the Colorado River. The preferential development of sand ramps on the west slopes of mountains along each path indicates that the eastward-moving, wind-driven sand was not restricted by topographic divides between separate drainage basins around the individual playas and valleys. Preliminary sediment analysis of selected samples shows that there are discrete associations of sand characteristics along the sand pathways, with an inferred possible relationship between the stabilized sands in Rice Valley (within the Bristol Trough path) west of the Colorado River and stabilized linear dunes on the Cactus Plain and La Posa Plain east of the Colorado River. Sand transport along the paths appears to have been episodic, based on multiple soil horizons present in several dissected sand ramps.

ACKNOWLDGEMENTS

Several people assisted during the collection and processing of the data presented here: J. Heisinger and A. Johnston provided valuable image processing assistance at the Center for Earth and Planetary Studies, National Air and Space Museum, and the Large Format Camera prints are courtesy of Ron Greeley, Dan Ball, and the Arizona State University Planetary Geology group. The authors are grateful for careful reviews of an earlier version of the paper by Ron Dorn and Andrew Bach. Tchakerian acknowledges the support of the donors of the Petroleum Research Fund, administered by the American Chemical Society (ACS-PRF 26124-G2).

REFERENCES

Antevs, E. (1962) Late Quaternary climates in Arizona. *American Antiquity*, v. 28, p. 193-198.

Bassett, A. M., and Kupfer, D. H. (1964) A geologic reconnaissance in the southeastern Mojave Desert. *California Division of Mines and Geology Special Report*, v. 83.

Benson, L. V., Currey, D. R., Dorn, R. I., Lajoie, K. R., Oviatt, C. G., Robinson, S. W., Smith, G. I., and Stine, S. (1990) Chronology of expansion and contraction of four Great Basin systems during the past 35,000 years. *Paleogeography, Paleoclimatology, Paleoecology*, v. 78, p. 241-286.

Blackwelder, E. (1933) Lake Manley: An extinct lake of Death Valley. *Geographical Review*, v. 23, p. 464-471.

Blackwelder, E. (1954) Pleistocene lakes and drainage in the Mojave region, southern California. In R.H. Jahns (ed.) *Geology of Southern California*. California Division of Mines Bulletin, v. 170, p. 35-40.

Blount, H. G., and Lancaster, N. (1990) Development of the Gran Desierto sand sea. *Geology*, v. 19, p. 724-728.

Blount, H. G., Smith, M. O., Adams, J. B., Greeley, R., and Christensen, P. R. (1990) Regional aeolian dynamics and sand mixing in the Gran Desierto: Evidence from Landsat Thematic Mapper images. *Journal of Geophysical Research*, v. 95, p. 15463-15482.

Breed, C. S., and Grow, T. (1979) Morphology and distribution of dunes in sand seas observed by remote sensing. In E. D. McKee (ed.) *A Study of Global Sand Seas*, U.S. Geological Survey Professional Paper 1052, p. 253-302.

Brown, W. J. (1989). *Late Quaternary Stratigraphy, Paleohydrology, and Geomorphology of Pluvial Lake Mojave, Silver Lake and Soda Lake Basins, Southern California*. M.S. thesis, University of New Mexico.

Chadwick, O. A., and Davis, J. O. (1990) Soil forming intervals caused by eolian sediment pulses in the Lahontan Basin, northwestern Nevada. *Geology*, v. 18, p. 243-246.

Dohrenwend, J. C. (1987) Basin and range. In W. L. Graf (ed.) *Geomorphic Systems of North America*, Centennial Special, v. 2. Boulder, Colorado, Geological Society of America, p. 303-342.

Dorn, R. I., Jull, A.J.T., Donahue, D. J., Linick, T. W., and Toolin, L. J. (1989) Accelerator mass spectrometry radiocarbon dating of rock varnish. *Geological Society America Bulletin*, v. 101, p. 1363-1372.

El-Baz, F., Breed, C. S., Grolier, M. J., and McCauley, J. F. (1979) Aeolian features in the western desert of Egypt and some applications to Mars. *Journal of Geophysical Research*, v. 84, p. 8205-8221.

El-Baz, F., and Maxwell, T. A., eds. (1982) *Desert landforms of southwestern Egypt: A basis for comparison with Mars*. NASA Contractor Report CR-3611.

Greeley, R., and Iversen, J. D. (1978) Field guide to the Amboy lava field, San Bernadino County, California. In R. Greeley et al. (eds.) *Eolian Features of Southern California: A Comparative Planetary Geology Guidebook*. Arizona State University p. 24-52.

Greeley, R., and Iversen, J. D. (1985) *Wind as a Geologic Process*. New York, Cambridge University Press.

Greeley, R., Arvidson, R. E., Elachi, C., Geringer, M. A., Plaut, J. J., Saunders, R. S., Schubert, G., Stofan, E. R., Thouvenot, E.J.P., Wall, S. D., and Weitz, C. M. (1992) Aeolian features on Venus: Preliminary Magellan results. *Journal of Geophysical Research*, v. 97, p. 13319-13345.

Jahns, R. H., ed. (1954) *Geology of Southern California*. California Division of Mines, Bulletin 170.

Laity, J. E. (1987) Topographic effects on ventifact formation, Mojave Desert, California. *Physical Geography*, v. 8, p. 113-132.

Laity, J. E. (1992) Ventifact evidence for Holocene wind patterns in the east-central Mojave Desert. *Zeitschrift für Geomporhologie*, v. 84, p. 73-88.

Lancaster, N. (1993) Kelso Dunes. *National Geographic Research and Exploration*, v. 9, p. 444-459.

Lancaster, N., Greeley, R., and Christensen, P. R. (1987) Dunes of the Gran Desierto sand-sea, Sonora, Mexico. *Earth Surface Processes and Landforms*, v. 12, p. 277-288.

Lancaster, N., Wintle, A. G., Edwards, S. R., Duller, G., and Tchakerian, V. P. (1991) Chronology of aeolian activity at Kelso Dunes: evidence from luminescence dating of dune sediments. *Geological Society of America Abstracts with Programs*, v. 23, p. 355.

Lancaster, N., Gaddis, L., and Greeley, R. (1992) New airborne imaging radar observations of sand dunes: Kelso Dunes, California. *Remote Sensing of the Environment*, v. 39, p. 233-238.

Leopold, L. B., Wolman, M. G., and Miller, J. P. (1964) *Fluvial Processes in Geomorphology*. W.H. Freeman & Co., San Francisco.

McFadden, L. D., Wells, S.G., and Jercinovich, M. J. (1987) Influences of eolian and pedogenic processes on the origin and evolution of desert pavements. *Geology*, v. 15, p. 504-508.

Merriam, R. (1969) Source of sand dunes of southeastern California and northwestern Sonora, Mexico. *Geological Society of America Bulletin*, v. 80, p. 531-534.

Miller, R. R. (1946) Correlation between fish distributions and Pleistocene hydrography in eastern California and southwestern Nevada, with a map of the Pleistocene waters. *Journal of Geology*, v. 54, p. 43-53.

Paisley, E.C.I., Lancaster, N., Gaddis, L. R., and Greeley, R. (1991) Discrimination of active and inactive sand from remote sensing: Kelso Dunes, Mojave Desert, California. *Remote Sensing of the Environment*, v. 37, p. 153-166.

Sagan, C., Veverka, J., Fox, P., Dubisch, R., Lederberg, J., Levinthal, E., Quam, L., Tucker, R., Pollack, J. B., and Smith, B. A. (1972) Variable features on Mars: Perliminary Mariner 9 television results. *Icarus*, v. 17, p. 346-372.

Sharp, R. P. (1966) Kelso dunes, Mojave Desert, California. *Geological Society of America Bulletin*, v. 77, p. 1045-1074.

Sharp, R. P. (1978) The Kelso Dune complex. In R. Greeley et al. (eds.). *Eolian Features of Southern California:* A Comparative Planetary Geology Guidebook. Arizona State University, p. 53-63.

Shelton, J. S., Papson, R. P., and Womer, M. (1978) Aerial guide to geological features of southern California. In R. Greeley et al. (eds.) *Eolian Features of Southern California: A Comparative Planetary Geology Guidebook*. Arizona State University, p. 216-249.

Smith, H.T.U. (1967) *Past versus present wind action in the Mojave Desert region, California*. U.S. Air Force Cambridge Laboratory Report AFCRL-67-0683.

Smith, R.S.U. (1982) Sand dunes in the North American Desert. In G.L. Bender (ed.) *Reference Handbook on the Deserts of North America*. Greenwood Press, Westport, Connecticut, p. 481-554.

Spaulding, W. G. (1991) A middle Holocene vegetation record from the Mojave Desert of North America and its paleoclimatic significance. *Quaternary Research*, v. 35, p. 427-437.

Tchakerian, V. P. (1991) Late Quaternary aeolian geomorphology of the Dale Lake sand sheet, southern Mojave Desert, California. *Physical Geography*, v. 12, p. 347-369.

Tchakerian, V. P., Zimbelman, J. R., and Williams, S. H. (1992) Transport of aeolian sediments across desert basins, California and Arizona. *Association of American Geographers Abstracts*, 88th Annual Meeting, p. 235.

Thompson, D. G. (1929) *The Mojave Desert Region, California: A Geographic, Geologic, and Hydrologic Reconnaissance*. U.S. Geological Survey Water-Supply Paper 578.

U.S. Geological Survey (1969a) Topographic map of the Needles area, California. Map NI 11-6, scale 1:250,000, Denver, CO.

U.S. Geological Survey (1969b) Topographic map of the Salton Sea area, California. Map NI 11-9, scale 1:250,000, Denver, CO.

Wells, S. G., McFadden, L. D., and Dohrenwend, J. C. (1987) Influence of late Quaternary climatic changes on geomorphic and pedogenic processes on a desert piedmont, eastern Mojave Desert, California. *Quaternary Research*, v. 27, p. 130-146.

Williams, S. H., Zimbelman, J. R., and Tchakerian, V. P. (1991) Evidence of aeolian sand transport across the Colorado River. *EOS,Transactions of the American Geophysical Union*, v. 72, p. 214.

Wilson, I. G. (1971) Desert sandflow basins and a model for the development of ergs. *Geographical Journal*, v. 137, p. 180-199.

Zimbelman, J. R., and Williams, S. H. (in preparation) Aeolian wind streaks: Geological and botanical effects on surface albedo contrasts.

6 FIELD METHODS IN A STUDY OF THE PROCESS-RESPONSE SYSTEM CONTROLLING DUNE MORPHOLOGY, SALTON SEA, CALIFORNIA

Kevin R. Mulligan
Department of Geography
Texas A&M University

ABSTRACT

To better understand the nature of aeolian processes, a field study was undertaken to investigate the interaction between wind flow, sand transport, and dune morphology. The study focused on a barchan located west of the Salton Sea in southern California. This paper describes the research design, study area, field methods, and instrumentation used in the study. To describe and analyze the wind events that occurred during the course of the field season, wind speed and direction were monitored on a continuous basis over a four-week period in late spring 1991. In response to these wind events, survey data and measurements from a grid of erosion stakes were both used to monitor changes in dune morphology. To help explain observed changes in dune morphology, an array of 40 anemometers was deployed to measure spatial variability in near-surface wind speed and the vertical structure of the wind field over the dune profile. One of the principal aims of the study was to develop a physically based numerical model of aeolian sand transport that could be used to simulate observed patterns of erosion and deposition over the dune profile.

INTRODUCTION

Aeolian sand dunes tend to develop wherever an abundance of loose sand is subject to strong and persistent wind. Whereas vegetation can play an important role in the growth and development of dunes, and have a strong influence on dune morphology, most active sand dunes are considered to be true aerodynamic bedforms (Bagnold 1941). Active dunes result from instability in the flow of air over a deformable boundary and dune morphology is directly related to the complex interaction and mutual adjustment that occurs between form and flow.

Unfortunately, our present understanding of this form-flow interaction is limited. Although it is widely accepted that dune morphology is a function of wind speed and direction, grain size, available sand supply, and upwind roughness, questions remain regarding the manner in which these variables control dune form and size. Of particular interest are the wind flow and sand transport processes that control the downwind profile of a transverse dune, insofar as an understanding these basic aeolian processes is prerequisite to understanding more complex dune morphology (Tsoar 1985).

Desert Aeolian Processes. Edited by Vatche P. Tchakerian. Published in 1995 by Chapman & Hall, London.
ISBN 978-94-010-6519-1

Although several field studies have been undertaken to examine the interaction between wind flow and dune morphology (Landsberg 1942, Olson 1958, Inman et al. 1966, Tsoar 1983, Livingstone 1986, Lancaster 1985, 1989, Mulligan 1988, Havholm and Kocurek 1988, Hesp et al. 1989, Burkinshaw et al. 1993, Burkinshaw and Rust 1993), until recently, most of this research has been constrained by limited instrumentation. Moreover, few field studies have focused specifically on the wind flow and sand transport processes that control the profile geometry of a barchan (Howard et al. 1978, Wiggs 1993), the simplest type of transverse dune. Thus, our understanding of form-flow interaction remains constrained by a paucity of quantitative field data. Without detailed measurements of wind flow and subsequent erosion and deposition, it is difficult to elucidate the interaction between wind flow, sand transport, and dune morphology (Watson 1987, Lancaster 1987).

To address this problem, a field study was designed to investigate the process-response system controlling dune morphology. In particular, the research focused on the wind flow and sand transport processes that control the profile geometry of a barchan. This paper describes the specific research objectives, the study area, field methods, and instrumentation. The results from this study will be explored in a subsequent paper that seeks to integrate field methods and instrumentation with the field data. The limitations inherent in the methodology are discussed in the context of our current understanding of aeolian processes.

The Process-Response System

Figure 1 is a conceptual model that illustrates the most important elements of the process-response system controlling dune morphology. In this model, a sand dune is viewed as being part of the interface between larger atmospheric and terrestrial systems. In the atmosphere, the magnitude and frequency of regional winds define the prevailing wind regime. As part of the terrestrial system, sand volume and grain size distribution define the potential source area.

At the atmosphere-terrestrial interface, the prevailing wind regime is modified by local upwind topography and upwind roughness. In a similar manner, topography and upwind roughness can also influence the available sand supply. In turn, the available sand supply and local wind regime define potential transport and, therefore, the rate of sand input to the dune system.

Once sand begins to accumulate, it creates velocity perturbations in the wind field. Dune morphology and the local wind regime both influence the flow field over a dune and, thus, the spatial pattern of near-surface wind speed and direction. In turn, this spatial pattern of near-surface wind flow controls spatial variability in surface shear stress, spatial variability in sand transport and, ultimately, spatial variability in the pattern of erosion and deposition. Finally, this spatial pattern of erosion and deposition is what constitutes a change in dune morphology.

If we consider this conceptual model as it applies to a barchan, there are

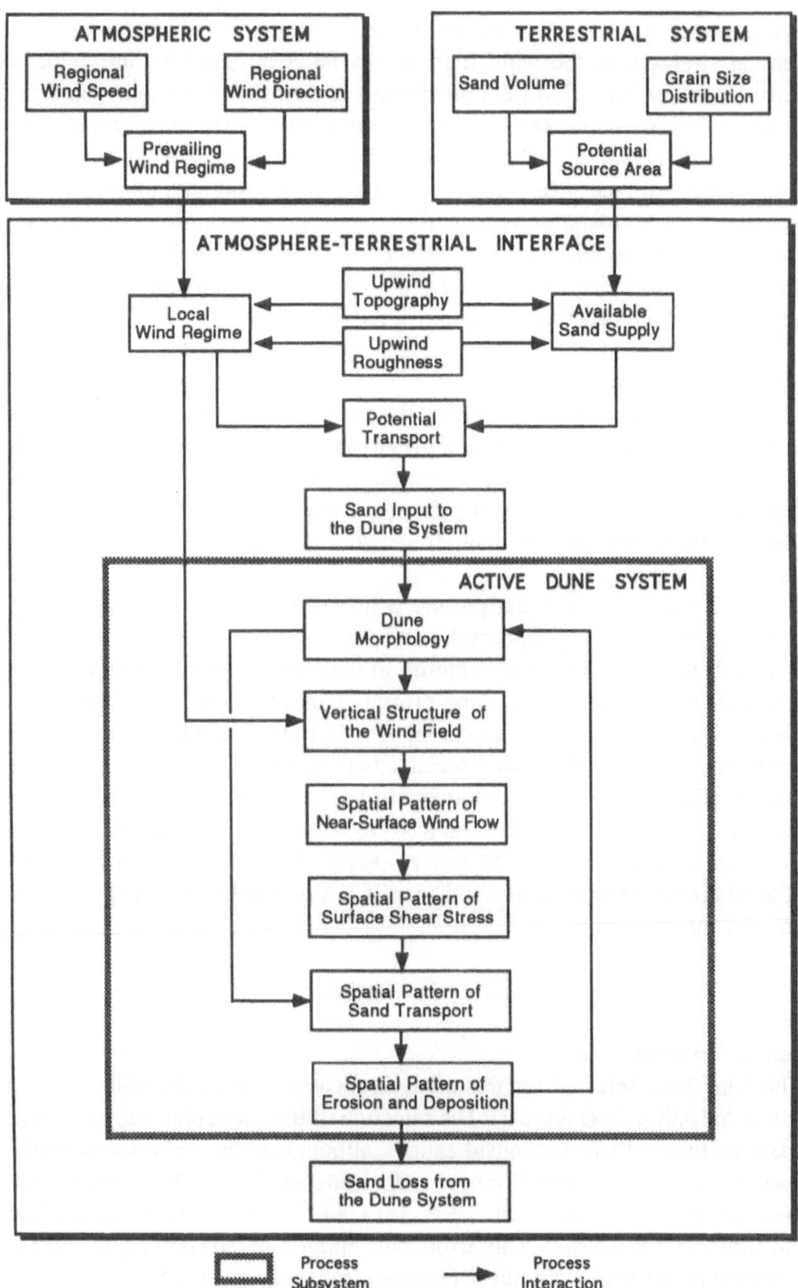

Figure 1. Conceptual model of the process-response system controlling dune morphology.

several important research questions that need to be addressed. First, what is the vertical structure of the wind field over a barchan, and how are velocity perturbations in this wind field manifested as downwind variability in near-surface wind speed? Second, given the wind field over the dune profile, how are downwind changes in near-surface wind speed manifested as downwind variability in surface shear stress, rates of sand transport and, ultimately, the spatial pattern of erosion and deposition? Lastly, given the profile of a barchan, and downwind variability in near-surface wind speed, is small-scale sediment transport theory adequate to model the observed patterns of erosion and deposition which constitute a change in the dune profile?

Research Objectives

To address these research questions, the field study was designed to collect detailed measurements of wind flow over a barchan and detailed measurements of the subsequent change in dune morphology. In many ways, the research design was strongly influenced by the conceptual model outlined in Figure 1. This model served as a general framework, in formulating five specific objectives of this study.

First, I describe and analyze the wind events that occurred during the course of the field season. Second, in response to the wind events, I describe and analyze the changes that occurred in dune morphology, with particular emphasis on changes in the dune profile. Third, in order to explain these observed changes in dune morphology, I describe and analyze the spatial variability in near-surface wind speed, again with particular emphasis on the dune profile. Fourth, in order to explain the downwind variability in near-surface wind speed, I describe and analyze the vertical structure of the wind field over the dune profile. Fifth, I develop a physically based numerical model of aeolian sand transport that could be used to simulate the observe patterns of erosion and deposition.

STUDY AREA

Salton Sea Dune Field

The dune field selected for this study is located west of the Salton Sea in southern California (Figure 2). The selection of this particular study area was based mainly on logistical considerations, although the barchans have been the focus of previous research (Rempel 1936, Long and Sharp 1964, Norris 1966, Howard 1977, Howard et al. 1978, Haff and Presti 1984, Walmsley and Howard 1985, Lee 1987). This affords the opportunity to compare the results obtained in this study with those obtained from previous work.

The Salton Sea barchans consist of individual and coalesced forms which migrate toward the east across a relatively flat and barren desert plain (Figure 3). Haff and Presti (1984) estimate the number of dunes to be approximately 70. Whereas most of the dunes exhibit the general form of barchans, or coalesced forms, relatively few dunes attain the classic symmetrical shape.

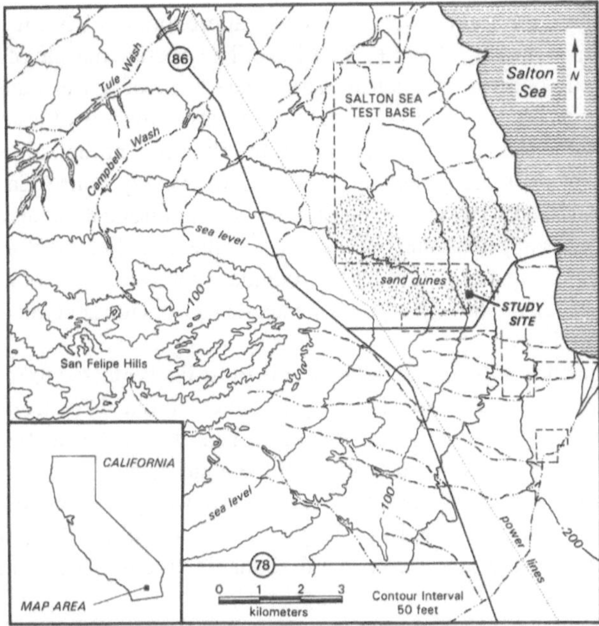

Figure 2. Location of the Salton Sea dune field.

Figure 3. Oblique aerial photograph of the Salton Sea barchans approximately six months before the main field season began. View is looking toward the northwest. The study dune is the isolated barchan located in the bottom center of the photograph.

The main dune field covers an area of approximately 15 km^2, most of which is located on the Salton Sea Test Base. The western margin of the dune field is roughly defined by a northwest-southeast line of power transmission towers located east of California Highway 86. From this line of transmission towers, the dune field extends eastward to the shoreline of the Salton Sea.

The surface of the desert is composed of weakly consolidated clay, mudstone, and sandstone beds of the Brawley formation and the source area for the sand extends approximately 40 km west of the dune field to Borrego and Clark valleys (Long and Sharp 1964). Potential sources of dune sand include poorly consolidated sedimentary beds, beach sands from former Lake Cahuilla, and fresh alluvium.

The existence of the dune field itself is most likely related to the lack of stream dissection in the area occupied by the dunes. Haff and Presti (1984) note that the terrain is gently convex and, therefore, runoff from the San Felipe Hills is channeled to either the north or south of the dune field (Figure 2). Haff and Presti (1984) also note that the terrain to the north, west, and south of the dune field is more heavily dissected with larger and deeper gullies. These observations suggest that the larger ephemeral channels preclude the formation of migratory dunes.

The climate of the dune field can be characterized as hot and arid. In summer, maximum daytime temperatures commonly exceed 40°C in the afternoon, with very low humidity. Rainfall in the region is very low. With the exception of Death Valley, and perhaps small parts of the Mojave Desert, the climate in the region is among the most arid in the United States. Potential evaporation is on the order of 180 cm per year and mean annual rainfall for most stations in the vicinity of the Salton Sea is on the order of 5 to 10 cm (Hely et al. 1966). Most of the precipitation occurs during winter as light rainfall, although infrequent heavy showers can occur in late summer as a result of convective storms.

Unfortunately, no long-term record of hourly wind speed and direction is available for the dunes, although climate summaries have been compiled for stations in the vicinity of the Salton Sea. Wind speed data from Thermal, located north of the Salton Sea, and Calipatria, located southeast of the Salton Sea, both show the greatest frequency of strong winds (>8.9 ms^{-1}) in April, May, and June. On the Salton Sea Test Base itself, monthly average wind speeds are available for a station at Sandy Beach (Haff and Presti 1984). These data also show the strongest winds in April, May, and June. Whereas none of these summary data can be used to describe or analyze the magnitude and frequency of sand-driving events, it is apparent that strong winds are most likely to occur in late spring and early summer, which is typical of desert stations in California.

In response to this climatic regime, vegetation is well adapted to xeric conditions. On the desert floor, creosote (*Larrea tridentata*) and white bursage (*Ambrosia dumosa*) are the dominant species. Winter ephemerals, such as the poppy (*Eschscholzia*), blue-bonnet (*Lupinus*), and evening primrose

(*Oenothera*), may also be abundant during the late spring (Crosswhite and Crosswhite 1982).

Description of the Field Site

A study site was selected within the dune field following an overflight of the area and preliminary ground reconnaissance in late October 1990, approximately six months before the main field season. The study site is located on the border of the Salton Sea Test Base, several hundred meters west of the road leading from Highway 86 to the shore of the Salton Sea (Figure 2). The dune is an isolated barchan, approximately 3.5 m in height and approximately 40 m from the windward toe to the brink of the slipface (Figure 3). Upwind of the dune, over a distance of 140 m, the terrain is relatively flat with little vegetation. The local source of sand is a complex set of dunes located 140 m directly west of the field site.

FIELD METHODS AND INSTRUMENTATION

From the wind speed summaries recorded in the vicinity of the Salton Sea, it is apparent that most of the dune activity should occur during late spring and early summer, during the months of April, May and June. To maximize the likelihood of obtaining data during one or more sand-driving events, most of the field work was undertaken over a six-week period in late spring, from 12 May to 23 June 1991.

Local Climate and Prevailing Wind Conditions

The local prevailing wind speed and direction were measured with a cup anemometer and wind vane to describe and analyze the wind events that occurred during the course of the main field season. These wind speed and direction sensors were mounted at a height of 10 m on a mast located 20 m upwind of the dune. In addition to wind speed and direction, ambient air temperature was also measured at a height of 10 m on the mast. The wind speed, wind direction, and temperature data were measured continuously from 20 May to 16 June.

At a small climate station located adjacent to the dune, additional background climate data were also measured throughout the field season. The background data included solar radiation, ambient air temperature at 1.5 m, and the temperature of the desert surface. A small rain gauge was also deployed but, as expected, no rainfall was recorded during the field season.

Surveyed Changes in Morphology

Conventional surveying techniques were used to measure changes in dune morphology and, in particular, the changes that occurred in the dune profile. It was evident, based on field observations, that the dune was aligned with respect to winds from 270° (all directional bearings are based upon true north). On 31 March, approximately six weeks before the main field season, a survey baseline was established 30 m upwind of the dune. The baseline was 70 m in length and oriented crosswind on a bearing of 0° (Figure 4). From the midpoint on this baseline, at 0 + 35 m, a single downwind profile was surveyed over the central axis of the dune. The profile was surveyed on a bearing of 90° to correspond with the presumed dominant wind direction.

During the main field season this downwind profile was resurveyed at weekly intervals: on 12, 19, and 27 May, and on 02, 09, 16, and 23 June. With these relatively frequent repetitive surveys, it was possible to monitor changes in the profile geometry as the dune advanced downwind. Following the main field season this profile was resurveyed on 15 March 1992, approximately one year after the first survey. From this last set of measurements it was possible to estimate the annual rate of dune movement.

In addition to these weekly repetitive surveys of the central dune profile, the entire field site was mapped on three separate occasions. These three surveys of the field site were used to develop base maps of dune topography and to measure changes in surface elevation over the entire dune. Fifteen downwind profiles were surveyed in order to map the field site. The profiles were spaced at 5-m intervals along the 70-m baseline, and each profile was surveyed 100 m downwind on a bearing of 90° (Figure 4). The entire field site was first surveyed on 27 May. A second survey was conducted one week later, on 2 June, following a very large sand-driving event. The final survey of the entire field site was done on 23 June, at the end of the main field season.

Measurements of Erosion and Deposition

Although the survey data are adequate to map changes in dune form on a weekly basis, it is difficult to relate these changes to individual wind events which often occur over a period of less than 12 hours. To provide more detailed data concerning changes in dune morphology, erosion stakes were used to monitor changes in surface elevation on a daily basis.

The erosion stakes were constructed from 50 cm lengths of 1/2-inch-diameter PVC pipe. During an initial survey of the dune, the erosion stakes were placed 35 cm into the sand. In this way it was possible to measure a maximum of either 35 cm of erosion or 15 cm of deposition. While there was often some minor scour of the sand surface at the base of erosion stakes, with a few exceptions, the measurements should be reliable to within ±0.5 cm. Moreover, the erosion stakes were releveled only when the dune was resurveyed. In this way it was possible to cross-check the measurements of cumulative erosion and deposition against the surveyed changes in surface elevation.

Initially, 10 erosion stakes were set into place when the dune profile was surveyed at the beginning of the field season, on 12 May. The erosion stakes were surveyed into place at 5-m intervals along the dune profile and changes in surface elevation were monitored on a daily basis. An additional 86 erosion stakes were surveyed into place when the entire dune was mapped on 27 May. These additional stakes were also placed at 5-m intervals along each survey line, thereby creating a 5-m by 5-m grid (Figure 4). In total, 96 stakes were used

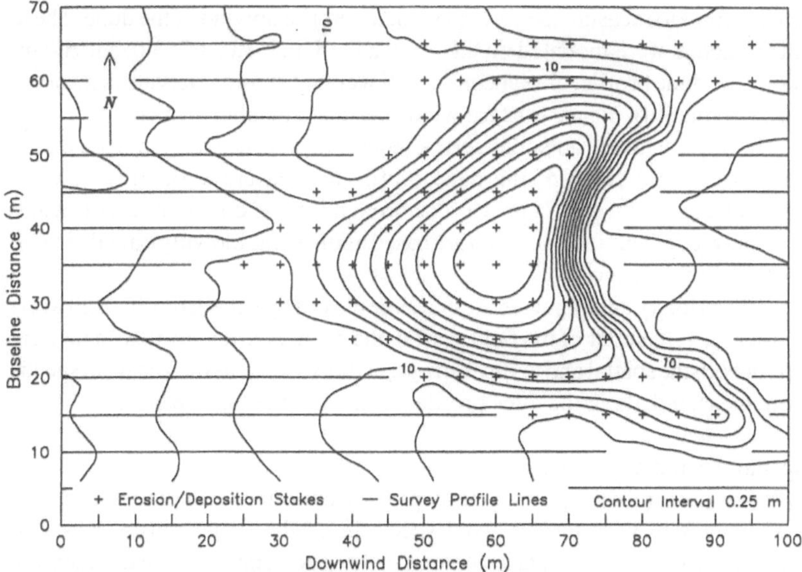

Figure 4. Location of survey lines and erosion stakes used to monitor changes in dune morphology.

to monitor erosion and deposition on a daily basis from 27 May to 16 June. Measurements were also made on 23 June to coincide with the last complete survey of the field season. Additional follow-up measurements were made on 7 July, 10 August, and 25 November 1991.

Deployment of Wind Sensors

Both the spatial variability in near-surface wind speed and the vertical structure of the wind field over the dune profile were measured in order to help explain the observed changes in dune morphology. Detailed measurements of near-surface wind speed were also needed as the primary data input to the simulation model. Considering these multiple objectives, it was necessary to prioritize the field work in terms of a deployment strategy for the wind sensors. During the course of the field season, three different anemometer arrays were deployed; the primary anemometer-array, the spatial anemometer-array, and the vertical anemometer-array.

The Primary Anemometer-Array

Insofar as the most important aspect of this study involves understanding the processes that control the profile geometry of a barchan dune, the primary anemometer-array was deployed during the first phase of the field season to measure the velocity profile upwind of the dune and spatial variability in near-surface wind speed along the dune profile. The primary array consisted of 16 anemometers. Figure 5 shows the deployment of the anemometers with respect to dune topography. In addition to the wind speed and direction sensors used to measure the prevailing wind, six anemometers were also mounted on the upwind mast to measure the vertical velocity profile upwind of the dune. These anemometers were mounted on the mast at 0.15, 0.5, 1.0, 1.7, 3.0, and 5.0 m. Along the dune profile, nine anemometers were set at 5-m intervals to measure spatial variability in near-surface wind speed at a height of 15 cm.

In addition to the anemometers, a second wind vane was deployed on the upper windward slope of the dune profile to measure variability in wind direction at a height of 0.75 m. The data from this wind direction sensor ensure that analyses of wind acceleration are based upon winds blowing directly in line with the dune profile.

The primary anemometer-array was deployed by itself from 20 to 29 May. During the remainder of the field season, the primary array formed an integral part of both the spatial and vertical arrays. In this way, near-surface wind speeds were measured along the dune profile throughout the field season.

Spatial Anemometer-Array

Whereas data from the primary array provide important information concerning changes in near-surface wind speed along the dune profile, it is also important to place these data in context. In the second phase of the field season, anemometers were deployed to measure spatial variability in near-surface wind speed over the entire stoss slope of the dune. To measure this spatial variability in near-surface wind flow, the primary array was expanded by adding 24 anemometers.

Figure 6 illustrates the deployment of the spatial array with respect to dune topography. In addition to the 16 anemometers which comprised the primary array, 12 anemometers were deployed on the northern flank of the dune and 12 anemometers were deployed on the southern flank. Along the central axis of the dune, the deployment of anemometers and the deployment of the wind vane remain unchanged. Figure 7 shows the spatial array of anemometers deployed over the stoss slope of the dune. This spatial array was deployed from 29 May to 8 June and, again, each of the anemometers was set to measure wind speed at a height of 15 cm.

Vertical Anemometer-Array

In the final phase of the field season, from 7 to 16 June, anemometers were redeployed to measure the vertical structure of the wind field over the dune profile. The principle aim was to develop a better understanding of the wind flow processes that control spatial variability in near-surface wind speed and, ultimately, sand transport along the central axis of the dune.

Figure 8 shows the deployment of anemometers in the vertical array. In addition to the mast located upwind of the dune, four masts were set at 10-m intervals along the dune profile. The first mast was set on the lower windward slope near the toe of the dune. The second mast was placed on the middle windward slope. The third mast was located on the upper windward slope near the dune crest and the fourth mast was placed approximately 2 m upwind of the brink.

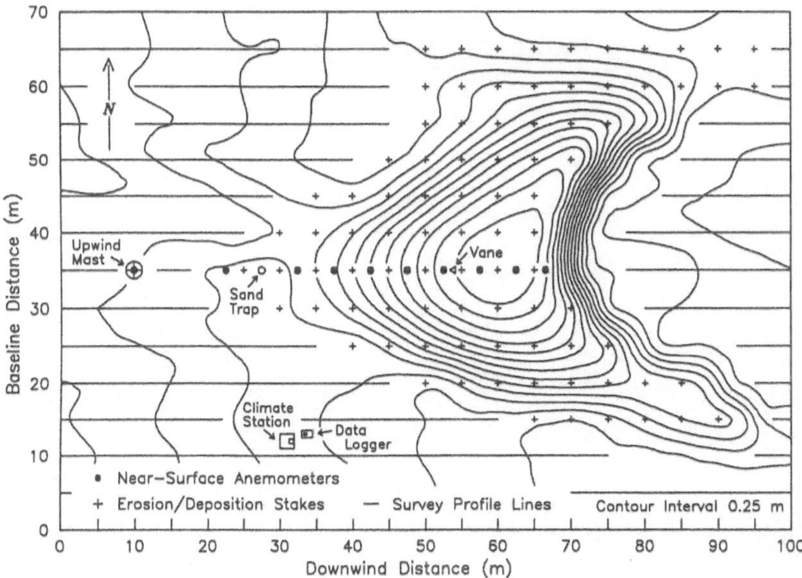

Figure 5. Deployment of wind sensors in the primary anemometer-array.

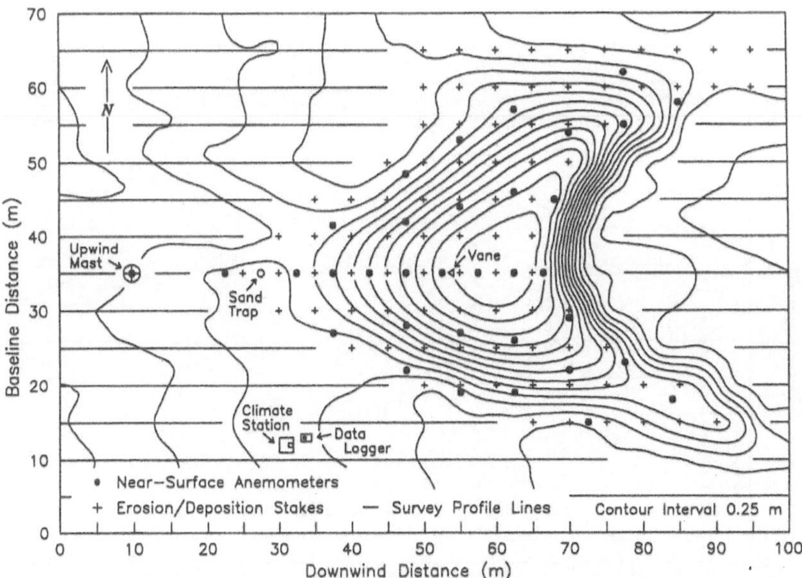

Figure 6. Deployment of wind sensors in the spatial anemometer-array. Note that the deployment of anemometers along the dune profile is the same as the deployment used in the primary array.

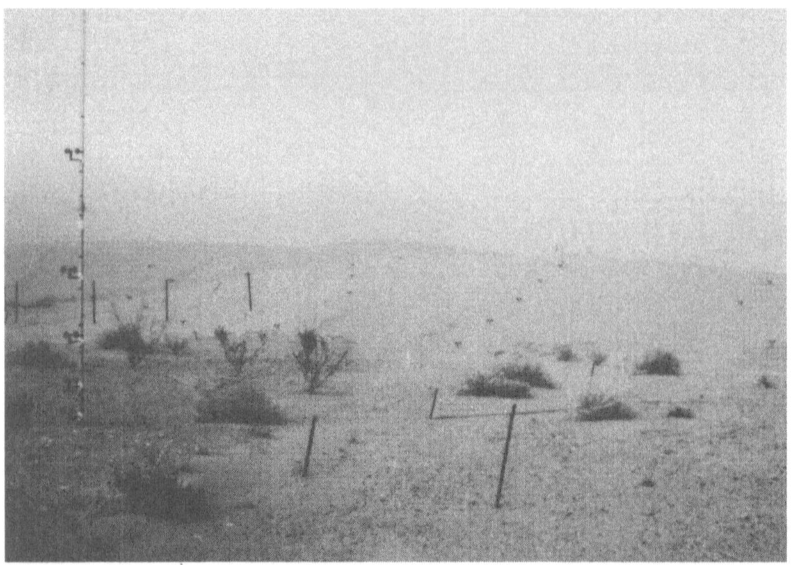

Figure 7. Anemometers deployed as part of the spatial array.

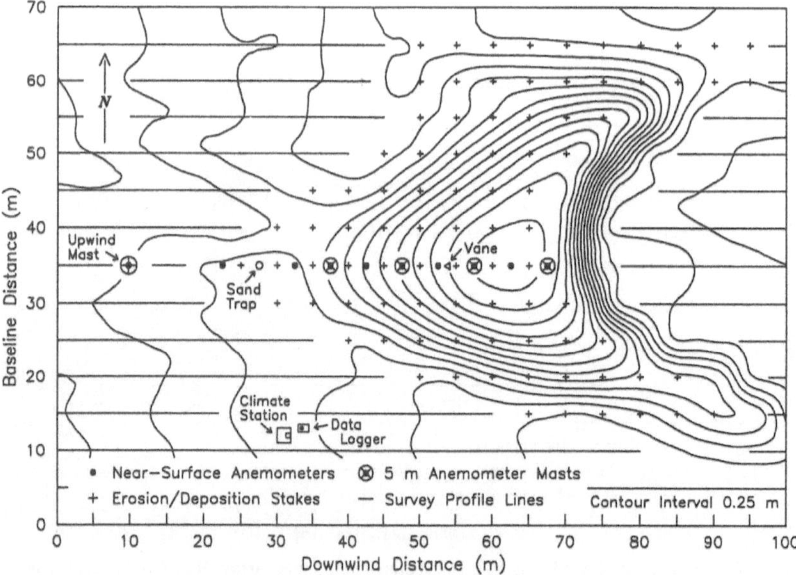

Figure 8. Deployment of wind sensors in the vertical anemometer-array.

Seven anemometers were mounted on each of the four masts located along the dune profile (Figure 9). Each mast was 5 m in height with anemometers mounted at 0.15, 0.40, 0.65, 1.0, 1.7, 3.0, and 5.0 m. Between each of the masts, a single anemometer was set to measure wind speed at a height of 15 cm. In addition, the wind vane deployed on the upper windward slope of the dune was maintained as part of the vertical array. Again, data from the wind vane ensure that analyses of wind flow over the dune profile are based upon winds blowing directly in line with the central axis of the dune.

Data for the Simulation Model

In addition to measuring the wind field over a dune, one of the main objectives in this study was to develop a physically based numerical simulation model of sand transport over the dune profile. The primary data required as input to the simulation model include near-surface wind speeds measured along the dune profile, the mean grain size of the sand, and the rate of sand input to the dune from upwind.

The near-surface wind data used as input to the simulation model were derived from anemometers located along the central axis of the dune. Insofar as all three anemometer arrays provide wind speed data measured at a height of 15 cm along the dune profile, input-data for the simulation model could be derived from any of the wind events that occurred during the course of the field season.

In the simulation model, it was also necessary to specify the mean grain size of the sand. To determine mean grain size, sand samples were collected at 5 m intervals along the dune profile. Each sample was approximately 50 g. The samples were collected by scraping the thin veneer of coarse-grain lag from the surface and scooping from the top 2 or 3 cm of sand. Each sample was then subjected to a mechanical sieve analysis at 1/4 ø intervals.

To measure the rate of sand transport input to the dune, a Leatherman-type sand trap was deployed approximately 10 m upwind. (Leatherman 1978). Above ground, the trap was 50 cm tall with a 1-cm-wide opening (Figure 10). To allow air flow through the trap, a 2.5-cm-wide slot was cut into the back of the trap and covered with a fine nylon mesh. Owing to the difficulty of continuously monitoring a sand trap, the data from the trap are intermittent samples.

Whereas the accuracy of sand trap data is often questionable, considering the problems associated with trap efficiency, this does not pose a significant problem in the analysis. Most of the sand transported over a dune is not derived from upwind, but rather, the sand is eroded from the dune surface itself. Nevertheless, the trap data provide a minimum rate of sand transport into the dune system.

Instrumentation

A total of 40 anemometers were used in this study to measure both spatial and temporal variability in wind speed. The sensors are rotating, three-cup an-

Figure 9. Anemometer masts deployed along the central axis of the dune as part of the vertical array.

Figure 10. Sand trap deployed upwind of the dune.

emometers manufactured by Trade-Wind® Instruments. The transmitter in each unit is a magnetic reed switch that produces a pulsed output signal directly proportional to wind speed. Prior to the field season, each of the anemometers was individually calibrated in the low-turbulence wind tunnel at the Iowa Institute of Hydraulic Research (Patel 1990). This particular wind tunnel is a large recirculating tunnel with a 1.5-m by 1.8-m rectangular test section. The seven-point calibration curves are all very similar and linear over the range of calibration, from 2 to 25 ms^{-1}. The threshold wind speed for the anemometers is less than 1.5 ms^{-1}.

As discussed previously, the anemometers were mounted in either of two configurations. If an anemometer was deployed to measure near-surface wind speed, it was mounted on a 30-cm section of PVC pipe that was inserted into the sand (Figure 11). If, on the other hand, anemometers were deployed to measure a wind velocity profile, the anemometers were mounted on a telescopic mast. The wind speed sensors were mounted on sections of PVC pipe secured to the mast with hose clamps (Figure 12).

In addition to these 40 anemometers, two wind vanes were used to measure variability in wind direction. The wind vanes are manufactured by Met One® Instruments and have a damping ratio 0:10. In the field, both wind vanes were oriented upon true north and the output signal calibrated to read from 0° to 360°.

Ambient air temperature was measured using type T thermocouples. One thermocouple was placed in a six-plate Gill radiation shield mounted on the upwind mast at a height of 10 m. At the small climate station adjacent to the dune, a second thermocouple was placed in a radiation shield mounted on a mast at a height of 1.5 m. Type T thermocouples were also used to measure the temperature of the desert surface at the climate station. Solar radiation was measured with a Li-Cor® silicon pyranometer.

Whereas the thermocouples are adequate to measure diurnal changes in ambient air temperature, these sensors do not provide the accuracy or precision necessary to analyze thermal gradients in the atmosphere. During strong winds, the effects of atmospheric stability or instability are assumed to be negligible and no attempt was made to directly measure temperature profiles.

Data Acquisition

Data from the wind speed, wind direction, temperature, and solar radiation sensors were recorded on magnetic cassette tapes using a Campbell Scientific® 21X datalogger. To accommodate the large number of wind speed data channels, the pulse-count capacity of the datalogger was expanded using five input modules, each with an eight-channel capacity. Whereas the use of a single datalogger required approximately 1500 m of cable to connect all of the anemometers, it was critical to have data recorded as synchronized measurements on single time clock.

All of the wind speed and wind direction data were recorded as both 1-minute and hourly average values. To measure wind speed, the total pulse count from each anemometer was recorded over a 60-second interval. These raw

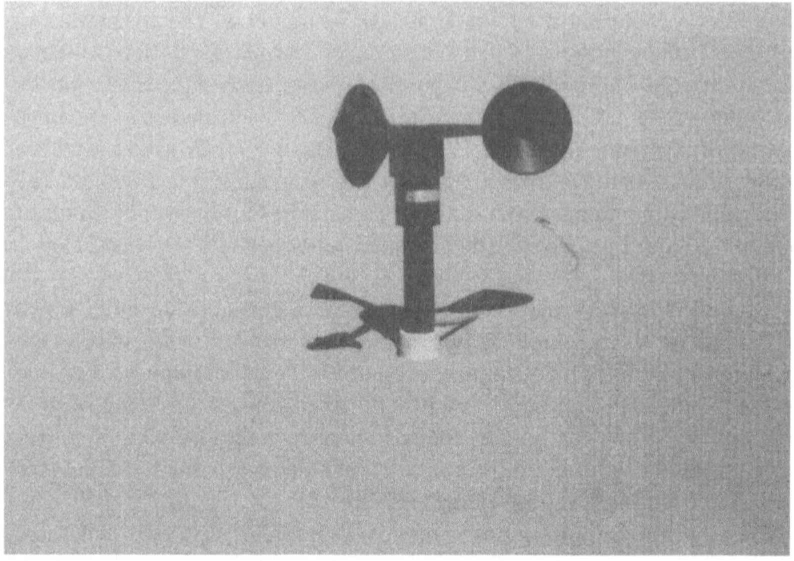

Figure 11. Anemometer deployed to measure near-surface wind speed.

Figure 12. Anemometer deployed as part of a wind velocity profile.

values were later converted to wind speeds using the individual anemometer calibrations. To measure wind direction, the output from each wind vane was sampled at 2-second intervals and 30 values were used to calculate the average wind direction and standard deviation over 1 minute. Temperature and solar radiation data were recorded as instantaneous values sampled once every 60 seconds. In all, approximately 2 million observations were recorded during the field season.

DISCUSSION

The field methods described in this paper represent an attempt to collect detailed measurements of wind flow over a dune and detailed measurements of the subsequent change in dune morphology. Whereas the field season was quite successful in terms of collecting these data, the analysis is far more tentative given our current understanding of aeolian processes.

Methodology Limitations

To relate measurements of wind flow with measurements of the subsequent change in dune morphology, we must recognize several assumptions inherent in the analysis. To link the various components of the process-response system, we assume: (1) that wind speed measurements are accurate, (2) that shear velocity or surface shear stress can be accurately derived from these wind speed measurements, (3) that sand transport can be accurately derived from these shear velocity or shear stress calculations, and (4) that variations in sand transport can be modeled to accurately reflect the observed patterns of erosion and deposition.

The first assumption concerns the accuracy of wind speed measurements. Whereas most researchers are aware that cup anemometers have a delayed response to changes in wind speed (MacCready 1966), this problem is rarely acknowledged in the dune literature. As in most previous research, this study assumes that wind speed data are reliable, although it is important to recognize this could be a significant source of error. Kagnov and Yaglom (1976) estimate that cup anemometers will tend to overestimate the actual wind speed by approximately 8% to 10%.

Assuming wind speed data are reasonably reliable, we make a second assumption concerning the accuracy of shear velocity or surface shear stress calculations. This is perhaps the largest source of error in any analysis of sand transport over dunes. Whereas shear velocity can be derived from logarithmic wind profiles, wind profiles over a dune are nonlogarithmic. To overcome this problem, it has been suggested that shear velocity can be estimated using a single measurement of wind speed close to the dune surface (Mulligan 1988). While this approach has obvious limitations (Gerety 1985), at the present time there is no better way to estimate shear velocity using conventional anemometry. Clearly, this problem needs to be addressed, insofar as the potential for large errors constrains efforts to model sand transport over a dune.

If we assume that shear velocity can be derived from near-surface wind speed measurements, the third assumption inherent in the analysis concerns the accuracy of sand transport calculations. In this case, there are a least three components to the problem. The first component involves the accuracy of sand transport equations. Over the years, several transport equations have been developed in wind tunnel experiments, but it remains unclear as to how well these equations work in a particular field situation. To field-test the accuracy of transport equations, researchers must normally rely on data from sand traps (e.g., Svasek and Terwindt 1974, Sarre 1988, 1989). Unfortunately, it is difficult to determine trap efficiency and, therefore, the results from this field-testing tend to remain inconclusive.

In a similar manner, we do not fully understand the direct effect of slope on calculations of sand transport. The most widely cited equation (Bagnold 1956) remains largely theoretical. Moreover, if we compare the predicted effects of slope with data from field experiments, the field data suggest that slope has a much greater effect on sand transport (Hardisty and Whitehouse 1988). Clearly, more field research is needed to fully resolve this issue.

Another problem that can affect calculations of sand transport is the lag time necessary for saltation to equilibrate following a change in wind speed. If there is a sharp increase in wind speed, presumably it takes time for the saltation layer to become fully saturated. Conversely, saltation may exceed the transport capacity of the wind following a sharp decrease in wind speed. This problem was discussed by Bagnold (1941, 1956), who believed that changes in the rate of sand transport could lag several meters behind a change in wind speed. In essence, Bagnold argued that spatial variability in sand transport over a dune is not in equilibrium with the wind. Again, this is an important question that needs to be resolved because, if true, a significant lag in sand transport may have a significant effect on transport calculations.

The fourth problem inherent in this type of analysis concerns the assumption that variations in sand transport can be modeled to accurately reflect the observed pattern of erosion and deposition. In this particular study, the modeling effort is concerned with the dune profile. Calculations of sand flux are volumetric assuming a 1-m width along the central axis of the dune. In the model, it is assumed that wind flow is directly in line with the dune profile and there is no convergence or divergence in sediment transport streamlines. Obviously, if wind direction is not in line with the dune profile, the model will fail to accurately predict the observed erosion and deposition. Similarly, if there is notable divergence or convergence in transport streamlines along the central axis of the dune, this could also be a potential source of error in the model.

Whereas each of the problems outlined above offers a significant challenge to overcome, perhaps the greatest concern is the possibility of canceling errors in the analysis. For example, there may be canceling errors in the model developed by Howard et al. (1978). In this model of sediment transport over a barchan, wind speeds were measured at a scaled height equal to 0.8 m above the dune surface. To calculate sand transport, a constant shear velocity was assumed below the measurement height, but subsequent research has shown that this is not a valid assumption. Numerous field studies have found that wind profiles are nonlogarithmic below 0.8 m and shear velocity must be calculated from wind speeds measured much closer to the dune surface. Yet the simulation model seemed to perform quite well. Unfortunately, this inconsistency suggests that canceling errors may be present in the model.

In essence, there are two problems that can arise when evaluating the performance of a numerical model. If a large discrepancy exists between field observations and model output, it is difficult to determine the specific source of error responsible for the discrepancy. Equally disconcerting is the possibility that model results will closely resemble field observations. In this situation, it is difficult to establish whether or not there are significant canceling errors in the analysis.

SUMMARY

The purpose of this paper has been to describe the research design, study area, field methods, and instrumentation used in a study of dune-scale aeolian processes. A conceptual model of the dune system was developed to serve as a guide in the research design. The research used this model as a general framework and focused on the wind flow and sand transport processes controlling the profile geometry of a barchan. More specifically, there were five objectives in this study.

The first objective was to describe and analyze the wind events that occurred during the course of the field season. An anemometer and wind vane were mounted on a 10-m mast placed 20 m upwind of the dune in order to measure the prevailing wind speed and direction. Wind speed and direction were monitored on a continuous basis over a four-week period, from 20 May to 16 June 1991.

In response to these wind events, the second objective was to monitor the changes that occurred in dune morphology. A single downwind profile was surveyed over the central axis of the dune in order to measure the changes in the dune profile as the dune advanced downwind. This downwind profile was resurveyed at regular intervals over a six week period, from 12 May to 23 June. In addition to these repetitive surveys of the dune profile, the entire dune was mapped on three separate occasions. To map the field site, 15 downwind profiles were surveyed over the dune. Whereas the survey data are adequate to monitor the overall changes in dune morphology, a 5-m by 5-m grid of 96 erosion stakes was used to measure erosion and deposition on a daily basis.

The third objective in this study was to measure spatial variability in near-surface wind speed over the dune in order to explain the observed changes in dune morphology. Two different anemometer arrays were deployed in order to measure this spatial variability. During the first phase of the field study, 16 anemometers were deployed to measure the velocity profile upwind of the dune and the spatial variability in near-surface wind speed along the dune profile.

During the second phase of the field study, the anemometer array was expanded to measure near-surface wind speed over the entire windward slope. In addition to the original 16 anemometers, 12 anemometers were deployed on the northern flank of the dune and 12 anemometers were deployed on the southern flank. In total 40 anemometers were used to measure the upwind velocity profile and spatial variability in wind speed 15 cm above the dune surface.

The fourth objective in the study was to measure the vertical structure of the wind field over the dune profile in order to help explain this spatial variability in near-surface wind speed. In the third phase of the field season the anemometer array was reconfigured to measure vertical velocity profiles in order to measure the vertical structure of the wind field. In addition to the velocity profile measured upwind of the dune, simultaneous velocity profiles were measured near the toe of the dune, on the middle windward slope, near the crest of the dune, and near the brink. At each location seven anemometers were mounted on a 5-m mast.

The last objective in this research was to develop a physically based numerical model to simulate sand transport and changes in the dune profile over time. The primary data input to the simulation model were near-surface wind speed measurements. These data were derived from anemometers placed along the central axis of the dune. To test the output from the model, calculated changes in the dune profile were compared to field measurements of erosion and deposition. These measurements of erosion and deposition were derived from survey data and the grid of erosion stakes.

The field methods described in this paper provide an integrated approach to understanding dune-scale aeolian processes. In future research concerned with the process-response system, every effort should be made to collect detailed measurements of both wind flow and the subsequent change in dune morphology. These data are essential for modeling, in a quantitative manner, the sand transport processes over a dune. While strongly advocating a field approach to the study aeolian processes, it is also important to recognize the limitations inherent in the methodology. Hopefully, these limitations will be addressed in future research.

ACKNOWLEDGMENTS

I gratefully acknowledge Vatche P. Tchakerian, John R. Giardino, Margaret F. Mulligan, and Lucia S. Barbato. Their assistance and efforts made this study possible.

REFERENCES

Bagnold, R. A. (1941) *The Physics of Blown Sand and Desert Dunes*. Methuen, London.

Bagnold, R. A. (1956) Flow of cohesionless grains in fluid. *Philosophical Transactions of the Royal Society of London*, Series A, v. 249, p. 235-297.

Belly, P. Y. (1964) *Sand Movement by Wind*. U.S. Army Corps of Engineers, Coastal Engineering Research Center, Technical Memo No. 1.

Burkinshaw, J. R., Illenberger, W. K, and Rust, I. C. (1993) Wind-speed profiles over a reversing transverse dune. In K. Pye (ed.) *The Dynamics and Environmental Context of Aeolian Sedimentary Systems.* Geological Society Special Publication No. 72, p. 25-36.

Burkinshaw, J. R., and Rust, I. C. (1993) Aeolian dynamics on the windward slope of a reversing transverse dune, Alexandria coastal dunefield, South Africa. In K. Pye and N. Lancaster (eds.) *Aeolian Sediments, Ancient and Modern.* Special Publication of the International Association of Sedimentologists, v. 16, p. 13-21.

Crosswhite, F. S., and Crosswhite, C. D. (1982) The Sonoran Desert. In G.L. Bender (ed.) *Reference Handbook on Deserts of North America.* Greenwood Press, Westport, Connecticut, p. 163-295.

Gerety, K. M. (1985) Problems with determination of u_* from wind velocity profiles measured in experiments with saltation. In O. E. Barndorff-Nielson, J. T. Møller, K. R. Rasmussen, and B. B. Willets (eds.) *Proceedings of International Workshop on the Physics of Blown Sand.* Department of Theoretical Statistics, Institute of Mathematics, University of Aarhus, Memoir 8, p. 271-300.

Haff, P. K., and Presti, D. E. (1984) *Barchan Dunes of the Salton Sea Region, California.* California Institute of Technology, Brown Bag Preprint Series in Basic and Applied Science, BB-16.

Hardisty, J., and Whitehouse, K.J.S., (1988) Evidence for a new sand transport process from experiments on Saharan dunes. *Nature*, v. 332, p. 532-534.

Havholm, K., and Kocurek, G. (1988) A preliminary study of the dynamics of a modern draa, Algodones, southeastern California. *Sedimentology*, v. 35, p. 649-669.

Hely, A. G., Hughes, G. H., and Irelan, B. (1966) *Hydrologic Regimen of Salton Sea, California.* U.S. Geological Survey Professional Paper 486-C.

Howard, A. D. (1977) The effect of slope on the threshold of motion and its application to the orientation of wind ripples. *Geological Society of America Bulletin*, v. 88, p. 853-856.

Howard, A. D., Morton, J. B., Gad-el-Hak, M., and Pierce, D. B. (1978) Sand transport model of barchan dune equilibrium. *Sedimentology*, v. 25, p. 307-338.

Hesp, P. A., Illenberger, W. K., Rust, I. C., McLachlan, A., and Hyde, R. (1989) Some aspects of transgressive dunefield and transverse dune geomorphology and dynamics, south coast, South Africa. *Zeitschrift für Geomorphologie*, Supplementband 73, p. 111-123.

Inman, D. L., Ewing, G. C., and Corliss, J. B. (1966) Coastal sand dunes of Guerrero Negro, Baja California, Mexico. *Geological Society of America Bulletin*, v. 77, p. 787-802.

Kagnov, E. I., and Yaglom, A. M. (1976) Errors in wind speed measurements by rotation anemometers. *Boundary-Layer Meteorology*, v. 10, p. 15-34.

Landsberg, H. (1942) The structure of the wind over a sand-dune. *Transactions of the American Geophysical Union*, v. 23, p. 237-239.

Lancaster, N. (1985) Variations in wind velocity and sand transport on the windward flanks of desert sand dunes. *Sedimentology*, v. 32, p. 581-593.

Lancaster, N. (1987) Reply: variations in wind velocity and sand transport on the windward flanks of desert sand dunes. *Sedimentology*, v. 34, p. 516-520.

Lancaster, N. (1989) The dynamics of star dunes: an example from the Gran Desierto, Mexico. *Sedimentology*, v. 36, p. 273-289.

Leatherman, S. P. (1978) A new aeolian sand trap design. *Sedimentology*, v. 25, p. 303-306.

Lee, J. A. (1987) A field experiment on the role of small scale gustiness in aeolian sand transport. *Earth Surface Processes and Landforms*, v. 12, p. 331-335.

Livingstone, I. (1986) Geomorphological significance of wind flow patterns over a Namib linear dune. In W. G. Nickling (ed.) *Aeolian Geomorphology.* Allen & Unwin, Boston, p. 97-112.

Long, J. T., and Sharp, R. P. (1964) Barchan-dune movement in Imperial Valley, California. *Geological Society of America Bulletin*, v. 75, p. 149-156.

MacCready, P. B. (1966) Mean wind speed measurement in turbulence. *Journal of Applied Meteorology*, v. 5, p. 219-225.

Mulligan, K. R. (1988) Velocity profiles measured on the windward slope of a transverse dune. *Earth Surface Processes and Landforms*, v. 13, p. 573-582.

Norris, R. M. (1966) Barchan dunes of the Imperial Valley, California. *Journal of Geology*, v. 74, p. 292-306.

Olson, J. S. (1958) Lake Michigan dune development 1: wind velocity profiles. *Journal of*

Geology, v. 66, p. 254-263.

Patel, V. C. (1990) *Design and Construction of the 1.8 m x 1.5 m Low-Turbulence Wind Tunnel.* Iowa Institute of Hydraulic Research, Limited Distribution Report No. 170.

Rempel, P. (1936) The crescentic dunes of the Salton Sea and their relation to vegetation. *Ecology*, v. 17, p. 347-358.

Sarre, R. D. (1988) Evaluation of aeolian sand transport equations using intertidal zone measurements, Saunton Sands, England. *Sedimentology*, v. 35, p. 671-679.

Sarre, R. D. (1989) Aeolian sand drift from the intertidal zone on a temperate beach: potential and actual rates. *Earth Surface Processes and Landforms*, v. 14, p. 247-258.

Svasek, J. N., and Terwindt, J.H.J. (1974) Measurement of sand transport by wind on a natural beach. *Sedimentology*, v. 21, p. 311-322.

Tsoar, H. (1974) Desert dunes morphology and dynamics, El-Arish (northern Sinai). *Zeitschrift für Geomorphologie*, Supplementband 20, p. 41-61.

Tsoar, H. (1983) Dynamic processes acting on a longitudinal (seif) sand dune. *Sedimentology*, v. 30, p. 567-578.

Tsoar, H. (1985) Profile analysis of sand dunes and their steady state significance. *Geographfiska Annaler*, v. 67A, p. 47-59.

Walmsley, J. L., and Howard, A. D. (1985) Application of a boundary-layer model to flow over an aeolian dune. *Journal of Geophysical Research*, v. 90, p. 10631-10640.

Watson, A. (1987) Comments: variations in wind velocity and sand transpoort on the windward flanks of desert sand dunes. *Sedimentology*, v. 34, p. 511-516.

Wiggs, G.F.S. (1993) Desert dune dynamics and the evaluation of shear velocity: an integrated approach. In K. Pye (ed.) *The Dynamics and Environmental Context of Aeolian Sedimentary Systems*. Geological Society Special Publication No. 72, p. 37-46.

7 BARCHAN DUNES OF THE SALTON SEA REGION, CALIFORNIA+

Peter K. Haff[1] and David E. Presti[2]
[1]*Department of Geology, Duke University*
[2]*Veterans Administration Medical Center, San Francisco*

ABSTRACT

Rates and directions of motion have been determined for 19 barchan dunes near the Salton Sea, California, for the period 1941 to 1981. Between 1941 and 1956 these barchans moved with an average speed of 15.5 m y^{-1} in the direction 89° T (T = from true north). During the period 1956-1963 the rate of motion increased to 24.4 m y^{-1}, and the direction shifted northward to 80° T. Finally, during the 18 years between 1963 and 1981, the rate of advance of these dunes declined to 16.2 m y^{-1}, and the bearing was again more southerly, 96° T. Fluctuations in speed and direction are superimposed on changes in the dunefield brought on by an apparent diminution of sand supply over the last 140 years. A decrease in the sand supply may be because of increasing maturation of an upwind, incised drainage pattern that interferes with sand flow and dune mobility, and to modern channelization associated with highway construction. At the downwind edge of the dune area, some barchans are entering the Salton Sea. They may ultimately become vegetated shoreline dunes. Members of the southernmost group of dunes appear to dissipate entirely before reaching the water. The overall picture is of a dunefield declining in area and population owing to a combination of adverse geological and cultural developments.

INTRODUCTION

We combine the results of previous studies with our own measurements to analyze the motion of individual barchan dunes over a 40-year-long period of time. Many studies of barchan dunes have been carried out previously with attention variously given to rates of movement, shape, and internal structure (Cornish 1897, Beadnell 1910, King 1918, Rempel 1936, Bagnold 1941, Finkel 1959, Norris and Norris 1961, Long and Sharp 1964, Norris 1966, McKee 1966). The present work was stimulated by the availability of data on individual barchan dunes in the Salton Sea region of California extending back several decades. These data deal principally with rates and directions of movement of individual barchans, with changes in dune heights, widths, and lengths, and with the long-term evolution of the dunefield.

Long and Sharp (1964) provided data covering the motion of a number of dunes in the Salton Sea region between 1941 and 1963. Dune positions

+Field work was carried out when the authors were in the Division of Physics, Mathematics and Astronomy and the Division of Biological Sciences, respectively, at the California Institute of Technology.

Desert Aeolian Processes. Edited by Vatche P. Tchakerian. Published in 1995 by Chapman & Hall, London.
ISBN 978-94-010-6519-1

recorded on U.S. Bureau of Reclamation maps in 1941 were found by Long in the winter of 1955-1956 to have changed substantially. New positions were recorded on the USGS 7.5° quadrangles Kane Springs NE (1956) and Kane Springs NW (1956). In 1963 Sharp made plane table surveys, locating the positions of 47 barchan dunes. Of these, 34 could be traced back to dunes shown on the 1941 maps.

During the 15 years between 1941 and 1956, the movement of 34 dunes averaged 15.3 m y^{-1} in the direction 89° T. In the years between 1956 and 1963, the average rate of movement increased to 25.0 m y^{-1}. The average direction of displacement over this period of time shifted about 9° to the north, with the direction of dune movement averaging 80° T. The acceleration of dune motion in the 1956-1963 period and the substantial shift in the average direction of motion were the most striking results of the study by Long and Sharp (1964). The present work extends the period of observation of individual dune motion to 1981.

In addition to the opportunity of developing a long time-based study of individual dunes, the Salton Sea site also offers a chance to glimpse some of the factors influencing the dunefield as a whole. There is evidence that the dunefield is, or has recently been, in a state of contraction, with active dunes covering a smaller area than they did earlier in this century. Some large dunes are currently starved for sand and may be dwindling in size. Areas where dunes were forming 20 years ago now show no evidence of new barchan forms, and in some sections dunes apparently die out with high probability long before reaching what otherwise would be their ultimate graveyard, the Salton Sea.

PHYSICAL SETTING

The Salton Sea dunefield (Long and Sharp 1964, Norris 1966) covers an area of roughly 15 km^2, about 13 km south of Salton City, Imperial County, California (Figure 1). Topographic coverage of the active field is provided by the USGS 7.5° quadrangles Kane Springs NW (1979) and Kane Springs NE (1956).

The dunes rest upon nearly flat, barren country composed of Pliocene sedimentary rocks which, to the east, are covered with a mantle of younger Salton Sink clays. The easternmost extension of the dunefield is truncated by the shore of the Salton Sea. From the shore, the dunefield extends westward about 6 km. The western boundary is marked approximately by the position of the Imperial Irrigation District powerline. The north-south width of the dune area, although variable, is about 3 km. From 1.5 to 2 km west of the powerline poles the San Felipe Hills rise abruptly to more than 70 m above the surrounding land.

The slope of the dunefield area is gentle, typical values being 10-25 m km^{-1} toward the north and east. Elevations range between 10 and 70 m below sea level. There is little stream dissection within the dunefield. Gully depths below the surrounding terrain are typically less than 1 m, with local

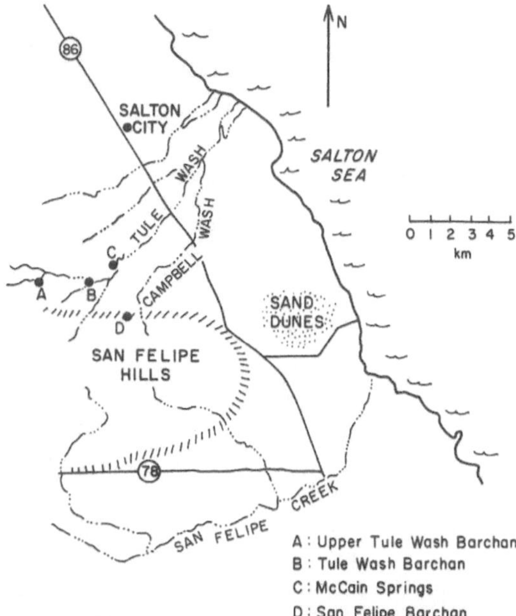

Figure 1. General location of the main Salton dunefield, labeled "SAND DUNES," The positions of several small dunes (A, D), of the remnants of the Tule Wash dune (B), and of McCain Springs (C) are also shown.

exceptions. Greater dissection occurs to the north, south, and west of the dune area. The dunefield terrain is gently convex, so that drainage channels from the San Felipe Hills trend either to the north, contributing to the Campbell Wash system, or to the south to join San Felipe Creek.

Vegetative cover is sparse and creosote bush is the most abundant large plant variety. On occasion it may be an important locus of dune formation. Rempel (1936) has discussed the relationship of dunes and vegetation in this area.

Long and Sharp (1964) argued that the Salton dunes are fed principally from sand derived from Borrego and Clark Valleys, 40 km to the west. This is consistent with observations of sand drifts and climbing and falling dunes located west of the study area. Rempel (1936) emphasized the importance of the old Lake Cahuilla shoreline as a potential source of sand.

DUNE FORMS

Figure 2 shows the positions in 1981 of all major dunes in the Salton group. Approximately 70 dunes were located by plane table survey. The smallest sand piles, generally lacking a slip face, were not recorded. The various dune forms existing in the Salton area may be described as barchans, distorted barchans, dune complexes, parabolic dune forms or vegetated dunes.

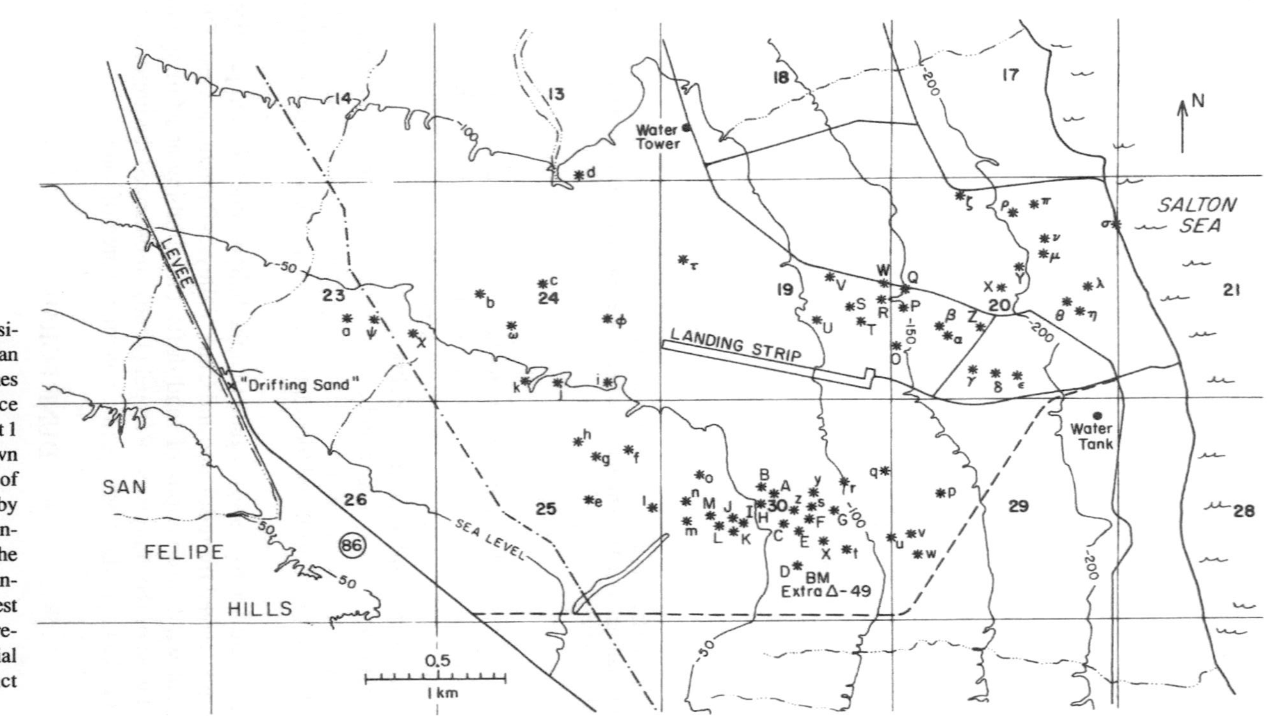

Figure 2. The positions (marked by an asterisk) of all dunes having a slipface greater than about 1 m high are shown as of the spring of 1981. The letters by each asterisk identify the dune. The dot-dashed line running from northwest to southeast represents the Imperial Irrigation District powerline.

Figure 3 shows a plan view and longitudinal cross-section of an idealized barchan. Relatively few of the Salton dunes approach the geometrical perfection illustrated in Figure 3. Dune i(13)[+], the large, isolated dune in the center of Figure 4, is a good example of the well-developed barchan shape. Failure to manifest the ideal form is often caused by the proximity of other dunes. Bagnold (1941, p. 212) has described how the presence of an upwind dune can lead to the formation of a "double barchan." Such a form, with its windward dune, is shown in the left center of Figure 5. If the spacing of dunes is sufficiently small, physical overlap distorts or destroys the barchan form. Figure 4 shows dune complexes built up of barchanoid shapes. If the overlap is minor, the individual barchans retain their identities. In some cases, substantial overlap and amalgamation lead to a dune complex that cannot be analyzed in terms of a distinct number of barchan dunes. An example of such merger is provided by the Algodones dunes, whose main sand mass has been described in terms of giant, coalescing, barchan forms by Norris and Norris (1961). In the Salton dunefield, little difficulty was encountered in connecting a given slipface to a definite dune.

Roughness of terrain can also affect dune form. However, except for minor bedrock knobs and drainage channels, the platform on which most of the Salton dunes move is smooth. There is little significant topographic modification of the larger dunes.

Asymmetry of horn lengths is common, although no general pattern was noted. Long and Sharp (1964) found instances of north horn elongation on some dunes and south horn elongation on others. It appears that subtle effects of topography, of wind speed and wind direction, and of sand supply must combine locally to produce the observed elongations. No insight into detailed causes was found in the present study.

The schematic illustration of a barchan cross-section in Figure 3 shows that the highest point, or crest, on the dune need not coincide with the highest point, or brink, on the slipface. In most dunes in the Salton group, the crest lies higher and several meters to windward of the brink. The heights of crest and brink were not measured separately, but in some cases the crest is as much as several meters higher. In a small number of dunes, the brink is the highest point on the dune.

Another departure from the idealized form of Figure 3 is the occurrence of a transverse asymmetry, in which the crest and brink are not aligned with what would otherwise be the longitudinal symmetry axis of the dune. Frequently, the crest is centrally located, while the highest point on the slipface is shifted toward one of the horns.

Parabolic dunes, like barchans, arise in regions of strong unidirectional winds, but, unlike barchans, they require anchoring vegetation to maintain their form (McKee 1966). Well-developed, parabolic dunes are U- or V-shaped with

---

[+]Greek and Latin letters refer to dunes identified in the present study. The numbers in parentheses refer to the nomenclature of Long and Sharp for the same dune.

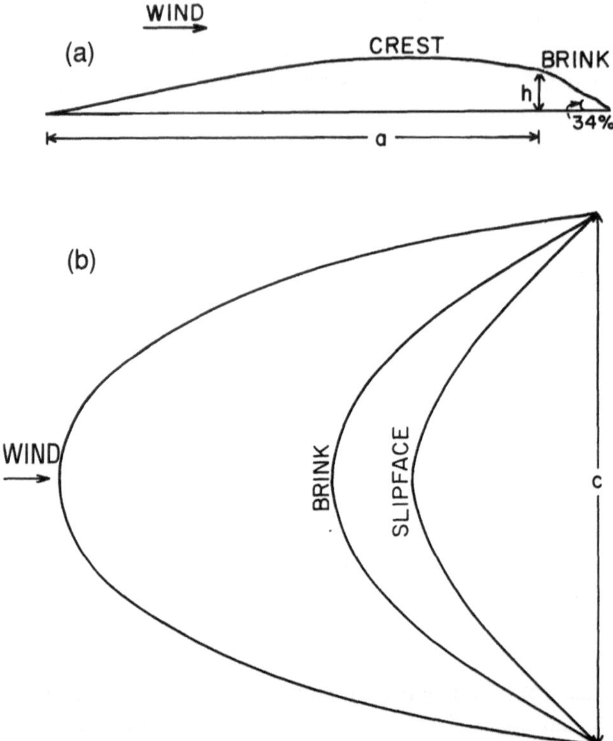

Figure 3. Cross section (a) and plan view (b) of an idealized barchan form. The dimensions a, c, and h were recorded for each dune in the Salton field. Where a slipface was absent the height of the crest was measured instead.

horns pointing in the upwind direction. They are blow-out features, with the trailing arms or horns anchored by vegetation. Slipfaces are often well-developed, so that the central portion avalanches forward in the manner of a barchan dune. Although parabolic forms are not uncommon in the Salton dunefield, neither are they the major expression of dune morphology. Barchan forms dominate the landscape.

In the eastern portion of the Salton dunefield, where vegetation is relatively more abundant, many dunes have trailing "tails," reminiscent of parabolic dune forms. The tails are anchored by vegetation such as desert buckwheat and various grasses. In some cases the tails are attached to what otherwise would be a well-formed barchan dune so that the dune has the appearance of a sea skate. Dune *m* is an example of a barchan with tails.

Elsewhere, sand piles without slipfaces show the effects of their movement by the presence of upwind tails. Most of these sandpiles are small, but some reach a respectable height. For example, dune *a* is 4 m in height with horizontal dimensions on the order of 100 m. This dune lacks a slipface, but the upwind tails suggest that the dune is mobile.

At several locations in the study area the dunes are vegetated and

Figure 4. This aerial view of the western part of the main Salton dunefield, looking north, was taken in 1958. Highway 86 and the Imperial Irrigation District powerlines are visible at the left hand edge of the figure. The large barchan at the center of the picture is designated "i" (13) in the text. It is being overtaken by the smaller dune "i" (12) immediately to windward (left). The numbers refer to the notation of Long and Sharp (1964); the letters are the notation of this paper. (Photo and copyright by John S. Shelton.)

Figure 5. This view shows the same area as that in Figure 4, but seven years later, in 1965. Note that dune "i" (12) has nearly caught up with dune "i" (13) and will soon merge with it. To the left of this pair, a large double barchan and its causative dune to windward are visible. Dune "d," discussed in the text, is located near the upper right-hand corner of the photograph. (Photo and copyright by John S. Shelton.)

apparently fixed. One string of fixed dunes is found along the shoreline in Section 17 (Figure 2), just north of benchmark "Mitchell" on the Kane Springs NE quadrangle. Other vegetated forms lie in Sections 28 and 29, south and east of the paved access road leading from State Highway 86 to the Salton Sea. Mesquite and tamarisk seem to play a major role in stabilizing these sand accumulations. It is interesting that many of the vegetated dunes are located below contour -200 feet on the Kane Springs NE quadrangle. The highest level of the Salton Sea following its initial filling in 1905 was -198 feet, attained in 1907 (Norris and Webb 1976, p. 166). Following the diversion of the Colorado River back to its original course in that year, the newly formed lake retreated, and dunes which had been recently submerged were uncovered (see plate 12B of MacDougal 1914). At least some of these were reactivated by the wind (Sykes 1937, p. 75). Between 1907 and 1925 the lake level was reduced through evaporation by about 15 m (Norris and Webb 1976, p. 167), an average of 80 cm y^{-1} over an 18-year period. This means that a (just) completely submerged dune 5 m high would require approximately six years from the time its crest first re-emerged above the level of the water until it was resting on dry land. The dune would therefore require about this same amount of time before it was able to move freely again. This period of enforced stability, coinciding with abundant moisture, would seem to be ideal for stabilizing vegetation to gain a foothold.

DUNE MOTION

Identification of Dunes

There are three periods for which data is available, 1941-1956 (15 years), 1956-1963 (7 years), and 1963-1981 (18 years). In 1963, Long and Sharp could trace 34 dunes back to 1941. In the years between 1963 and 1981, 15 dunes were lost either because they dissipated or fused with other dunes, or because it was impossible to make a correspondence with reasonable certainty between dune locations shown on the 1963 and 1981 surveys. There are only 19 dunes for which a history is available for the entire 40-year span. While some dunes indeed have dissipated or merged, it is probable that many of the missing dunes are mapped on Figure 2.

Average Rates of Motion

One of the conclusions reported by Long and Sharp (1964) was an increase in average dune speed for the period 1956-1963. With only two exceptions, every dune surveyed in 1963 that could be traced back to 1941 had undergone a substantial increase in rate of motion. The average rate R for the 1941-1956 period, for all dunes, was 15.3 m y^{-1}. For the 1956-1963 period the rate was $R = 25.0$ m y^{-1}, a 60% increase. Citing two possible causes for the change, an increase in the velocities or duration of the sand-moving winds and an increase in the sand supply, Long and Sharp (1964) favored the latter. They argued that much of the sand in the Salton dunes can ultimately be traced to Borrego and

Clark Valleys. Increased cultivation of these valleys might have led to an enhancement of the easterly sand flow. The idea is that a higher sand flux impinging on the dunes would beget more sand in motion, with the result that the forward motion of the dune would be accelerated.

During the years between 1963 and 1981, the dunes have once again been moving at a more leisurely average pace. Again, with only two exceptions, the average speed of each dune during the 1963-1981 period is substantially less than the average speed of that dune during the time between 1956 and 1963. In fact, in most cases, the average rates of movement during the periods 1941-1956 and 1963-1981 are comparable on a dune by dune basis. The average rate in the most recent time period is $R = 16.2$ m y^{-1}.

Long-term records in the form of monthly average wind speeds at the "Sandy Beach" locality on the Salton Sea for the period 1948-1979 were kindly made available by the Imperial Irrigation District office in Imperial, California. These records indicate that mean wind velocities were somewhat higher during the period of enhanced dune motion than either before or afterward. For the period 1948-1955 the year-averaged speed was 6.5 km h^{-1}, for 1956-1963, 6.9 km h^{-1}, and for 1964-1978, 5.9 km h^{-1}. These are gross averages and cannot be used to calculate sand fluxes quantitatively, but they suggest that the period 1956-1963 was one of increased wind velocity.

Direction of Motion

Another piece of evidence supporting the notion that the wind regime was a prime factor in the observed long-term changes in average dune motion is provided by the data on directions of dune movement. Directional information was compiled for the three intervals 1941-1956, 1956-1963, and 1963-1981. The course of dune movement is illustrated in Figure 6. In the figure the kink represented by the 1956-1963 leg can be clearly seen. The nearly due eastward average movement for the period 1941-1956 shifted 9° to the north for the 1956-1963 interval. Then, over the 18-year period 1963-1981, the average motion has been 6° south of east. If a shift in the direction of sand-moving winds is required to explain the data, it would not be surprising if the magnitude and frequency of sand driving winds were also subject to change over the same period of time.

Dune Size and Its Influence on Dune Motion

For each of the dunes mapped in 1981 (Figure 2), three characteristic linear dimensions were measured. These were a, the distance from brink to toe, c, the distance between the tips of the two horns, and h, the maximum height of the slipface (Figure 3). The distances a and c were measured by pacing. The horn to horn distance c is more accurately determined than a, because the positions of the horn tips are better defined than the position of the upwind toe.

The maximum height of the slipface, h, is the easiest dimension to measure accurately, and probably the dimension with the greatest direct influence on dune motion. Values of h were measured with a stadia rod and telescopic alidade. Where no slipface existed, the height of the crest was measured.

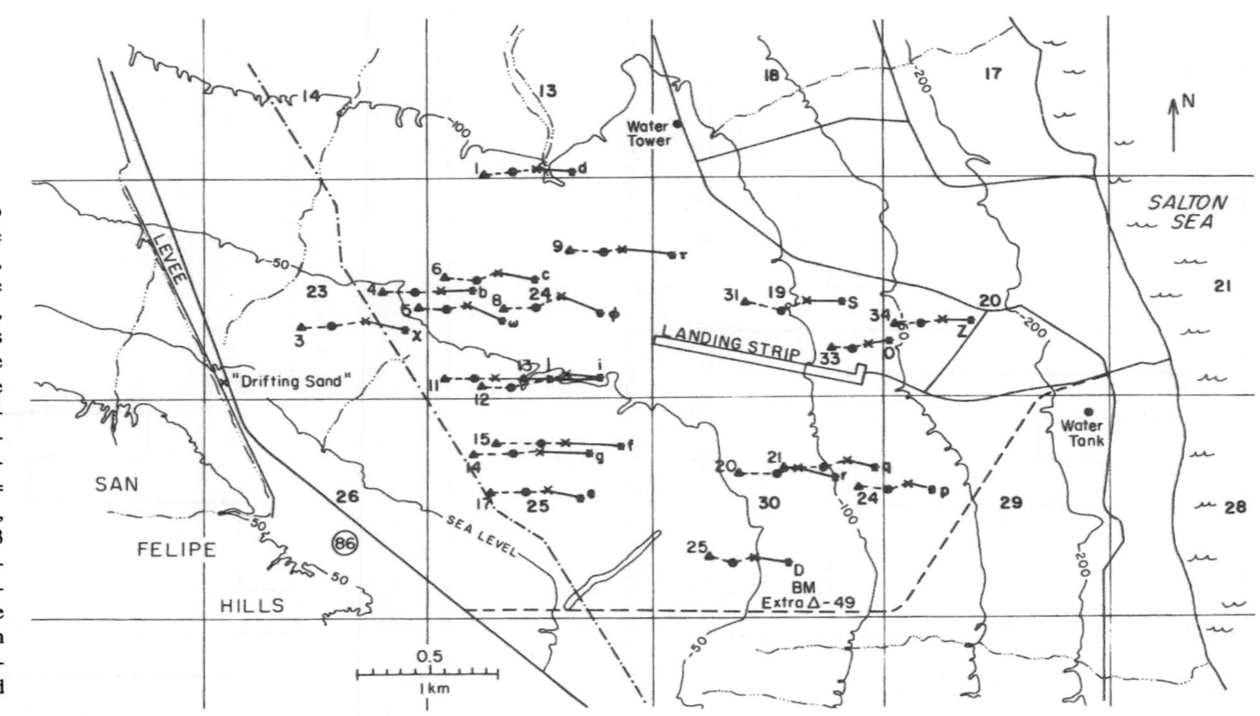

Figure 6. This map is similar to the one in Figure 2, but it shows the 1941, 1956, 1963, and 1981 positions of all dunes whose identity can be traced over the 40-year period. Triangles = 1941 positions, circles = 1956 positions, crosses = 1963 positions, and asterisks = 1981 positions. Note the northward jog in the motion between 1956 and 1963.

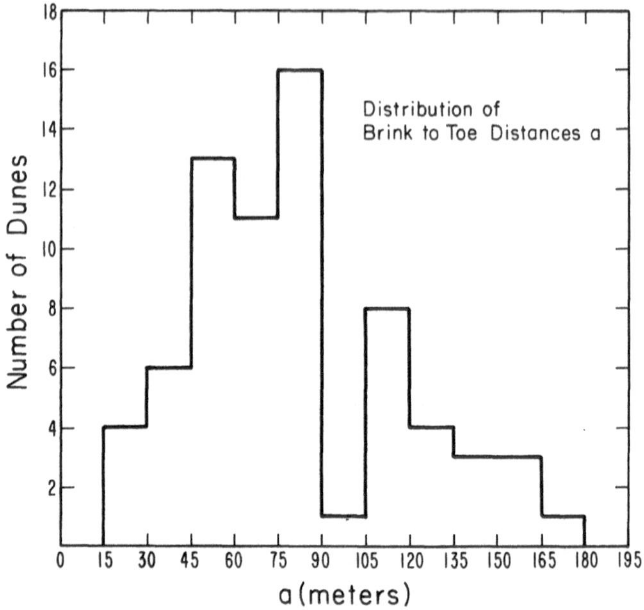

Figure 7. Histogram to illustrate the 1981 measured distribution of toe to brink distances, *a*.

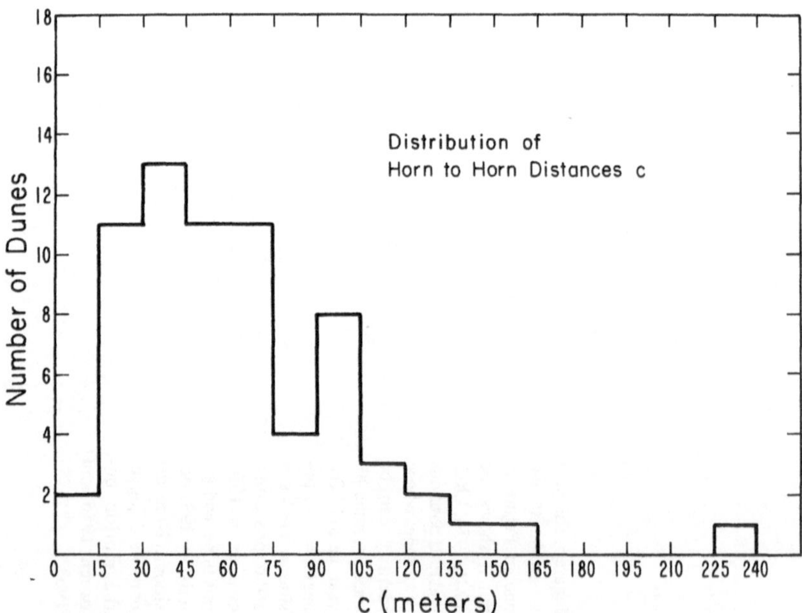

Figure 8. Histogram to illustrate the 1981 measured distribution of horn to horn distances, *c*.

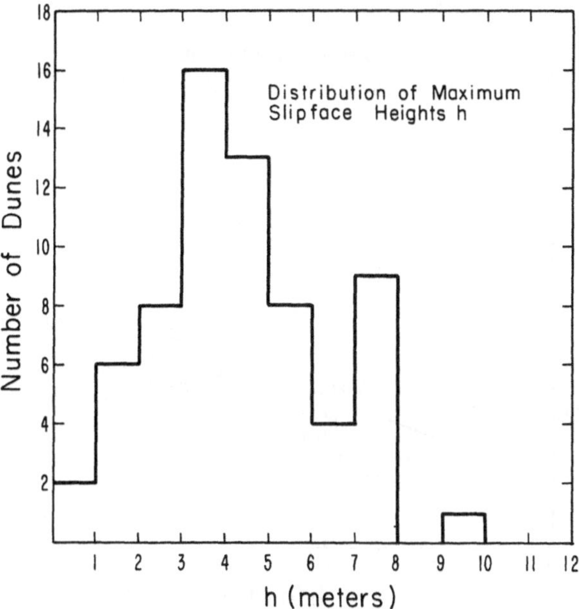

Figure 9. Histogram to illustrate the 1981 measured distribution of maximum slipface heights, *h*.

The average dimensions of a dune in the Salton group in 1981 were $a = 79$ m, $c = 63$ m, and $h = 4.3$ m. These averages include all dunes in the field having slipfaces or crests more than about 1 m high. Figures 7, 8, and 9 show histograms of the 1981 dimensions for a, c, and h, respectively, to illustrate the distribution of values used to calculate the averages. Comparison of our measurements in 1981 with those of Long and Sharp in 1963 points to a modest but general decline in the size of the Salton barchans. On average, in the 18 years between 1963 and 1981, the horn to horn distance, c, decreased by 28% and the average slipface height, h, decreased 23%. The values for the toe to brink distance, a, show an increase of 30%. Because of difficulties in precisely defining the location of the toe, values of a are more uncertain than for c or b. The general impression is one of contraction of dune size. These results are consistent with evidence, discussed below, indicating an attenuation of the sand supply.

Because the height of the slipface, h, is a principal factor in governing dune motion, many authors have studied the relation between h and rate of dune movement R. Bagnold (1941) argued that for dunes whose size is not changing rapidly, these two quantities are inversely proportional, $R^{-1} = \kappa h$, with κ a constant depending on the value of the sand flow. This relation reflects the observation that small dunes move more rapidly than large dunes. In Figures 4 and 5, taken in 1958 and 1965, respectively, the small barchan i(12), located near the center of the field, is seen to be overtaking the large dune i(13). Data

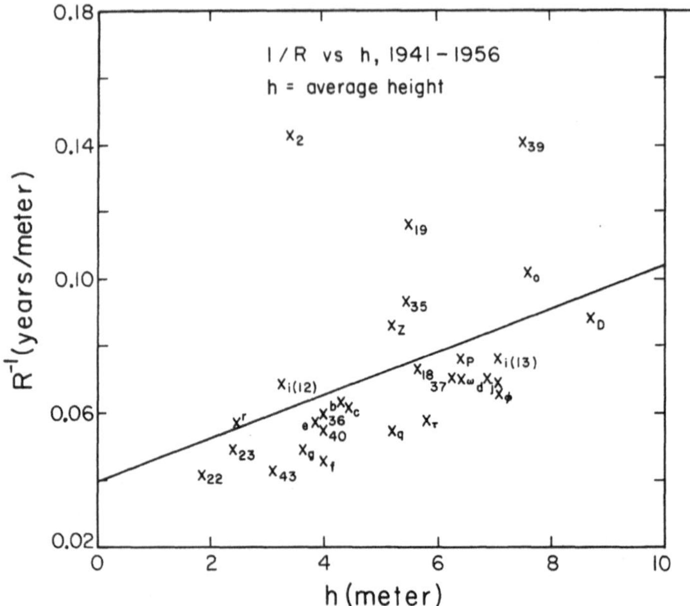

Figure 10. The inverse rate of dune motion R^{-1} is plotted versus slipface height h for the period 1941-1956. For each dune, the height h is the average of the 1941 and 1956 slipface heights for that dune. The letter and number labels identify the dune. By 1981 the dunes identified by numbers only had lost their identity or dissipated. Positions of lettered dunes are shown in Figure 6. The straight line is a least-squares fit to the data. In the units shown on the axis, the equation of the line is $R^{-1} = 0.00646\,h + 0.0394$.

on dune motion acquired in Peru by Finkel (1959) and at the Salton site by Long and Sharp (1964) shows that h and R are related by equations of the form $R^{-1} = \kappa_1 h + \kappa_2$, where κ_1 and κ_2 are empirical constants. The additive constant was needed because the smaller dunes did not move as rapidly as the simplest interpretation of Bagnold's form of the equation predicted.

Figures 10, 11, and 12 show dune motion as a function of average slipface height, over the period 1941-1981. Although individual dunes underwent large fluctuations in height, the changes in the average height were more modest. A slight increase in average dune height, detected in 1963, reversed itself so that in 1981 average dune heights were slightly smaller than at any time since 1941.

Figure 10 plots $1/R$ vs. h for 29 dunes whose heights could be determined on both the 1941 and 1956 maps. The value of h used in the construction of the figure is the average h determined for each dune from the 1941 and 1956 maps.

The straight lines in Figures 10, 11, and 12 represent least-squares fits to the data points. The speed-up of dunes in the 1956-1963 interval is clearly displayed by the smaller slope of the line in Figure 11.

According to Bagnold (1941), if the flux of sand impinging on the toe is neglected, then $R^{-1} = \rho h/q(h)$, where ρ is the sand bulk density and $q(h)$ the flux

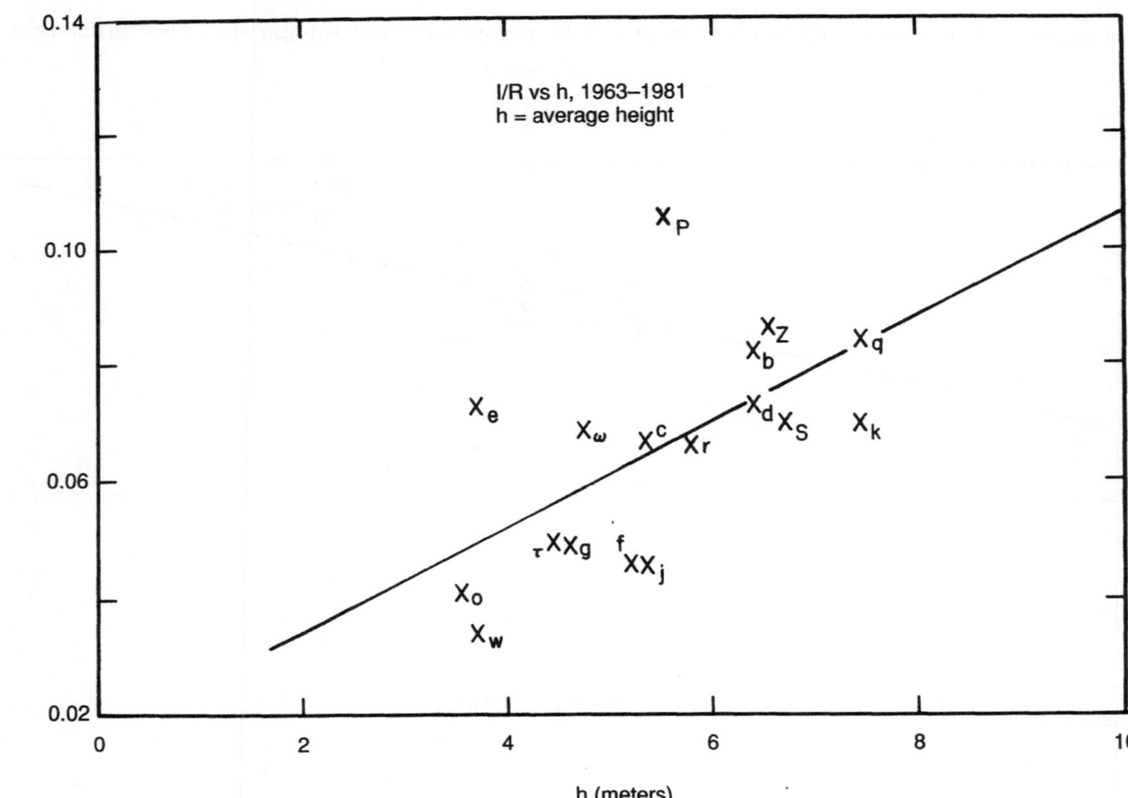

Figure 12. Same as Figure 10, except that the data is for the 1963-1981 period. Dune slipface heights are averages of 1963 and 1981 values. The equation of the line is $R^{-1} = 0.00899\,h + 0.0162$.

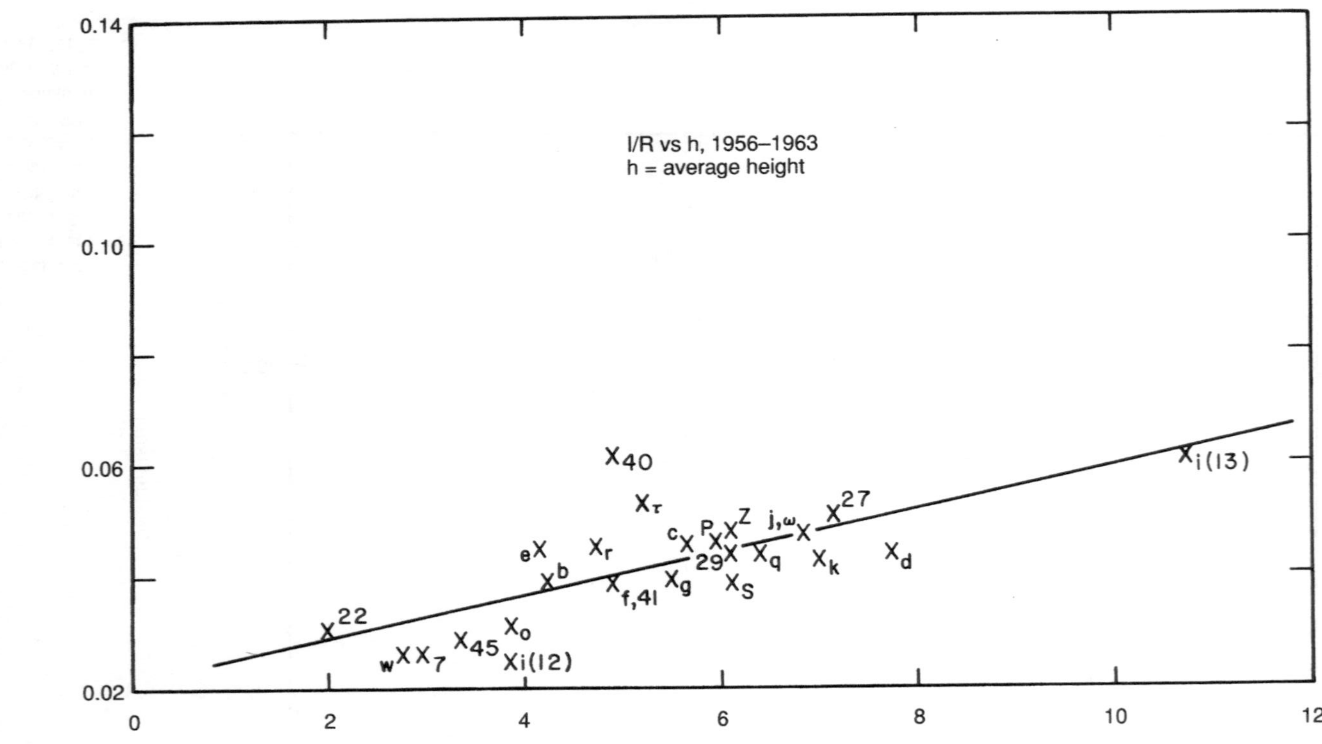

Figure 11. Same as Figure 10, except that the data is for the 1956-1963 period. Dune slipface heights are averages of 1956 and 1963 values. The equation of the line is $R^{-1} = 0.00379\ h + 0.0213$.

of sand over the slipface. The empirically determined equation representing dune motion in the Salton field, $R^{-1} = \kappa_1 h + \kappa_2$, can be cast into the Bagnold form by writing

$$\frac{1}{R} = \frac{\rho h}{q_0}\left(1 + \frac{\kappa_2 q_0}{h\rho}\right)$$

where

$$q_0 = \frac{\rho}{\kappa_1}$$

Then we can define

$$q(h) = \frac{q_0}{1 + \left(\frac{\kappa_2 q_0}{\rho h}\right)} = \frac{q_0}{1 + \left(\frac{\kappa_2}{h\kappa_1}\right)}$$

as an empirical expression for the flux of sand captured by the slipface. This quantity is plotted in Figure 13. For small h, $q(h) = \rho\, h/\kappa_2$, where k_2^{-1} is the velocity of the smallest dunes. Here q is proportional to h, and the speed of small dunes is essentially independent of their size. For large dunes, $q(h)$ becomes equal to the constant $q_0 = \rho/\kappa_1$.

Comparisons with Measurements Elsewhere Rate of Barchan Movement
For the Salton barchans, R averaged approximately 17 m y^{-1} over 40 years. Beadnell (1910) measured the rate of advance of five barchan dunes in the Kharga Oasis in the Libyan Desert over a year's time in 1908. During this period the dunes averaged a displacement of about 17 m, with actual values ranging between 10 and 19 m. Finkel (1959), measuring the motion of scores of barchans over a three-year period in the northern Peruvian desert, found rates of movement averaging 15.4 m y^{-1}. The average height of Beadnell's dunes was ~ 12-13 m, the average height of Finkel's ~ 3-4 m, while the height of Salton dunes for which rate measurements are available averaged about 5 m. If Bagnold's relation between rate of movement, R, and sand flux, q, is valid in each of these dunefields, then the sand flux q must be greater for Beadnell's dunes and less for Finkel's, than for dunes at the Salton site. In turn, this implies that the speeds and/or duration of typical sand-moving winds must be greater for Beadnell's dunes and less for Finkel's dunes.

Barchan Dimensions
The dimensions of the Salton dunes are not large compared to those of barchan dunes in many other locations. Beadnell's dunes were much higher. Bagnold (1941) estimates the maximum height of barchans in the Libyan Desert to be around 30 m, and the maximum length and width to be about 400 m. In 1981, the greatest dune height at the Salton site was that of dune "O," 9.8 m. In 1963, dune 13 was 12.2 m high. Presently the greatest horn distance is 227 m (dune O); dune 29 was 252 m across in 1963. These measured dimensions are all

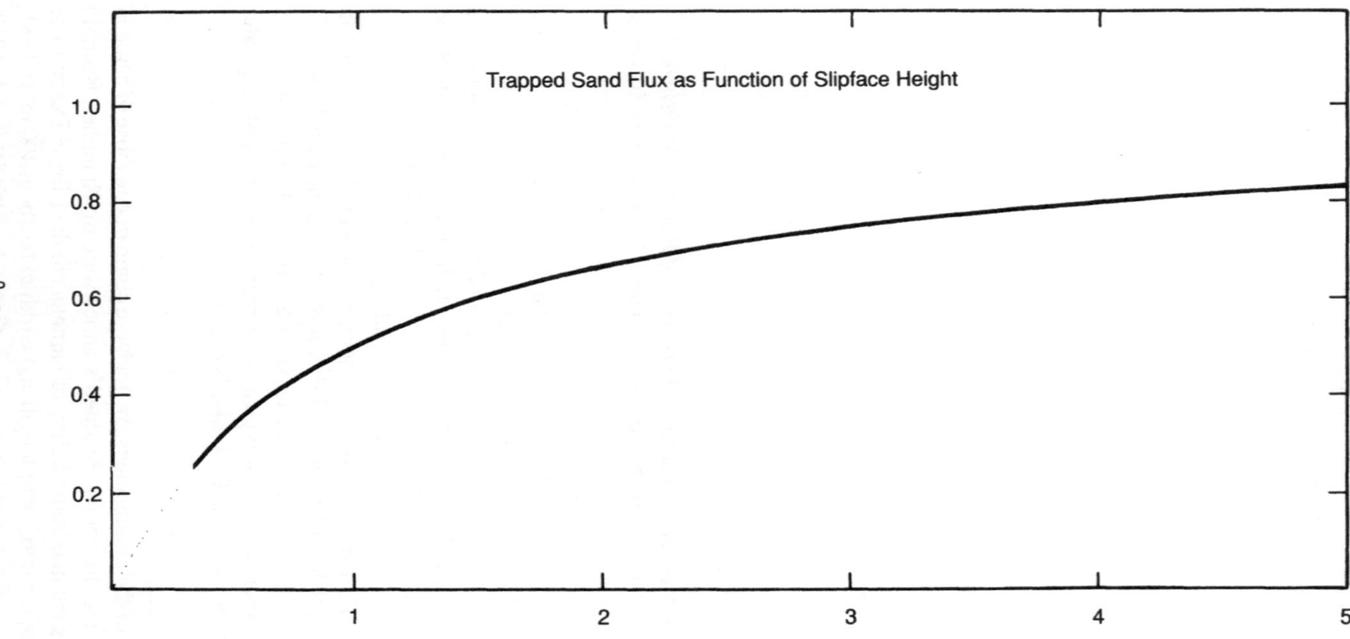

Figure 13. An empirical expression for the trapped sand flux at the dune brink is plotted versus slipface height. Sand flux $q(h)$ is expressed in units of q_0 and h is expressed in units of $\frac{\kappa_2}{\kappa_1}$.

considerably less than those which a barchan may potentially attain under more favorable conditions. In the 1930s, Russell (1932) noted that some of the crescents were 1000 feet (305 m) from horn to horn, with maximum heights "in the neighborhood" of 30 feet (9 m). Only one barchan, dune O, presently approaches this size in its lateral dimensions. At about the same time Rempel (1936) reported a dune 19 m high at the crest and 137 m between the horns. This dune would have been more than twice as high as dune O. The observations suggest that the size of the largest dunes has decreased over the last 50 years or so. There is no evidence that the dunes ever approached the maximum dimensions possible for the barchan form.

THE EVOLUTION OF THE SALTON DUNEFIELD

Sand Supply in the Tule Wash Area

Comparison of the present state of the Salton dunefield and of the Tule Wash area north of the San Felipe Hills (Figure 1) with the descriptions of early observers suggests that the sand flux through these areas is now less than in former times. As recently as the 1500s the Salton basin contained a freshwater lake standing at about 12 m above sea level (Waters 1983). As this body of water, Lake Cahuilla, evaporated, the gently sloping and undissected lake bottom was exposed. A supply of beach sand remained in the vicinity of the 12 m contour, and strong westerly winds presumably blew then as now. These conditions would have been ideal for the formation of barchan dunes. With time, the lake beds became dissected by deep gullies, making it increasingly difficult for dunes to survive. In the early 1900s, three large barchans were present in the Tule Wash area (Mendenhall 1909, Norris 1966). By 1981, only traces of one of these dunes, the Tule Wash barchan, remained. The destruction of this dune had been predicted by Norris (1966) based on its approaching encounter with the deep middle fork of Tule Wash.

Although the terrain north of McCain Springs and upwind from the Salton dunefield (Figure 1) is relatively free of sand today, Blake (1858) noted the occurrence of sand "drifts" while approaching the springs from that direction in 1853. Upon leaving McCain Springs and heading around the San Felipe Hills toward Carrizo Creek, Blake again noted the presence of parallel ridges and drifts of sand up to 10 feet in height. At the present time barchan dunes and other significant sand accumulations are sparse or absent near the San Felipe Hills. In comparison with modern conditions, it seems likely that active sandforms were more prevalent 140 years ago in the region west of the Salton dunefield. Deep channels with steep walls, like Tule Wash, have probably been effective traps, not only for migrating barchan dunes, but for loose blowing sand as well.

Sand Supply at the Western Edge of the Salton Area

Qualitative evidence from the last 50-60 years suggests that a diminution of sand supply is presently afflicting at least part of the western portion of the main Salton dunefield. Aerial photographs taken by John S. Shelton in 1958 and

1965 (Figures 4 and 5) show three, and possibly four, barchan, or barchanoid, dunes west of the Imperial Irrigation District powerline (see Figure 2 for powerline location). Long and Sharp (1964) observed that "a few small barchans and one of medium size at the western edge of the area have formed since 1956." When the field work for the present paper was carried out in 1981, no barchans existed west of the powerline except for several small dunes outside the main Salton dunefield. The only significant sand deposits in the Salton area at the present time in the vicinity of the powerline poles are dunes a, ψ, χ (Figure 2). None of these dunes has the barchan shape and none has a slipface. Furthermore, with the exception of dune χ, there are presently no dunes within ~ 700 m east of the powerline. However, in the 1940s and 1950s there were at least eight moderate-sized to large dunes located in this area (Figure 6). It seems likely that the peculiar jog in the alignment of the powerline (Figures 2 and 6) was because of the fact that when the powerline was constructed in the 1940s or early 1950s, the preferred straightline path was obstructed by a number of sand dunes of significant size. (Barchan dunes in the eastern part of the dunefield have overrun transmission and telephone lines, e.g., dune O in Figure 14 and dune σ.)

Other indirect evidence of more abundant sand supply and deposition in the past is afforded by observations made in the 1930s. Both Russell (1932) and Rempel (1936) cite dune dimensions that are greater than any found today. Rempel also recorded that "120 to 130 barchans were counted." In 1981 the total number of dunes was about 70, and not all of these were barchans.

In 1981, during periods of strong winds from the west, observations in the vicinity of dune d (Figure 2) suggested that that dune was receiving a negligible flux of sand from the upwind direction. At a time when saltating sand was streaming across the surface of the dune and over the brink, there was nowhere any sign of sand impinging the dune from windward. The dune surface was an isolated island of furiously saltating sand in an otherwise static landscape.

Considered as a whole, the evidence suggests a diminution of the sand supply to the western edge of the dune field. One cause of this diminution is probably the construction of north-south trending flood control levees immediately west of Highway 86. Some of these levees (Figure 2) were in place as early as 1941, according to the USBR maps, and might explain the growing paucity of dunes in the swath along the powerline. The levees are low embankments that were bulldozed into place to protect the highway from runoff draining from the San Felipe Hills. The levees themselves are only a meter or so high, and have gently rounded profiles, thus presenting little obstacle to the passage of wind blown sand. However, water running down the many small washes transected by the levee system is deflected northward. The outwash of many channels combine behind the levees to form a potentially substantial stream of water, with the levees as the east bank. Over the past four decades or more a substantial channel, 2-4 m deep and 2-3 m wide, has been cut into the ground just west of the levee. This transverse feature probably intercepts a large portion of any eastwardly saltating sand. Further west, deep

Figure 14. The view is to the north from the top of dune O, the tallest Salton dune in 1981.

natural gullies formed since the desiccation of Lake Cahuilla perform the same function (Norris 1966).

Sand Supply in the Eastern Part of the Salton Area

In the region east of the landing strip (Figure 2) the mantle of sand upon the ground between the dunes is generally thicker than elsewhere, probably because large dunes to the west serve as reservoirs of material for dune-building in the downwind direction.

In the Salton dunefield the spacing of barchans is the reverse of the usual pattern described by Bagnold (1941, p. 218). He illustrates how at the upwind end of a typical belt of barchans, the dunes are closely packed, even overrunning each other, while in the downwind direction they become increasingly well separated. The dunes in the present study are well-separated at the upwind edge of the field but become more crowded to the east.

This spatial pattern is what might be expected from a group of dunes that are slowly dying out. As the source of sand gradually fails, smaller upwind dunes merge with larger dunes, or dissipate entirely. Large upwind barchans march on as best they can under increasingly adverse conditions, stripped each year of a portion of their volume by sand-driving winds. Downwind, large and small barchans alike creep ahead, thriving even in lean years on sand stripped from their upwind cousins, and showing as yet little effect of the desolation overtaking them from behind.

Fate of Dunes on the Eastern Border of the Salton Dunefield

Extrapolated trajectories of present-day Salton dunes connect across a distance of 40 km to the vicinity of the northern tip of the Algodones dune chain on the eastern side of the Salton Sink. Although there would have been nowhere near enough time for an individual barchan to cross the entire sink between the most recent fillings of the trough, it is possible that at one time a belt of dunes, and saltating sand, formed a direct connection to the much older (Sharp 1979) Algodones chain. Sykes (1937, p. 60) noted that a substantial amount of sand once occupied the lower portions of the sink, now underwater.

All dunes in the Salton field are now moving on courses that will take them eventually into the Salton Sea. At present rates of motion, the dunes farthest to the west will cross the present shoreline of the Salton Sea in about 250-300 years. In 1981, there was one large dune (σ) actually entering the water (Figure 15). The leading point of the northern horn had been washed away. In places the slipface descended directly into the water, while in other places a narrow beach, a few meters in breadth, separated the base of the slipface from the water.

The dune itself was damp on the surface for a height of about 60 cm above the water. At the base, where the slipface descended into water, a small scarp about 15 cm high had formed in the wet sand. Sand avalanches starting high up on the slipface cascaded down over the scarp. Wave action immediately began to round and smooth the new deposits into conical fans extending out underwater from the foot of the scarp.

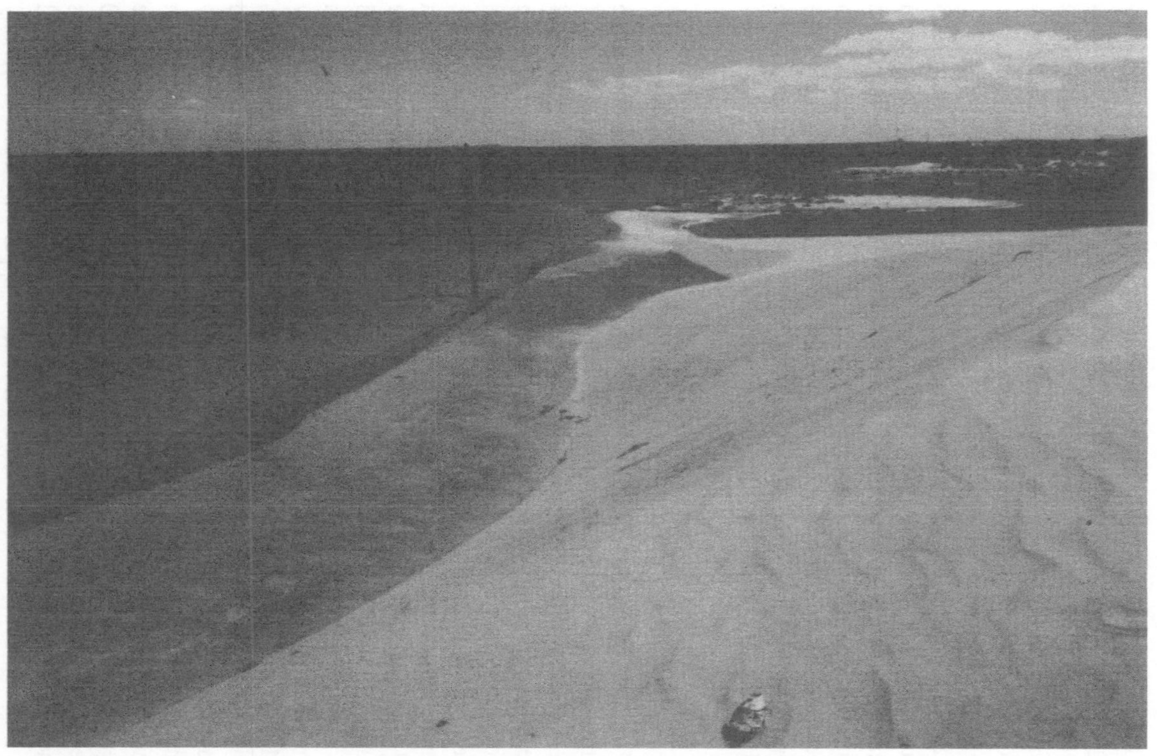

Figure 15. This large dune (σ) is about to enter the Salton Sea. The view is toward the south. Object in foreground is a fish.

In time, the dune will probably be beveled by the wind into a damp, low-lying deposit of sand sitting directly on the shore. There are many low, vegetated sand deposits in this area that are likely old dunes stabilized by the waters of the sea. Some may date from the filling of the Salton Sink. Others may be younger dunes which were never submerged, but became stuck and then vegetated near the shoreline.

South of the landing strip and east of the beach access road in Section 29 (Figure 2) there is an absence of active dunes. Some hillocks in this region appear to be former dunes artificially stabilized by burial under bulldozed layers of clay. At least 33 active dunes in Sections 25 and 30 are presently on trajectories that would eventually carry them into this empty quarter. The absence of large dunes here is unexplained. Perhaps dunes in this area, once used as a military base, were eliminated by bulldozing.

SUMMARY

The present Salton dunefield appears to be the remnant of a once more extensive and active barchan field. The decrease in the number and size of dunes is probably owing to both natural and human causes. The evaporation of Lake Cahuilla initially provided an adequate sand supply and an underlying smooth platform, which, together with strong and nearly unidirectional winds, represented an ideal environment for barchan formation. With time, channelization of the platform began to frustrate the transport of wind-blown sand to the east, and to impede the progress of all but the largest dunes. By the beginning of the 20th century, only a few large dunes remained in the upwind area west of present Highway 86. Flooding of the Salton Trough drowned an extensive train of barchans marching across the bottom of the sink, so that by the 1930s, only an isolated band of somewhat more than a hundred dunes survived on the higher undissected terrain east of the San Felipe Hills. Flood control operations along Highway 86 may have further decreased sand supply from the west. The western boundary of the dunefield has moved nearly 1 km eastward of its position in the 1940s and 1950s, and dunes on the western edge appear to be starved for sand. In 1981, the dunefield contained only about 70 dunes. All dunes are on trajectories that will eventually carry them to the shores of the Salton Sea. There they will be destroyed by wave action or transformed into low vegetated hillocks. Some dunes have apparently been artificially stabilized by covering with clay layers, while the absence of any significant barchans south of the landing strip suggest dune elimination by bulldozing. The Salton dunefield appears headed for eventual extinction as a result of a variety of natural and human causes.

ACKNOWLEDGMENTS

Field data and notes were generously provided by R. P. Sharp. We would like to acknowledge helpful comments, suggestions and assistance from R. Schwarm,

J. S. Shelton, and T. A. Tombrello. Thanks to K. Mulligan for comments on an earlier version of the manuscript. Finally, we would like to thank the U.S. Navy for permission to work on the Salton Sea Test Base. This work was supported in part by the National Science Foundation, Grant No. EAR-89-15983.

REFERENCES

Bagnold, R. A. (1941) *The Physics of Blown Sand and Desert Dunes.* Chapman & Hall, London.

Beadnell, H.J.L. (1910) The sand-dunes of the Libyan Desert. *Geographical Journal,* v. 35, p. 379-395.

Blake, W. P. (1858) *Report of a Geological Reconnaissance in California.* H. Bailliere, New York.

Cornish, V. (1897) On the formation of sand-dunes. *Geographical Journal,* v. 9, p. 278-309.

Finkel, H. J. (1959) The barchans of southern Peru. *Journal of Geology,* v. 67, p. 614-647.

King, W.J.H. (1918) Study of a dune belt. *Geographical Journal,* v. 51, p. 16-33.

Long, J. T., and Sharp, R. P. (1964) Barchan dune movement in Imperial Valley, California. *Geological Society of America Bulletin,* v. 75 p. 149-156.

MacDougal, D. T., ed. (1914) *The Salton Sea.* Carnegie Institution of Washington, Publication 193, Washington, D.C.

McKee, E. D. (1966) Structures of dunes at White Sands National Monument, New Mexico (and a comparison with structures of dunes from other selected areas). *Sedimentology,* v. 7, p. 1-70.

Mendenhall, W. C. (1909) *Some Desert Watering Places in Southeastern California and Southwestern Nevada.* USGS Water-Supply Paper 224.

Norris, R. M., (1966) Barchan dunes of Imperial Valley, California. *Geological Society of America Bulletin.* v. 72, p. 292-306.

Norris, R. M., and Norris, K. S. (1961) Algodones Dunes of Southeastern California. *Geological Society American Bulletin.* v. 73, p. 605-620.

Norris, R. M., and Webb, R. W. (1976) *Geology of California.* John Wiley, New York.

Rempel, R. J., (1936) The crescentic dunes of the Salton Sea and their relation to the vegetation. *Ecology.* v. 17, p. 347-358.

Russell, R. J., (1932) Land forms of San Gorgonio Pass, Southern California. *University of California Publications in Geography,.* v. 6, p. 23-121.

Sharp, R. P. (1979) Intradune flats of the Algodones Chain, Imperial Valley, California. *Geological Society American Bulletin,* v. 90, p. 908-916.

Sykes, G. (1937) *The Colorado Delta.* Carnegie Institution of Washington, Publication 460, Washington, D.C.

Waters, M. R. (1983) Late Holocene lacustrine chronology and archaeology of ancient Lake Cahuilla, California. *Quaternary Research,* v. 19, p. 373-387.

8 AEOLIAN MODIFICATIONS OF GLACIAL MORAINES AT BISHOP CREEK, EASTERN CALIFORNIA

Andrew J. Bach
Department of Geography
Arizona State University

ABSTRACT

Aeolian landforms and deposits are important sources of paleoclimatic data, especially when associated with other climate controlled landforms. Aeolian activity has modified Pleistocene glacial moraines at Bishop Creek, California, through the deposition of dust, tephra, and sand on moraine surfaces and the abrasion of till boulders. Many moraine surfaces are mantled by 1 to 20 cm of Holocene age aeolian dust. The magnitude of the aeolian modifications to the moraines appear to vary over small distances in response to the complex topography. Dust influx influences pedogenesis and rock weathering significantly. The spatial variation of dust deposits is superposed on the chronosequence of glacial moraines and confounds expected weathering and pedologic trends. Ventifact populations formed ~15-22 ka and ~60-65 ka reveal that aeolian erosive activity is associated with full and retreating glacial conditions. Before chronosequence research is conducted on dryland landforms, site to site variability in aeolian influence must be analyzed to assess constancy in affects.

INTRODUCTION

The significance of aeolian activity at the dryland/alpine interface can be found in a wide number of research themes. Geomorphic evidence for past aeolian activity includes stabilized sand dunes, aeolian dust mantles, and ventifacts covered with rock varnish. These aeolian landforms can provide information on the geographical extent, duration, and timing of episodes of enhanced aridity and changes in circulation patterns over time. Because wind action is often enhanced by dry conditions, the existence of aeolian deposits, ventifacts, and yardangs have long been regarded as indicators of aridity. However, aeolian activity is also enhanced in glacial environments, where strong (katabatic) winds exist and abundant fine sediment is supplied by the glacial system (Neuman 1993). Aeolian processes are sensitive to changes in both atmospheric parameters and surface conditions, and are influenced by changes in wind strength and direction, vegetation cover, and sediment availability. Spatial and temporal changes in aeolian activity in response to both local and regional change can influence the deposition and erosion rates of landforms in arid lands, including glacial moraines.

Abundant evidence for aeolian activity is found along the eastern margin of the Sierra Nevada, California, where Pleistocene glacial deposits are found today in an arid environment. Similar settings can be found in Tibet (Derbyshire

Desert Aeolian Processes. Edited by Vatche P. Tchakerian. Published in 1995 by Chapman & Hall, London.
ISBN 978-94-010-6519-1

et al. 1991), Africa (Rosquist 1990), Mexico (White 1990), Hawaii (Dorn et al. 1991), in the South American Andes (Hastenrath 1971), Colorado Front Range (Muhs et al. 1992), Big Horn Mountains, Wyoming (Sharp 1949), Iceland (Antevs 1928), and the glaciated ranges in the Great Basin (Zielinski and McCoy 1987).

The purpose of this paper is to document the nature and recurrence of aeolian modifications to the terminal moraine complex at Bishop Creek, California. The moraines have sand, dust and tephra on their surfaces, and the boulders in tills show evidence of aeolian abrasion. A preliminary aeolian history is constructed by assigning relative, calibrated, and numerical ages to mapped ventifacts. An assessment of the impact of aeolian deposits on weathering and pedogenesis is reviewed. Recognizing the spatial and temporal heterogeneity of aeolian modifications in the eastern Sierra Nevada, and in similar settings elsewhere, is important for chronosequence studies of post-depositional features on glacial moraines.

PHYSICAL SETTING

The presence of stabilized sand dunes, ventifacts, and dust deposits indicate that aeolian activity has been widespread in the eastern Sierra Nevada during the late Quaternary (Blackwelder 1929, Bateman 1965, Marchand 1970). The study area is a late Pleistocene terminal moraine complex at Bishop Creek, located in east-central California about 11 km WSW of Bishop (Figure 1). The study area covers approximately 16 km^2 and ranges in elevation from about 1600 m to 2100 m. Terminal moraine relief averages 150 m above the piedmont, and locally exceeds 500 m along the incised creek. The moraines are composed of homogeneous glacial till derived from a basin underlain by 90% granodiorite and 10% metamorphic rock (Bateman 1965). The moraines are sparsely vegetated with mountain mahogany *(Cercocarpus ledifolius)*, bitterbrush *(Purshia tridentada),* and sage brush *(Artimesia tridentada)*.

Present Climate

The Sierra Nevada crest roughly delimits a climatic boundary between the cool, moist Pacific air masses to the west, and the seasonally extreme temperatures and dry continental conditions of the Basin and Range to the east. Bishop Creek, on the eastern side of the crest, lies within the rainshadow. Precipitation data collected at nearby Bishop indicate dry summers and wet winters with an average annual precipitation of 144 mm (Powell and Klieforth 1991). Precipitation is slightly higher over the study area because of its higher elevation. Average annual temperature is 13.3°C, with January the coldest month (2.9°C) and July the warmest month (24.9°C) (Powell and Klieforth 1991).

The primary factor determining regional wind flow is the location and strength of the semi-permanent Pacific high pressure cell, and a desert surface

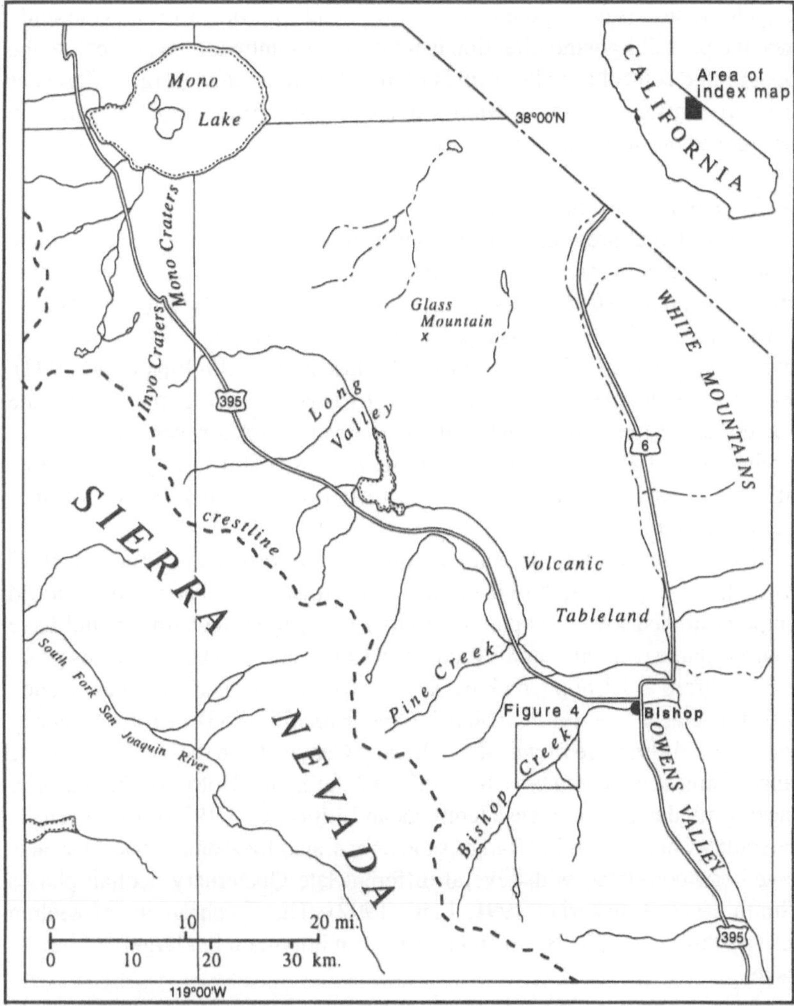

Figure 1. Location map for the study area and main volcanic centers.

low pressure cell over southern Nevada. During months of high sun, the desert surface low is most intense while to the west the Pacific high is strongest. This results in prevailing westerly winds with some southerly flow bringing moisture from the Gulf of California. In autumn and winter, the pressure gradient is often weaker and synoptic winds have lower speeds. Wind data are lacking within the study area, although at Bishop the prevailing direction is southerly, reflecting the topographic influence of Owens Valley (Figure 2a). A secondary wind direction is northerly to northwesterly, largely resulting from the passage of cyclonic storms during the winter (Powell and Klieforth 1991). In the study area, the prevailing wind direction is westerly to southwesterly, reflecting the topographic control of valley winds in the canyon. Locally, large differences in microclimate (including wind velocity) result from hummocky terrain superposed on subparallel lateral moraines.

Evidence of Paleoclimatic Conditions

The Bishop Creek area experienced significant and repeated changes in climate during the Quaternary period. Ice caps and valley glaciers in the Sierra Nevada advanced and retreated at least four times during the late Pleistocene (Bursik and Gillespie 1993). At Bishop Creek, there is evidence for as many as seven major glacial advances during the last 200 ka (Phillips et al. 1992). Owing to its unique tectonic and glacio-hydrologic setting, the Pleistocene glaciers at Bishop Creek changed their lower courses several times, preserving at least 12 distinct terminal moraines. In contrast, only two or three different Pleistocene stages have been recognized in other canyons in the Sierra Nevada.

During glacial periods, temperatures may have been as much as 10°C lower than at present, based on plant macrofossils. Because of reduced temperature and lower evapotranspiration rates, more moisture would have been available (Jennings and Elliott-Fisk 1993). Higher velocity winds probably occurred at Bishop Creek because of stronger westerly flow associated with full-glacial conditions (Manabe and Brocolli 1985). Strong katabatic winds would also have flowed down Bishop Canyon from the ice field atop the range. Three arid phases have been inferred during the Holocene from nearby packrat midden assemblages (Jennings and Elliott-Fisk 1993). Elsewhere in the southwestern Basin and Range, sand ramps, stabilized dunes, and ventifacts have been correlated with several different late Quaternary aeolian phases (Smith 1967, Tchakerian 1991, Laity 1992). The synchroneity of aeolian activity throughout the Basin and Range is an important but largely uncertain issue.

METHODS

The potential impact of aeolian processes was assessed through field examination of soils and ventifacts. Fifteen soils were sampled on moraine crests in areas determined to be free of excessive erosion or deposition. Morphological descriptions of soil horizons and stratigraphic units were made in the field and samples taken for laboratory analysis. Soil profiles were described using the nomenclature outlined by Birkeland (1984). Particle-size analysis of the fine fraction and soil pH measurements followed methods outlined by Singer and Janitzky (1986).

Granitic boulders were examined for aeolian abrasion. Well-formed grooves or flutes on relatively flat-lying surfaces were measured with a brunton compass. Ventifacted surfaces were collected for rock varnish analysis. A ratio

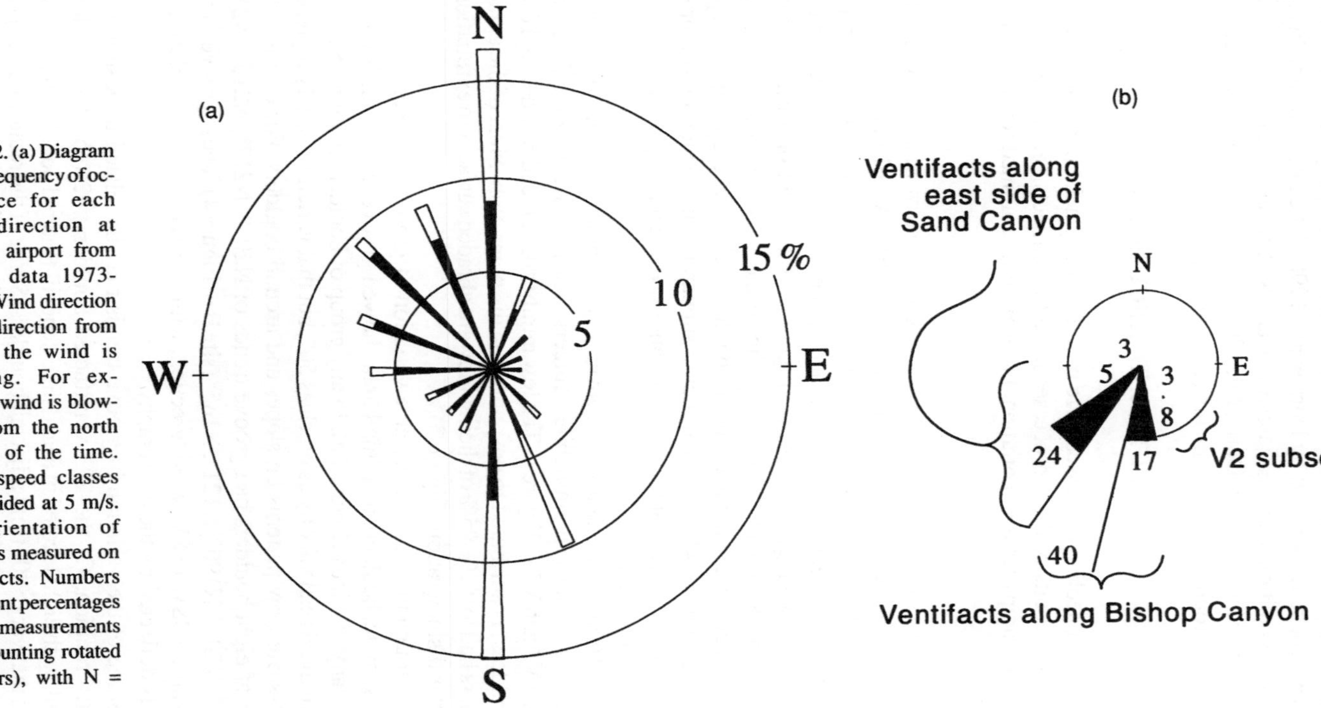

Figure 2. (a) Diagram of the frequency of occurrence for each wind direction at Bishop airport from hourly data 1973-1991. Wind direction is the direction from which the wind is blowing. For example, wind is blowing from the north 16.7% of the time. Wind speed classes are divided at 5 m/s. (b) Orientation of grooves measured on ventifacts. Numbers represent percentages of total measurements (not counting rotated boulders), with N = 231.

of $(K^+ + Ca^{2+})/Ti^{4+}$ was measured on cleaned scrapings of rock varnish following methods outlined by Dorn et al. (1990). The rate at which this ratio decreases is calibrated using radiocarbon and K-Ar ages from landforms in the region (calibration curve in Dorn et al. 1990). Cation-ratios were measured by electron microprobe measurements (wavelength dispersive).

RESULTS AND DISCUSSION

A number of aeolian processes have modified the glacial moraines at Bishop Creek. The results of these processes include ventifacts, tephra deposits, and dust. Sand ramps are also common, but were not examined for this study.

Ventifacts

Many boulders on glacial moraines along the eastern Sierra Nevada show evidence of wind abrasion (Blackwelder 1929). At Bishop Creek, most ventifacts are formed on granitic boulders 1-2 m in diameter, with some ranging up to 5 m in diameter. Evidence for aeolian abrasion includes facets, pitting, fluting and grooving, and frosting (Figure 3). Pits occur on surfaces at high angles to the wind (60°-120°) and indicate the windward side of boulders. Grooves occur on the top and sides of boulders. Grooves are U-shaped, open-ended depressions which are parallel to subparallel in orientation (Figure 3). Ventifacts are most abundant on moraine crests and slopes facing former positions of glaciers.

Wind trends were measured from groove alignments on flat-lying surfaces of 273 large, relatively immobile boulders. These boulders come from areas marked V2 and V3 in Figure 4. The low number of abraded boulders (<10% of all boulders) is attributed to post-abrasion disintegration. Most of the abraded boulders had lost 5%-95% of their windward abraded surfaces to exfoliation or granular disintegration (Figure 3).

Although there is considerable variation of groove alignment within the entire sample, boulders in a small area showed groove trends consistent with down canyon winds (Figure 2b). In any group of boulders (within 15 m^2), the groove trends can vary by as much as 50°, but that variation largely represents micro-scale flow patterns on slopes and around boulders. For example, one group of eight boulders had groove trends of N207°, N217°, N215°, N214°, N206°, N203°, N190°, N121°, N185° (site B in Figure 4). These grooves reveal a prevailing SSW to SW wind direction, with increasing southerly flow as the wind is deflected by the topography.

Some of the variation in groove trends may be related to post-abrasion boulder rotation. Of the 273 ventifacted boulders, 42 (15%) were clearly rotated, with eight located in a landslide deposit. Rotated boulders have groove trends that vary >40° from adjacent boulders. Some of the rotated boulders had abraded surfaces on their lower ends. Although boulder rotation can complicate paleo-circulation reconstruction, the general lack of significant rotation in the

Figure 3. Grooved granitic boulder at site B. The areas on both sides of the photo have been eroded since time of abrasion.

study area implies that the boulders and the moraines have been relatively stable over the last 15-20 ka.

Analysis of rock varnish on ventifacts at Bishop Creek suggest that they are relict (Dorn, this volume). Cation-ratio ages of varnishes on ventifacts fall into three age classes (Table 1). Ventifacts in V1 are ~60-65 ka, in V2, ~19-22 ka, and in V3, ~15-20 ka (Table 1, Figure 4). These data provide the first corroboration for Blackwelder's (1929) hypothesis that the ventifacts in the eastern Sierra Nevada formed during a glacial epoch, when strong winds armed with fine glacial detritus abraded boulders. Whether ventifaction was a result of increased wind speeds, increased sediment supply, or a combination of both remains to be determined.

The V1 ventifacts (Figure 4) formed when katabatic winds blew down Sand Canyon off the retreating Tahoe-age glacier (marine oxygen-isotope stage 4, ~65 ka maximum, Phillips et al. 1992). Fine detritus from the recently deglaciated valley floor most likely provided the sediment for abrasion (Figure 5a). Changes in glacier position from Sand Canyon to the present-day lower Bishop Canyon between the two glacial periods, isolated the V1 ventifacts from further aeolian abrasion. During the Tioga-age full glacial (stage 2, ~21 ka maximum, Phillips et al. 1992), ventifacts were formed in area V2 (Figure 5b). At this time, the boundary-layer in Bishop Canyon would have been the glacier surface at a similar elevation to area V2. Thus the winds would have a

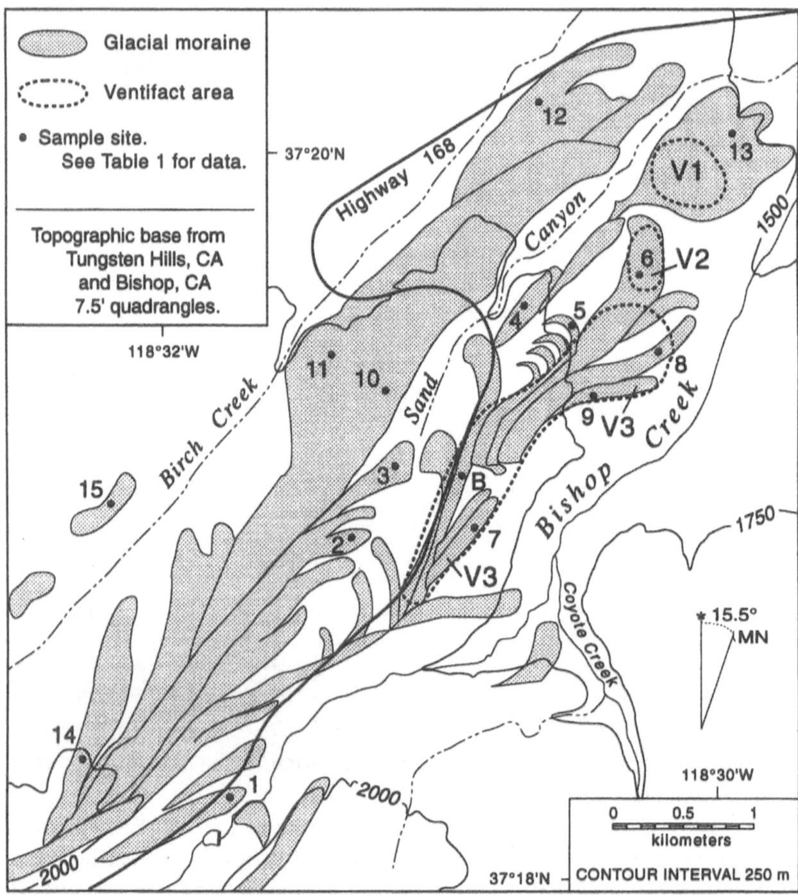

Figure 4. Map of Bishop Creek late-Pleistocene terminal moraine complex showing sample sites and ventifact areas. Upper Bishop Canyon continues in SW trend from lower left corner of map.

direct course over intervening topography into area V2 as indicated by SSE groove trends (Figure 2b). Winds from the retreating Tioga-age glacier would have flowed down the deglaciated Bishop Canyon. However, they could no longer reach the V2 ventifacts. Thus, wind abrasion ceased and rock varnish began to form. On the other hand, these same winds abraded boulders in area V3, along Bishop Canyon and Sand Canyon (Figure 5c). Boulders along Sand Canyon would have been abraded during each ventifaction event, as indicated by the best-formed flutes. Rock varnish cation-ratios suggests that the last abrasion in area V3 occurred ~14-15 ka.

Aeolian Deposits

Tephra is found near the surface of several soil pits, both as a reworked unit and mixed in the soil. The 'tephra-layer' is an Av soil horizon with a thickness of 1

Table 1
Cation-ratio ages for rock varnish formed on ventifacts at Bishop Creek.

V1		V2		V3	
Cation-ratio	Age	Cation-ratio	Age	Cation-ratio	Age
3.63	65	4.29	20	4.39	17
3.65	63	4.33	19	4.44	16
3.67	61	4.35	18	4.45	15
3.68	60	4.38	17	4.47	15
		4.24	22	4.51	14
		4.25	22	4.29	20
		4.27	21	4.32	19
		4.29	20	4.34	19
		4.30	20	4.35	18
		4.33	19	4.39	17

See Dorn et al. (1990) for calibration curve. Ages in thousand years (ka).

to 10 cm with an average of about 5 cm (Table 2). The tephra-layer is very pale brown (10YR 7/3) and contains vesicular pores. It has a friable blocky structure, which easily crushes to fines. The tephra-layer has a silty loam texture with 30%-45% silt and 4%-10% clay. Silt and clay percentages are much higher than the soil parent material or B-horizons (Table 2). The tephra layer at some sites is considerably reworked and contains arkosic sand.

Several volcanic centers in the region could have contributed ash to the study area: the Mono and Inyo Craters in Mono Basin, Glass Mountain near Long Valley, the Volcanic Tableland northeast of Bishop Creek, and the Coso area, 150 km to the south (Figure 1). The Mono/Inyo Craters have had several dozen silicic eruptions during the last 35,000 years, with the majority occurring during the Holocene (Sieh and Bursik 1986). Chemical analysis of tephra from one site suggests a mixture of several tephras from both the Inyo and Mono Craters (Andrei Sarna-Wojcicki, personal communication, 1994). Ages of the tephras range from as young as 525 years before present, to as old as 7020 years before present (Andrei Sarna-Wojcicki, personal communication, 1994). Since most of the Mono/Inyo tephras are similar in chemical composition, the age of the tephra layer is best characterizes as mid-to-late Holocene in age.

Pits excavated on moraine crests reveal distinctive surface layers which overlie the tephra-layer. The term "cover-bed" is used for surficial layers on soils with an origin different from bedrock parent material (Kleber 1992). Sources for post-tephra cover-bed include: (a) grus eroded from upslope clasts, (b) organic matter, (c) younger tephra, (d) aeolian dust, (e) material reworked

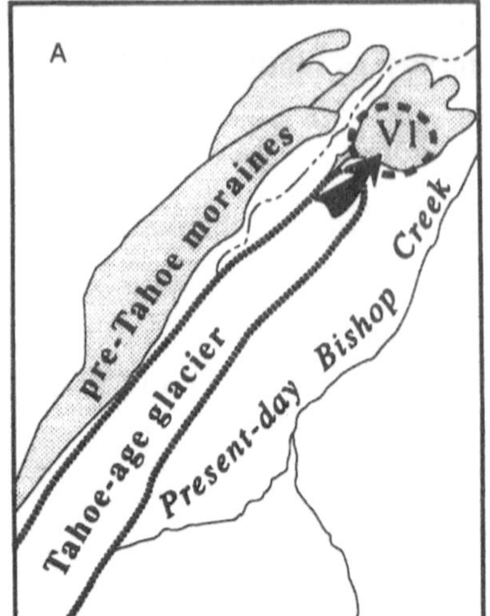

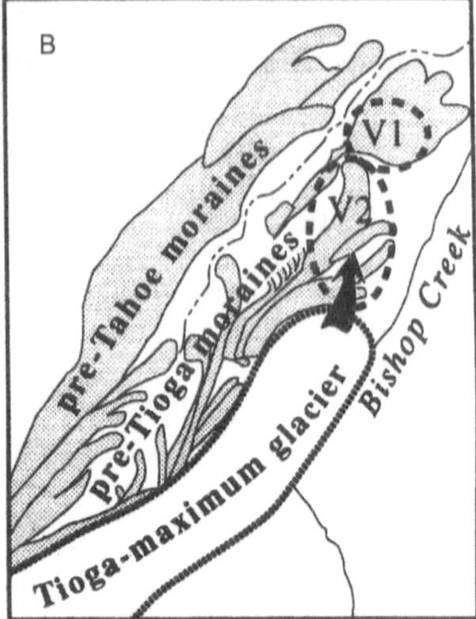

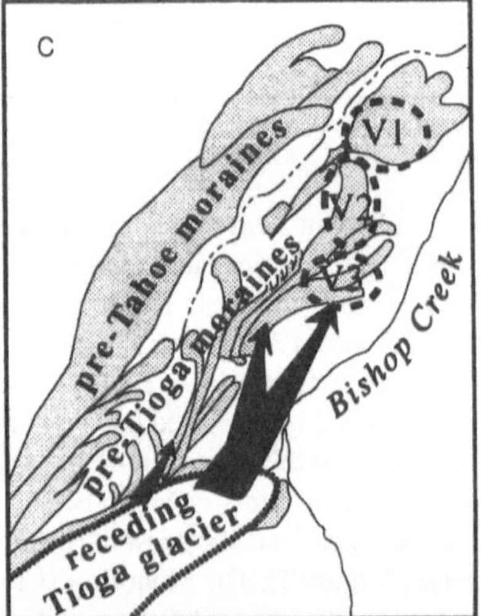

Figure 5. Hypothesized sequence of ventifact formation at Bishop Creek. (a). Study area about 60-66 ka. Tahoe-age glacier sat between a large moraine complex and the present-day Bishop Creek. Katabatic winds drained off glacier and blasted the foreglacial area. (b). Study area about 20-22 ka. The Bishop Creek glacier changed its course since Tahoe time (see a). At 20 ka the glacier sat in the present-day canyon of Bishop Creek. Katabatic winds blasted boulders in area V2. Next, the glacier began to retreat. (c). Study are about 14-16 ka. The glacier has retreated part way up canyon. Katabatic wind would blast the lower part of area V2, the topography protected the northern part of area V2 from further abrasion. Rock varnish did not begin to grow in area V3 until abrasion ceased ~14~16 ka.

Table 2
Characteristics of soils and cover-beds. Each site represents a morphostratigraphically older deposit.

Site	Relative age (see Bursik and Gillespie 1993)	Surface layer depth (cm)	Surface layer color (dry)	Surface layer texture *	Tephra thickness (cm)	Tephra texture *	Material Accumulating in upper B-horizon† CaCO$_3$	Si	Clay	Depth of soil (cm)
1		11.5	1OYR 5/3	SL (13/3)	8.5	SiL (30/9)	+			66
2	Tioga	4	2.5Y 5.5/3	SL (22/6)	7	SiL	+	+		68
3		8.5	10YR 7/3	SL	6	SiL			++	91
4		4	10YR 6/3	SL (21/4)	4	SiL (45/10)	+			85
5	Tahoe	4	10YR 6/2.5	L	10	SiL	+		+	89
6		3	10YR 5/3	SL (11/3)	6	SiL (44/4)	+	+	+	91
7		6	10YR 5.5/3	SL	4	SiL	++	++	+	114
8		6	10YR 5/3	SL	O††	-	++		+	107
9		3	2.5Y 5/3	SL	6	SiL	+		++	91
10		0**	-	-	13	L (18/8)	+		+	89
11	pre-Tahoe	0**	-	-	13	L	+		++	84
12		7	10YR 5/3	L	4	SiL	+		++	112
13		4	10YR 6/3	SL	7	SiL	+	+	+	147
14		6	10YR 5/2	SL	O††	-			++	137
15	Sherwin	4	10YR 5/3	SL	4	SiL	++	+	++	91

* SL= sandy loam, L= loam, SiL= silty loam, numbers in parantheses are percentages of silt/clay determined by particle size analysis (Singer and Janitzky 1986).

** Lack of surface layer indicate surfical erosion during the Holocene Loam textures of ash layer indicated mixing.
† Relative accumulation in B-horizons: + slight, ++ moderate, +++ significant
†† Lack of tephra indicates wind regime at time of deposition did not allow ash deposition, or that surfical erosion has occurred.

by overland flow, (f) *in situ* weathering, including frost shattering, and (g) bioturbation. Slope transportation by overland flow is minimal, as sampled sites are on thin-crested (1-3 m wide) moraines with gentle down-valley slopes (1°-12°). Grus in the cover-bed is eroded from granitic clasts lying on the moraine surface. Rates of bioturbation are undocumented in the Bishop region, but ants and ground squirrels may contribute a significant amount of material to the surface layer (e.g., Scott 1951, Johnson 1991).

The post-tephra cover-bed varies from 1 to 11.5 cm in thickness, with an average of about 5 cm (Table 2), and has a brown color (10YR 5/3) owing to the accumulation of organic matter in the A-horizon. It is separated from the till by a disconformity, often marked by the tephra, and has a different texture. The cover-bed is sandy loam in comparison and has about 5% silt and 2% clay more than the unweathered till (Table 2). The silt and clay are probably aeolian in origin owing to the fact that granitic tills in the Sierra Nevada contain very little fines.

The soil formed in till below the cover-beds varies in morphologic characteristics (Table 2). Relative weathering studies (e.g., Burke and Birkeland 1979) have distinguished only three glacial stage deposits: Tioga, Tahoe, and pre-Tahoe deposits, but not the additional deposits proposed by Phillips et al. (1992). Both carbonate and clay-films are found in the same B-horizons of some soils (Table 2). Both properties cannot have formed simultaneously since carbonate enrichment inhibits clay flocculation (Birkeland 1984). A reasonable explanation is that clay translocation occurred during the more humid late Pleistocene, and carbonate accumulation during the more arid phases of the Holocene (e.g., Jennings and Elliott-Fisk 1993).

GEOMORPHIC IMPLICATIONS

Dust Fallout Rates

Estimates of the dust fallout rates are speculative at this point due to poor age control. Based upon the young 525-year age for the underlying tephra, long-term dust fall rates vary from 89 to 340 g m^{-2} a^{-1} between sites, with an average of 148 g m^{-2} a^{-1} at Bishop Creek (Table 3). The older 7020-year age for the tephra gives dustfall rates from 6.6 to 25.4 g m^{-2} a^{-1} with and average of 11 g m^{-2} a^{-1} (Table 3). Either the dust flux into Bishop Creek is unusually high, the tephra is older than 525 years, or the input of other material into the surface layer is considerable. The dust influx in the region has been extremely high over the last 60 years because of the desiccation of Owens Lake, 100 km down wind (Gill and Cahill 1992). Short-term accumulations have not been measured in the study area, but are available for various localities in the southwestern United States (Table 3). However, it is difficult to separate dust flux from anthropogenic sources from natural inputs in these rates. A striking feature in Table 3 is the spatial variability of the dust fluxes. Such spatial variations are also seen in the dust deposits at Bishop Creek because of heterogeneous wind flow across a complex topography.

Table 3
Dust fall rates at Bishop Creek and other locations in the southwestern U.S.A.

Location	Dust Accumulation Rate (gm a)	Length of Record	Reference
Bishop Creek (young age estimate)	148 (89-340) *	average for ~525 years	this study
Bishop Creek (old age estimate)	11 (6.6-25.4)	average for ~7020 years	this study
White Mountains, Ca	1.4-15	4 months	Marchand (1970)
San Clemente Island, east side	24	one year	Muhs (1983)
San Clemente Island, west side	31	one year	Muhs (1983)
San Clemente Island, west side	28	average for ~2800 years	Muhs (1983)
Tempe, Az	54	one year	Pewe et al. (1981)
Las Cruces, New Mexico	9.3-123.8	one year	Gile and Grossman (1979)

* Assumed bulk density of surface layer is 1.55 gcm, age control on underlying tephra (see text).

Pedogenesis

Since the pioneering work of Marchand (1970), airborne dust as a soil-forming process has received increased attention (Machette 1985, Litaor 1987, Wells et al. 1987, Reheis 1990, Chadwick and Davis 1990). Soil characteristics may be strongly influenced by aeolian cover-beds. In some cases, cover-beds may be more important in pedogenesis than the underlying parent material (Kleber 1992). Clay minerals, $CaCO_3$, and silica in some soil B-horizons are not necessarily weathering products, but aeolian in origin (Colman 1982, Machette 1985, Chadwick et al. 1987b).

Some of the Bishop Creek soils are slightly to firmly cemented with silica (Table 2). Once cemented, the soil becomes less permeable and pedogenesis may be slowed. An important source of silica is rhyolitic ash, although the silica may also come from the solution of siliceous minerals like feldspar (Berry 1987, Chadwick et al. 1987b, Harden et al. 1991). Repeated wetting and drying slightly devitrites tephra, which is illuviated through the soil in solution, and is precipitated as opal-A (Chadwick et al. 1987a,b). Electron microprobe and scanning electron microscope (SEM) observations of soil peds from silicous B-horizons, reveal opal bridging sand grains and coating ped faces, clasts and pores much like clay films. Electron microprobe and EDX analysis with the SEM show the cementing agent to be pure silica. The distribution of silica in the soils shows no clear trend related to age or climate (Table 2), but may reflect a paleo-tephra distribution.

Some late Quaternary soils in dryland environments have shown increased $CaCO_3$ with increasing age (Gile and Grossman 1979, Machette 1985). Soils in the study area have accumulations of $CaCO_3$, but do not show a strong age-related trend (Table 2). The presence of $CaCO_3$ in the same horizons as clay-films suggests that the influx of $CaCO_3$ may be relatively recent.

Rock Weathering

Aeolian activity can affect rock weathering rates by: (a) direct abrasion of the rock surface, (b) increased mechanical weathering by wedging, and (c) the formation of case hardening that protects the underlying rock. Aeolian abrasion can mimic glacial polishing, especially if the weathering is not well developed as seen in Figure 3. Since the number of fresh, glacially polished boulders on moraines has been used to differentiate moraines into different glacial stages, aeolian abrasion must be recognized in order to accurately evaluate rock weathering parameters (e.g., Sharp 1969, Burke and Birkeland 1979, Bursik and Gillespie 1993).

Salts and expandable clay minerals accelerate mechanical weathering of surface clasts (Mustoe 1983). Wind-blown salts and expandable clay minerals deposited in micro- and macro-rock cracks will expand upon wetting, contributing to granular disintegration or spallation (Figure 6). Intense salt-spalling and granular disintegration from high dust influx can impart a "false age" appearance to granitic boulders in chronosequence studies. Dust deposition in rock microfractures can result in case hardening, especially in medium-to coarse-grained granites with abundant microfractures (Conca and Rossman 1982, Pye 1986). Case hardening preserves rock surfaces, giving a false impression of rock weathering rates (Bursik and Gillespie 1993).

Impact on Chronosequence Studies of Glacial Moraines

Observations and measurements of soil development and rock weathering characteristics have long been used to determine the relative ages of glacial moraines (e.g. Blackwelder 1931, Sharp 1969, Burke and Birkeland 1979, Bursik and Gillespie 1993). For more accurate relative weathering results, a chronosequence must exist, where all factors of soil development other than time are minimized (e.g., climate, biota, parent material, topography) (Jenny 1941). The relative amount of post-depositional modification between deposits (e.g., differences in soil development and rock weathering) reflect the time since moraine deposition (Birkeland 1984). As demonstrated in this paper, aeolian modifications to glacial moraines can occur over a variety of spatial and temporal scales, increasing or decreasing the apparent relative age of a landform. In order for a chronosequence in drylands to be properly evaluated, aeolian deposits and abrasion must be considered.

Aeolian processes can play a key role in modifying glacial moraine deposits and thus affecting the relative dating of the moraines. Aeolian processes adversely affect pedogenesis and rock weathering, the "traditional"

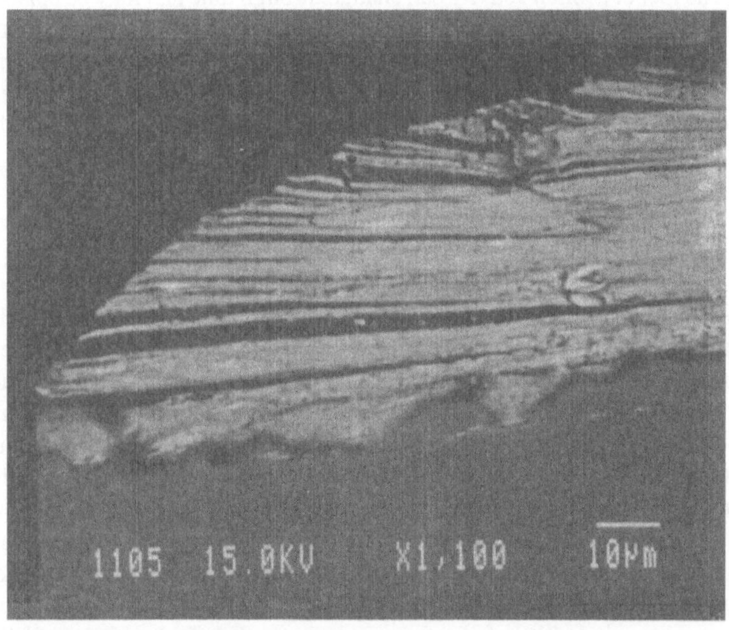

Figure 6. SEM micrograph of biotite grain on abraded surface of till boulder in ventifact area V3 (Figure 4). Biotite shows typical fraying weathering pattern. Clay minerals and quartz grains can be seen in the largest cracks. While some of this material may come from the weathered biotite, most has been blown in by wind.

methods of dating. However, ventifaction can be a valuable tool in new methods of surface-exposure dating, which includes rock varnish and cosmogenic isotope techniques (e.g., Dorn and Phillips 1992). Abrasion resets the rock varnish "clock" by removal, but does not remove enough rock to affect cosmogenic dating. Boulder rotation can influence cosmogenic dating by changing boulder geometry (e.g., Nishizumi et al. 1993). When rotated, the formally shielded part of the boulder may be moved to the top, where samples are typically taken. Recognition and use of ventifacts as indicators of stable boulders would greatly improve the reliability of cosmogenic sampling and dating.

Aeolian activity generally confounds the relation between moraine age and soil development at Bishop Creek. Aeolian erosion and deposition are heterogeneous across the complex moraine topography. For example, one site has 11.5 cm of Holocene dust at the surface, while 100 m away, a boulder has a paleo-soil line 70 cm above the present surface. The soil development reflects this heterogeneity by not displaying age related trends in silica or $CaCO_3$. To prevent misinterpretations, numerous sampling sites should be used to assess the micro-scale variability imposed by aeolian inputs.

The loci of aeolian activity can change through time and may be episodic at different time scales. Changes in geomorphic processes, landforms, circulation patterns, temperature, and/or precipitation may allow a region to change from a depositional sink to an erosional. Regions or sample sites which may not show evidence of aeolian activity today may have been affected by the wind in the past (i.e., fossil ventifacts).

CONCLUSIONS

Geomorphic evidence at Bishop Creek indicates that aeolian processes have varied over space and time. Deposits of dust, tephra, and sand on moraine surfaces occur next to wind-abraded boulders. The deposition of an average of 5 cm of aeolian dust during the mid-to-late Holocene indicates that the study area is currently a depositional sink. Yet, ventifacts that formed ~60-65 ka and ~14-22 ka suggest enhanced aeolian abrasional activity during glacial periods. Unstudied sand ramps in the area should provide further temporal information for aeolian processes (e.g., Tchakerian 1991).

Ventifacts at Bishop Creek are found near terminal moraines. Rock varnish dating suggests that the ventifacts formed at the same time as glaciers occupied those terminal positions. Furthermore, ventifact groove orientations indicate that winds blew from the same terminal positions. Since aeolian abrasion at Bishop Creek appears to have been episodic and related to glacial conditions, ventifacts may be used as time markers in glaciated regions. The association of ventifacts and glacial termini provides a basic means for establishing a check for dating glacier position, whereby ages of the ventifacts agree with ages of related glacial deposits. Ventifacts can also be used to assess the stability of boulder geometry for cosmogenic studies which can provide independent age control.

Currently, Bishop Creek is a depositional sink for dust with an average flux between 11 and 148 g m^{-2} a^{-1} during the Holocene. This influx varies spatially in response to wind velocity over the complex terrain and may not be representative of longer periods of time. Material deposited at the soil surface contributes to the chemical and physical make-up of the soil and may enhance soil formation at a site. The spatial variability of silica and $CaCO_3$ seen in soils at Bishop Creek may relate to patterns of aeolian deposition, rather than climate or age (Table 2). Aeolian dust can enhance rock weathering by mechanical wedging of crystals (Figure 6) and hence contribute to the overall weathering budget of an area.

Whereas great care is taken in chronosequence studies to control climate, parent material, topography, biota, and time, aeolian inputs have not received proper attention and study. In dryland environments, however, aeolian activity is often important, albeit episodic. The impact of aeolian processes on the chronosequence of glacial moraines at Bishop Creek, eastern California, points out the importance of dust flux in the study of weathering processes in arid lands.

ACKNOWLEDGMENTS

Supported in part by NSF grant SES-89-00401 (to Elliott-Fisk) , NSF grant SES-89-00402 (to Dorn), GSA grant 5111-93 (to Bach) and the White Mountain Research Station (to Bach). Wind data purchased and compiled under NSF grant EAR-9204648 to Nick Lancaster. Thanks for field assistance by and discussions with Debbie Elliott-Fisk, Ginger Schmid, Ron Dorn, Fred Phillips, and Doug Van Lare. Rock varnish was sampled and analyzed by Ron Dorn, and chemical analysis of tephra by C. Meyer and A. Sarna-Wojcicki. I thank Vatche Tchakerian, Ron Dorn, and an anonymous reviewer for their helpful comments on the manuscript.

REFERENCES

Antevs, E. (1928) Wind deserts in Iceland. *Geographical Review,* v. 18, p. 675-676.

Bateman, P. C. (1965) *Geology and Tungsten Mineralization of the Bishop District, California,* U.S. Geological Survey Professional Paper, 470.

Berry, M. E. (1987) Morphological and chemical characteristics of soil catenas on Pinedale and Bull Lake moraine slopes in the Salmon River Mountains, Idaho. *Quaternary Research,* v. 28, p. 210-255.

Birkeland, P. W. (1984) *Soils and Geomorphology.* Oxford University Press, Oxford.

Blackwelder, E. (1929) Sandblast action in relation to the glaciers of the Sierra Nevada. *Journal of Geology,* v. 37, p. 256-260.

Blackwelder, E. (1931) Pleistocene glaciation in the Sierra Nevada and Basin Ranges. *Geological Society of America Bulletin,* v. 42, p. 865-922.

Burke, R. M., and Birkeland, P. W. (1979) Reevaluation of multiparameter relative dating techniques and their application to the glacial sequence along the eastern escarpment of the Sierra Nevada, California. *Quaternary Research,* v. 11, p. 21-51.

Bursik, M., and Gillespie, A. R. (1993) Late Pleistocene glaciation of Mono Basin, California. *Quaternary Research,* v. 39, p. 24-35.

Chadwick, O. A., and Davis, J. O. (1990) Soil-forming intervals caused by eolian sediment pulses in the Lahontan basin, northwestern Nevada. *Geology,* v. 18, p. 234-246.

Chadwick, O. A., Hendricks, D. M., and Nettleton, W. D. (1987a) Silica in duric soils: I. A depositional model. *Soil Science Society of America Journal,* v. 51, p. 975-982.

Chadwick, O. A., Hendricks, D. M., and Nettleton, W. D. (1987b) Silica in duric soils: II. Mineralogy. *Soil Science Society of America Journal,* v. 51, p. 982-985.

Colman, S. M. (1982) Clay mineralogy of weathering rinds and possible implications concerning the sources of clay minerals in soils. *Geology,* v. 10, p. 370-375.

Conca, J. L., and Rossman, G. R. (1982) Case hardening of sandstone. *Geology,* v. 10, p. 520-523.

Derbyshire, E., Yafeng, S., Jijun, L., Benxing, Z., Shijie, L., and Jingtai, W. (1991) Quaternary glaciation of Tibet: The geological evidence. *Quaternary Science Reviews,* v. 10, p. 485-510.

Dorn, R. I., Cahill, T. A., Eldred, R. A., Gill, T. E., Bach, A. J., and Elliott-Fisk, D. L. (1990) Dating rock varnishes by the cation ratio method with PIXE, ICP, and the electron microprobe. *International Journal of PIXE,* v. 1, p. 157-195.

Dorn, R. I., and Phillips, F. M. (1992) Surface exposure dating: Review and critical evaluation. *Physical Geography,* v. 12, p. 303-333.

Dorn, R. I., Phillips, F. M., Zreda, M. G., Wolfe, E. W., Jull, A.J.T., Donahue, D. J., Kubik, P. W., and Sharma, P. (1991) Glacial chronology of Mauna Kea, Hawaii, as constrained by surface-exposure dating. *National Geographic Research,* v. 7, p. 456-471.

Gile, L. H., and Grossman, R. B. (1979) *The Desert Project Soil Monograph.* Washington D.C.: U.S. Department of Agriculture, Soil Conservation Service.

Gill, T. E., and Cahill, T. A. (1992) Playa-generated duststorms of Owens Lake. In C. A. Hall, Jr.,

V. Doyle-Jones, and B. Widawski (eds.) *The History of Water: Eastern Sierra Nevada, Owens Valley, White-Inyo Mountains.* p. 63-73.

Harden, J. W., Slate, J. L., Lamothe, P., Chadwick, O., Pendall, E., and Gillespie, A. (1991) Soil formation on the Trail Canyon alluvial fan, Fish Lake Valley, Nevada. In *1991 Pacific Cell Friends of the Pleistocene Guidebook, Fish Lake Valley, California-Nevada.* p. 139-160.

Hastenrath, S. L. (1971) On the Pleistocene snow-line depression in the arid regions of the South American Andes. *Journal of Glaciology,* v. 10, p. 255-267.

Jennings, S. A., and Elliott-Fisk, D. L. (1993) Packrat midden evidence of late Quaternary vegetation change in the White Mountains, California-Nevada. *Quaternary Research,* v. 39, p. 214-221.

Jenny, H. (1941) *Factors of Soil Formation.* McGraw-Hill, New York.

Johnson, D. L. (1991) Ants as key geomorphic agents in the evolution of biomantles. *1991 Annual Meeting Abstracts, Association of American Geographers,* p. 97.

Kleber, A. (1992) Periglacial slope deposits and their pedogenic implications in Germany, *Paleogeography, Paleoclimatology, Paleoecology,* v. 99, p. 361-371.

Laity, J. E. (1992) Ventifact evidence for Holocene wind patterns in the east-central Mojave Desert. *Zeitschrift f r Geomorphologie,* v. 84, p. 73-88.

Litaor, M. I. (1987) The influence of aeolian dust on the genesis of alpine soils in the Front Range, Colorado. *Soil Science Society of America Journal,* v. 51, p. 142-147.

Machette, M. N. (1985) Calcic soils of the southwestern United States. In D. L. Weide (ed.) *Soils and Quaternary Geology of the Southwestern United States.* Geological Society of America, Special Paper 203, p. 1-22.

Manabe, S. and Brocolli, A. J. (1985) The influence of continental ice sheets on the climate of an ice age. *Journal of Geophysical Research,* v. 90, p. 2167-2189.

Marchand, D. E. (1970) Soil contamination in the White Mountains, eastern California. *Geological Society of America Bulletin,* v. 81, p. 2497-2506.

Muhs, D. R., Benedict, J. B., and Evans, J. (1992) Sources of probable aeolian sediments on late Quaternary alpine moraines, Colorado Front Range: Evidence from trace element geochemistry. *AMQUA Program with Abstracts,* p. 73.

Mustoe, G. E. (1983) Cavernous weathering in the Capitol Reef Desert, Utah. *Earth Surface Processes and Landforms,* v. 8, p. 517-526.

Nishizumi, K., Kohl, C., Arnold, J., Dorn, R., Klein, J., Fink, D., Middleton, R., and Lal, D. (1993) Role of *in situ* cosmogenic nuclides ^{10}Be and ^{26}Al in the study of diverse geomorphic processes. *Earth Surface Processes and Landforms,* v. 18, p. 407-425.

Neuman, C. M. (1993) A review of aeolian transport processes in cold environments. *Progress in Physical Geography,* v. 17, p. 137-155.

Phillips, F. M., Zreda, M. G., and Elmore, D. (1992) Late Quaternary glacial history of the Sierra Nevada from cosmogenic ^{36}Cl dating of moraines at Bishop Creek, California. *EOS,* v. 73, p. 186.

Powell, D. R., and Klieforth, H. E. (1991) Weather and Climate. In C. A. Hall, Jr. (ed.) *Natural History of the White-Inyo Range, Eastern California-Nevada.*University of California Press, Berkeley, p. 3-29

Pye, K. (1986) Mineralogical and textural controls on the weathering of granitoid rocks. *Catena,* v. 13, p. 47-57.

Reheis, M. C. (1990) Influence of aeolian dust on the major-element chemistry and clay mineralogy of soils in the northern Bighorn Basin, U.S.A. *Catena,* v. 17, p. 219-248.

Rosquist, G. (1990) Quaternary glaciations in Africa. *Quaternary Science Reviews,* v. 9, p. 281-297.

Scott, H. W. (1951) The geological work of the mound-building ants in western United States. *Journal of Geology,* v. 59, p. 173-175.

Sieh, K. and Bursik, M. (1986) Most recent eruption of Mono Craters, eastern central California. *Journal of Geophysical Research,* v. 91(B), p. 12,539-12,571.

Singer M. J., and Janitzky, P. (eds.) 1986. *Field and laboratory procedures used in a soil chronosequence study.* United States Geological Survey Bulletin 1648.

Smith, H.T.U. (1967) *Past versus present wind action in the Mojave Desert region, California.* Air

Force Cambridge Research Laboratories Report 67-0683.
Sharp, R. P. (1949) Pleistocene ventifacts east of the Big Horn Mountains, Wyoming. *Journal of Geology*, v. 57, p. 175-195.
Sharp, R. P. (1969) Semiquantitative differentiation of glacial moraines near Convict Lake, Sierra Nevada, California. *Journal of Geology*, v. 77, p. 68-91.
Tchakerian, V. P. (1991) Late Quaternary geomorphology of the Dale Lake sand sheet, southern Mojave Desert, California. *Physical Geography*, v. 12, p. 347-369.
Wells, S. G., McFadden, L. D., and Dohrewned, J. C. (1987) Influence of late Quaternary climatic changes on geomorphic and pedogenic processes on a desert piedmont, eastern Mojave Desert, California. *Quaternary Research*, v. 27, p. 130-146.
White, S. F. (1990) Quaternary glacial stratigraphy of Mexico. *Quaternary Science Reviews*, v. 5, p. 201-206.
Zielinski, G. A., and McCoy, W. D. (1987) Paleoclimatic implications of the relationship between modern snowpack and late Pleistocene equilibrium-line altitudes in the mountains of the Great Basin, Western U.S.A. *Arctic and Alpine Research*, v. 19, p. 127-134.

9 ALTERATIONS OF VENTIFACT SURFACES AT THE GLACIER/DESERT INTERFACE

Ronald I. Dorn
Department of Geography
Arizona State University

ABSTRACT

Ventifacts are found on many glacial moraines in drylands, where aeolian abrasion is currently absent. This paper examines new methods for studying fossil ventifacts through surficial alteration. These include the development of rock coatings, thickening of weathering rinds, and the buildup of cosmogenic nuclides. Three areas are chosen to examine surficial alteration on fossil ventifacts: Mauna Kea in Hawaii, and Bishop Creek and Mono Basin in the eastern Sierra Nevada, California. Cosmogenic nuclides are valuable tools in fossil ventifact studies in that they provide a maximum-possible age for the ventifact by indicating the age of the glacial deposit. However, they cannot be used to indicate when the abrasion took place. In contrast, cation-ratio dating of rock varnish, layering of rock varnish, and the development of weathering rinds on ventifacted boulders provide data on when abrasion occurred. Ventifacts at Bishop Creek and Mauna Kea formed during glacial periods. Aeolian abrasion and subsequent boulder-surface alteration are rarely considered by glacial geomorphologists who use relative dating methods to assess the ages of glacial deposits.

INTRODUCTION

Aeolian erosion is normally associated with warm deserts, where it was first recognized and emphasized in the scientific literature (see review by Laity 1994). Aeolian abrasion is also common in modern and fossil glacial environments. Glacial ventifacts have been found on moraines adjacent to active glaciers and in glacio-fluvial settings, where strong katabatic winds abrade rocks with abundant pro-glacial sand and dust. Ventifacts are present in a wide number of cold-climate locations, for example, Iceland (Antevs 1928), northern Canada (McKenna-Neuman and Gilbert 1986), Antarctica (Hall 1989), South America (Czajka 1972), and Sweden (Schlyter 1991).

Fossil ventifacts associated with Pleistocene alpine glacial systems have been known for decades (e.g., Blackwelder 1929, Powers 1936, Sharp 1949), but have received little attention owing to the paucity of methods available to study them. Most Quaternary techniques require a stratigraphic context, but these ventifacts have remained at the surface. Yet, the alteration of the surfaces of ventifacts provides a means by which to assess the history of aeolian processes at a site. Laity (1991), using secondary electron microscopy, concluded that the weathering of ventifact surfaces can be used as a relative dating tool to assess the cessation of abrasion. Three different types of surficial alteration occur after aeolian abrasion ceases: buildup of cosmogenic nuclides, weather-

Desert Aeolian Processes. Edited by Vatche P. Tchakerian. Published in 1995 by Chapman & Hall, London. ISBN 978-94-010-6519-1

ing of the ventifact, and development of rock coatings on the abraded surface.

The purpose of this study is to assess whether different types of surficial alterations can yield useful data to aeolian geomorphologists. Six types of surficial alterations were analyzed: the buildup of ^{36}Cl, thickening of weathering rinds, layering in rock varnish, cation-ratio dating of rock varnish, development of oxalate coatings, and the growth of silica glaze. These surficial alterations are examined from three sites: the former ice cap on top of Mauna Kea, Hawaii (Dorn et al. 1991), a terminal moraine complex at Bishop Creek in Owens Valley, California (Zreda 1993), and moraines of the Mono Basin in eastern California (Sharp and Birman 1963).

SURFICIAL ALTERATION OF VENTIFACTS

Cosmogenic Nuclides

Cosmogenic nuclides (e.g., ^{3}He, ^{10}Be, ^{26}Al, ^{36}Cl) build up over time in minerals that are found in the upper meter of the Earth's surface. This technique has the potential to contribute to geomorphology (Dorn and Phillips 1991, Beck 1993) as an absolute dating tool for glacial deposits, bedrock slopes, sand dunes, basalt flows, paleo-beach ridges, and meteorite impact craters (Kurz 1986, Phillips et al. 1990, Nishiizumi et al. 1993, Zreda 1993).

The use of cosmogenic nuclides is explored in this paper through an analysis of ^{36}Cl ages on ventifacts in Dorn et al. (1991) and Zreda (1993). ^{36}Cl builds up in rocks because of the interactions of cosmic rays with atoms in minerals. The rate of accumulation is a function of altitude, geomagnetic latitude, rock chemistry, geometry of exposure to cosmic rays, time, and the cosmic ray flux. The cosmic ray flux does not vary (cf. radiocarbon production), but long-term averages integrate short-term fluctuations. For radionuclides like ^{36}Cl, the length of dwelling time at the Earth's surface must include the effects of both buildup and decay. For more detailed analysis consult Zreda (1993).

Weathering Rinds

After aeolian abrasion ceases, the newly created surface of the ventifact starts to weather. The net effect of biochemical weathering is a loss of mass in the outer skin of the rock. This change can be seen visually as a "weathering rind" around the outer rim of the ventifact. Weathering rinds on surficial deposits have been used as a dating tool in Quaternary research (e.g., Chinn 1981). The basic rationale involves the progressive thickening of rinds over time, provided climate, organisms, lithology, boulder topographic position, development of rock coatings, and surface microtopography are held constant and boulder erosion does not occur.

The study of weathering-rind thickness has yet to be applied to ventifacts. This is surprising, since the presence of ventifact polish supports an important assumption that is often glossed over in chronometric studies of weathering rinds: that no spalling of the weathering rind has occurred. Aeolian polish indicates a lack of erosion, thus the assumption that the "weathering clock" has not been reset by erosion. One possible reason why rinds have not been explored in ventifact studies is their poor development. Weathering rinds associated with ventifacts are much thinner than rinds on cobbles in soils of the same age, and much harder to measure accurately, owing to the fact that lithologies that preserve aeolian polish are often more resistant to weathering. Also, soil is a biogeochemical environment that typically enhances weathering.

There is a cost/benefit trade-off in deciding which method to use in the measurement of weathering rinds on ventifacts. Field measurements are the easiest, but often impossible because many rinds are too thin to visually "see," let alone measure thickness. Rinds can be identified more readily in thin section, viewed through a petrographic microscope at >30x magnification, but boundaries between the rind and unweathered rock are often unclear. In this paper, weathering rinds on ventifacts were analyzed with an electron microscope in order to accurately measure rind thicknesses. While the higher cost of using electron microscopy limited the number of samples, there is a much greater confidence in the accuracy and precision of the measurements.

Rinds were measured from nine different boulders at each site. Samples were collected from the tops of the boulders in order to minimize the role of water runoff and water collection. Only aplite samples were used in the Bishop Creek study, while in the Mauna Kea site, basalt. Cross-sections normal to the rock surface were polished, coated with carbon, and then imaged with back-scatter electron microscopy (BSE), following methods outlined by Krinsley and Manley (1989). As a result of the net loss of mass of chemically weathered minerals, weathered rinds in BSE appear more porous than the underlying unweathered rock. In this study, rind thickness was measured normal to the rock surface at 500 evenly spaced points (50 μm apart) along 2.5 cm of rind length on each boulder. Since rock coatings also affect weathering rinds, measurements were limited to places where coatings had not yet grown over the ventifact surface.

Development of Rock Coatings

After aeolian abrasion ceases, the newly polished ventifact surface starts to develop coatings. There are many different types of accretions: growth of organisms (lichens, fungi, cyanobacteria, moss, algae), iron films, phosphate crusts, carbonate, sulfate, polish smears, dust films, anthropogenic pigment, case hardening with carbonate and silica, manganiferous rock varnish, silica glaze, and oxalate-rich crusts. The biogeochemical environment determines whether a coating grows or erodes, how it grows, and how fast it grows.

Each coating has the potential to supply information on the post-abrasion history of the ventifact, but only three accretions are examined here: manganiferous rock varnish, silica glaze, and oxalate-rich crusts. Rock varnish is a paper-thin coating of manganese and iron hydroxides with intermixed clay minerals; it is analogous to a brick wall: clays are the "bricks" while the oxides are analogous to the mortar. Silica glaze is a lustrous accretion that is clear to orange in color, typically less than 1 mm and is dominated by amorphous SiO_2. A typical silica glaze found on glacial ventifacts would consist of ~70%-95% amorphous SiO_2, ~5%-20% amorphous Al_2O_3, and sometimes enough iron (~1%-10%) to give it an orange color. Oxalate-rich crusts are typically composed of a mixture of calcium oxalate and amorphous silica in roughly similar amounts. The calcium oxalate derives from lichens.

Microstratigraphy of Rock Varnish

Rock varnish contains layers, visible in both light and electron microscopes, that accumulate over time as a sedimentary deposit. Perry and Adams (1978) first observed continuous orange (Mn-poor) and black (Mn-rich) layers in rock varnish and argued that the layers were laterally continuous and reflected some unknown type of environmental change. Dorn (1990) and Jones (1991) established that these Mn:Fe microlaminations are most likely caused by fluctuations in alkalinity experienced on rock surfaces. When alkalinity levels are high (e.g., from deposition of aerosols deflated from margins of saline playas), varnish chemistry is not enriched in manganese and has an orange appearance. Manganese enrichment occurs with near neutral conditions. The time needed to develop a distinctive layer is typically on the order of 10^3 to 10^4 years (Dorn 1990), because varnish grows slowly in drylands.

Mn:Fe microlamination sequences in varnishes have not previously been reported for ventifacts. The ventifacts with Holocene cation-ratio and radiocarbon ages, reported by Laity (1991, 1994) have relatively thin layers (<20 μm) of manganese-poor varnish. This is characteristic of the Mojave Desert, California, where Holocene varnishes have developed in a more arid, alkaline environment. In contrast, Pleistocene varnishes in western North America typically have one or more cycles of Mn-poor and Mn-rich layers, because older varnishes have experienced paleoalkalinity fluctuations (Dorn 1990). This study assesses whether it is possible to use varnish layering to assign relative ages to Pleistocene ventifacts.

Cation-Ratio Dating of Rock Varnish

Cation-ratio (CR) dating is a method that assigns relative or calibrated ages to rock varnishes (Dorn 1989). Since its discovery (Dorn 1983), the trend of decreasing CRs with known age has been duplicated by five different groups around the world (e.g., Glazovskiy 1985, Pineda et al. 1988, 1990, Zhang et al. 1990, Bull 1991, Whitney and Harrington 1993). The ratio of cations of (K+Ca)/Ti decreases with age, because of the leaching of the more mobile potassium and calcium ions with time (Dorn and Krinsley 1991).

A least-squares semilog regression calibration can be constructed called a "cation-leaching curve," if CRs are measured at sites with known exposure ages. Figure 1 illustrates cation-leaching curves for study sites examined in this study. Calibrated ages are assigned to cation ratios, providing minimum-

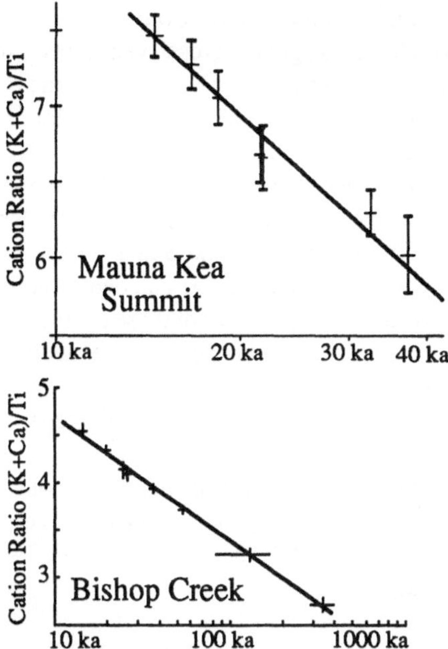

Figure 1. Cation-leaching curves for Bishop Creek, eastern California and Mauna Kea, Hawaii. The calibrations for Bishop Creek are from radiocarbon dates on subvarnish organic matter and K-Ar ages (Phillips et al., in preparation). The calibrations for Mauna Kea are from radiocarbon dates on subvarnish organic matter (Dorn et al. 1991, 1992).

limiting ages for the cessation of aeolian abrasion. Differences among curves in Figure 1 are from variations in the chemistry of airborne fallout and environmental factors. Details on the method of cation-ratio dating used here are provided in Dorn et al. (1990).

Although the general decline in cation ratios with age has been replicated, there are controversies with the technique. These are reviewed in detail in Beck (1993) and are beyond the scope of this paper. However, two general comments merit attention.

First, it is essential to select the right type of material for analysis. There are dozens of different types of rock varnishes. Many controversies have arisen because researchers have collected varnishes that are inappropriate for dating. For example, several varnish researchers specify that they collect the "darkest" and "smoothest" varnishes. These may not be necessarily the best choices because varnishes typically start out in rock joints. These "crack varnishes" are then exposed to the viewing eye by spalling. Crack varnishes do not yield information on when a geomorphic event occurred (e.g., flooding, glaciation), and form in a geochemical environment that makes them unsuitable for CR dating (Dorn 1990).

Second, any Quaternary dating results (even radiocarbon) should be "cross-checked" against other information. In this study, the cation-ratio results are compared with other types of ventifact alterations.

Development of Silica Glaze and Oxalate Coatings
Coatings other than rock varnish grow on fossil ventifacts. Two of these studied include silica glaze (Curtiss et al. 1985) and accretions of calcium oxalate ($CaC_2O_4.H_2O$) (Jones et al. 1981, Del Monte and Sabbioni 1987). Silica glaze and oxalates encapsulate organic matter that can be removed from under these coatings and radiocarbon dated (Dorn et al. 1991, 1992, Nobbs and Dorn 1993). The carbon in oxalate can be radiocarbon dated (Watchman 1991), and analyzed for its ^{13}C composition (Zak and Skala 1993). Both coatings show some evidence of layering that may reveal paleoenvironmental information. For the purpose of this study, the presence of these coatings is used to infer that the ventifact had been fossilized, because active aeolian abrasion would remove these soft coatings.

Silica glaze and oxalate-rich coatings were collected from the top surfaces of morainal boulders that were ventifacted. After collection, they were prepared for the laboratory by placing them in an epoxy mold, polishing a cross-section normal to the surface, coating the polished surface with a thin layer of carbon, and placing the specimens under an electron microscope for examination with BSE (cf. Traquair 1986).

RESULTS AND DISCUSSION

Ventifacts from glacial moraines were collected from three different environmental settings with different sets of rock-surface alterations. Bishop Creek offers one of the most arid settings for glacial moraines in North America (see Bach, this volume). Mauna Kea, above the trade-wind inversion, offers a chance to evaluate tropical rock varnish and weathering rinds on a basaltic lithology. Mono Basin moraines have ventifacts, but the environment does not preserve rock varnish, although other types of rock coatings are preserved.

Bishop Creek, Sierra Nevada, California
Along Bishop Creek in the eastern Sierra Nevada, at least seven major glacial advances are preserved from the last 150,000 years (Zreda 1993, Phillips et al., in prep.). Previously deposited morainal boulders were subject to aeolian abrasion, resulting in ventifacts ranging in size from ~1/3 m diameter to >2 m diameter (Figure 3).

Three different types of alterations were examined on the ventifacts from three positions at Bishop Creek (Figure 2): rock varnish CR dating, rock varnish microlaminations, and weathering rind thickness. The ventifacts sampled for dating in this study were from large boulders on moraine crests in order to coordinate alteration results with ^{36}Cl dating of the same boulders. In position V1 in Figure 2, the ^{36}Cl ages (from Zreda 1993) reveal that the moraines probably belong to marine oxygen isotope stage 6 (~150-160 ka). Morainal boulders in position V2 yield ^{36}Cl ages in the EoWisconsin, marine oxygen isotope stage 5b (~90 ka). Morainal boulders in position V1 have ^{36}Cl

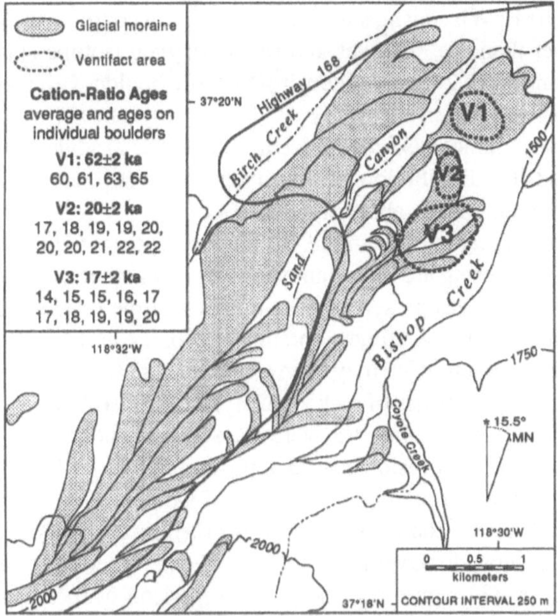

Figure 2. Ventifact sampling areas at Bishop Creek, eastern California (base map by A. Bach).

Figure 3. Aplite ventifact used for weathering rind study, sampled from population V1 at Bishop Creek. Note that the abraded surface is eroding, leaving a patchy appearance that provides further evidence that abrasion is not active at present.

ages also in the EoWisconsin, stage 5d (~110 ka). In contrast, the CR ages are much younger than the ^{36}Cl ages and rest in oxygen isotope stages 4 (early Wisconsin) and stage 2 (late Wisconsin). Specific boulder CR ages and group averages are presented in Figure 2.

Bach (this volume) describes the sequence of glacial advances and retreats that produced the distribution of ventifacts in Bishop Creek. The glacier supplied the sediment for abrasion (sand, and possibly loess) and katabatic drainage wind energy. According to Bach, glacial position, boulder position, and boulder weathering are important variables in whether a previous episode of aeolian abrasion was preserved. The position of the glacier changed over time, thus allowing the ventifacts abraded ~60-65 ka at V1 to be preserved. Although the ventifacts at V2 may have also been abraded ~60 ka, the CR ages only record the development of rock varnish since the last abrasion event ended by ~17 ka. The ventifacts at V2 were then also abandoned as the glacier retreated, and aeolian abrasion ceased at position V3 by ~14 ka (Figure 2).

The CR ages are supported by analysis of varnish layers and weathering rind thickness. The development of rock varnish layering separates the ventifacts into two age groups; the layering at V1 is considerably more complex than at positions V2 or V3. Figure 4 presents a comparison of rock varnish layers developed at V1, with those at V2 and V3. The ventifacts from V1 show two Mn-poor layers separated by a thick Mn-rich layer. In contrast, the ventifacts from V2 and V3 show only a thin Mn-rich layer at the bottom. The thick Mn-rich layer on V1 probably corresponds to glacial events, whereas Mn-poor layers correspond with interglacial periods. The microstratigraphic sequence in Figure 4 suggests the following. After aeolian abrasion ceased at V1, a more alkaline environment persisted and fostered the development of an Mn-poor varnish. Then, a less alkaline period occurred. Towards the end of this wetter phase, ventifact abrasion ceased and a thin Mn-rich layer was deposited during the more moist phase in the latest Pleistocene (cf. Jennings and Elliott-Fisk 1993). The Mn-poor surface layer was deposited on both V2 and V3 during the drier Holocene (cf. Jennings and Elliott-Fisk 1993).

The thickness of weathering rinds provides further support for these interpretations. Figure 5 illustrates BSE images of rinds from positions V1 and V2. On nine aplite ventifacts from V3, rind thicknesses averaged 0.9 mm with a standard deviation of 0.5 mm, statistically indistinguishable from the nine ventifacts from V2 (1.1 ± 0.6 mm). In contrast, nine aplite ventifacts from V1 yielded thicknesses of 3.2 ± 1.4 mm, a statistically distinct and older population (F-test, $p < 001$). In summary, greater rind thicknesses from V1 are consistent with the older CRs and more layers from this site.

In all cases, the ^{36}Cl ages are significantly older than what the CR ages, and other evidence of rock surface alteration, would suggest. This is not a contradiction in that the moraines were accumulating ^{36}Cl long before aeolian abrasion took place. This is because cosmic rays penetrate into boulders: the absorption mean-free-path in most rocks is about 50-60 cm deep in most rocks (Nishiizumi et al. 1993). In other words, aeolian abrasion of the upper few

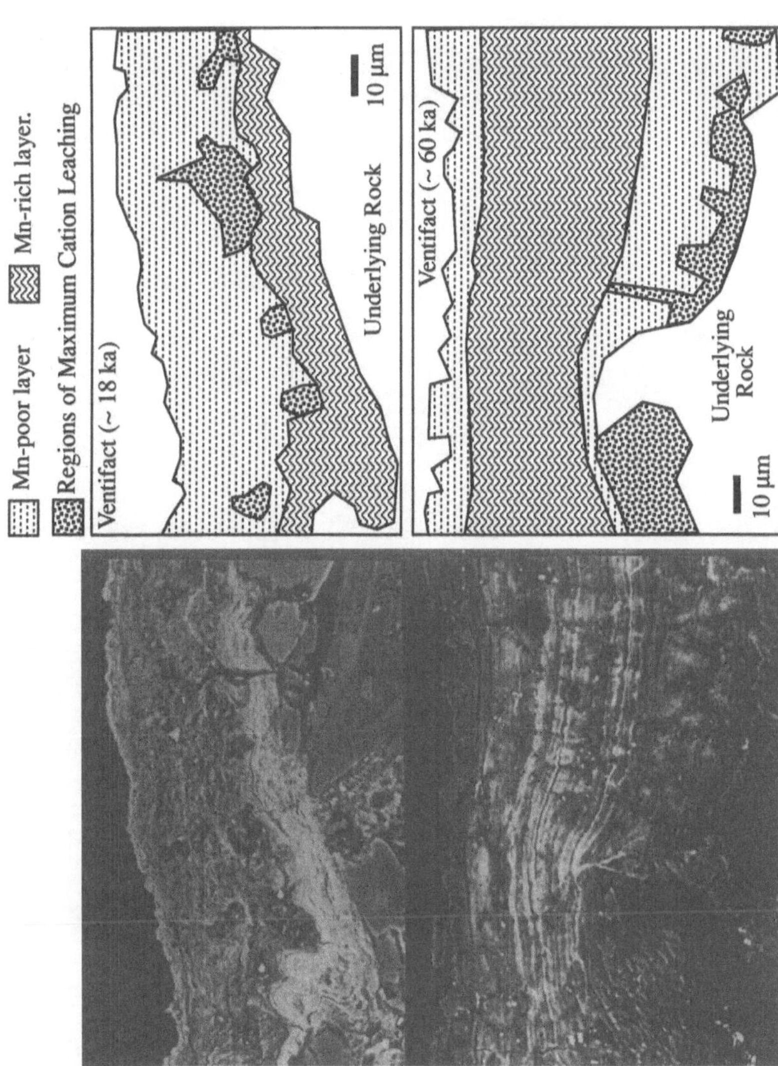

Figure 4. Typical varnish microlamination sequence on ventifacts from V1 (upper image and map) and V2 (lower image and map), viewed by backscatter electron microscopy. Brighter regions on BSE images have a higher average atomic number. Manganese-rich layers are brighter because there is more manganese and iron. In contrast, the silica-rich rock is typically darker. Areas of maximum cation leaching (cf. Dorn and Krinsley 1991) are darker and more porous because leaching removes mobile cations (e.g., K, Ca), as well as Mn and Fe.

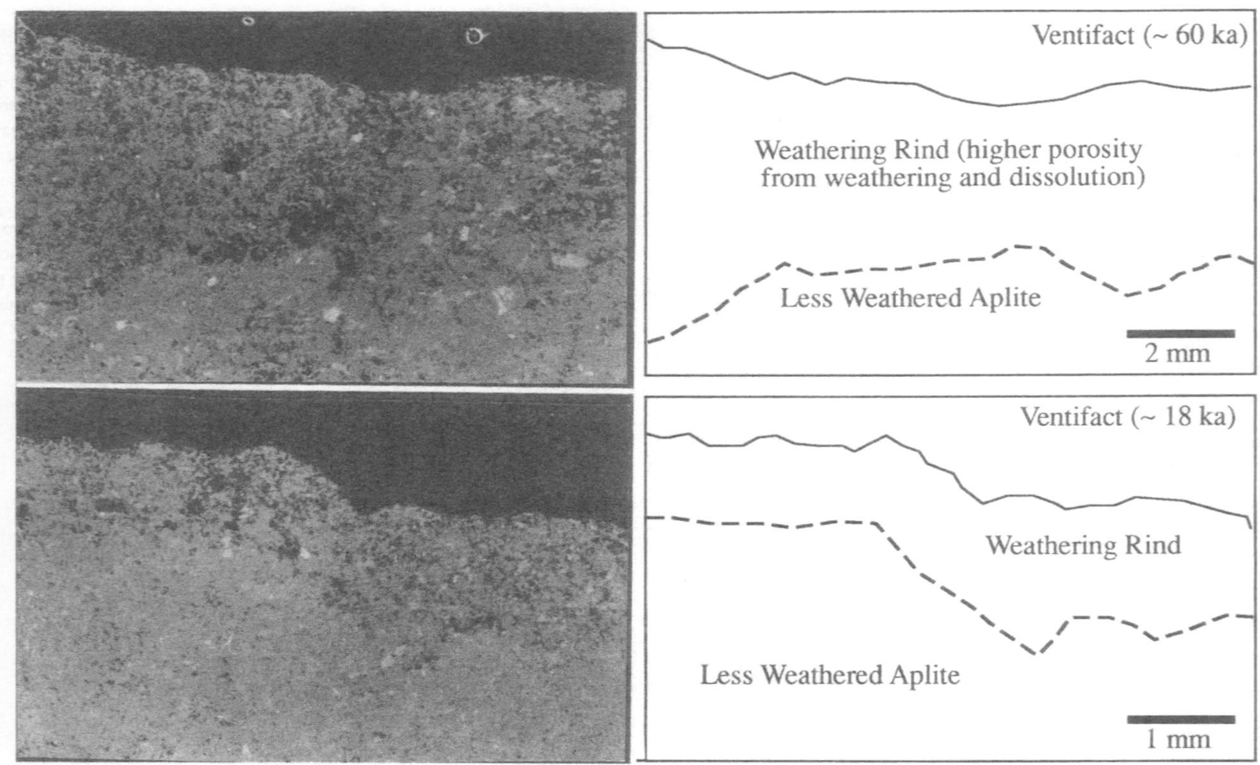

Figure 5. BSE images of weathering rinds developed on unvarnished ventifacts on V1 and V2. Note how rind thickness varies laterally; this explains the large standard deviations reported in the text.

centimeters of a boulder would not appreciably change its cosmogenic exposure age. This indicates that cosmogenic isotopes can be of assistance in studying ventifacts, by providing maximum-limiting ages for any subsequent abrasion events. Because boulders were deposited during glacial events preceding ventifaction, cosmogenic dating cannot be used to date the aeolian abrasion events themselves.

Mauna Kea, Hawaii

Mauna Kea was glaciated at least three times in the past (Figure 6). According to surface exposure dating of Mauna Kea tills with ^{36}Cl, rock varnish CRs, and radiocarbon (Dorn et al. 1991), ice caps grew during marine oxygen isotope stage 6 (Pohakuloa till), stage 4 (Waihu till), and stages 3 and 2 (Makanaka). An ice cap probably rested over the top of Mauna Kea for much of the period from at least 68 ka until about 15 ka, when the ice cap ablated completely.

There is abundant evidence for aeolian activity associated with the glaciation of Mauna Kea. Dunes (Porter 1979) and loess (S. C. Porter, personal communication 1993) can be found around the flanks of Mauna Kea. Several fields of ventifacted morainal boulders (e.g., Figure 7) occur on Mauna Kea. These were abraded in the past, but are now fossilized as indicated by coatings of silica glaze and rock varnish, as well as weathering rinds. Samples were collected from two sites to assess the potential of tropical rock varnish and weathering rinds to yield information on the history of aeolian abrasion: site 14, older Makanaka till with a ^{14}C exposure age of >37 ka; and site 8, a deposit of Waihu till with a ^{36}Cl age ~65 ka.

Evidence from varnish CRs, varnish layering, and weathering rind thickness is consistent with the hypothesis that these two ventifact groups were last abraded during the last glacial maximum ~22-20 ka. (1) CR dating of boulders from sites 8 and 14 revealed statistically indistinguishable age populations of 22 ± 2 ka and 20.5 ± 3 ka. (2) The development of microlaminations is limited on both sites to a layer of Mn-rich varnish under Mn-poor varnish. In comparison, an unventifacted boulder of the Waihu till has a much more complex series of microlaminations (Figure 8). (3) The weathering rind thicknesses of the basalt ventifacts are also statistically indistinguishable, with nine boulders from site 8 and nine boulders from site 14 having rinds of 6.3 ± 3.5 mm and 6.8 ± 3.7 mm.

Development of Silica Glaze and Oxalate-Rich Crusts on Mono Basin Moraines

The Mono Basin of eastern California witnessed pronounced aeolian activity throughout the late Quaternary. Piedmont glaciers, volcanic activity, and fluctuating lakes provided aeolian sediments, leading to the formation of dunes and loess deposits. Aeolian activity is common at present, although glacial katabatic winds were most likely more intense in magnitude and frequency.

The glacial moraines in the Mono Basin have abundant ventifacts, but the type of surficial alteration differs from Bishop Creek and Mauna Kea in that

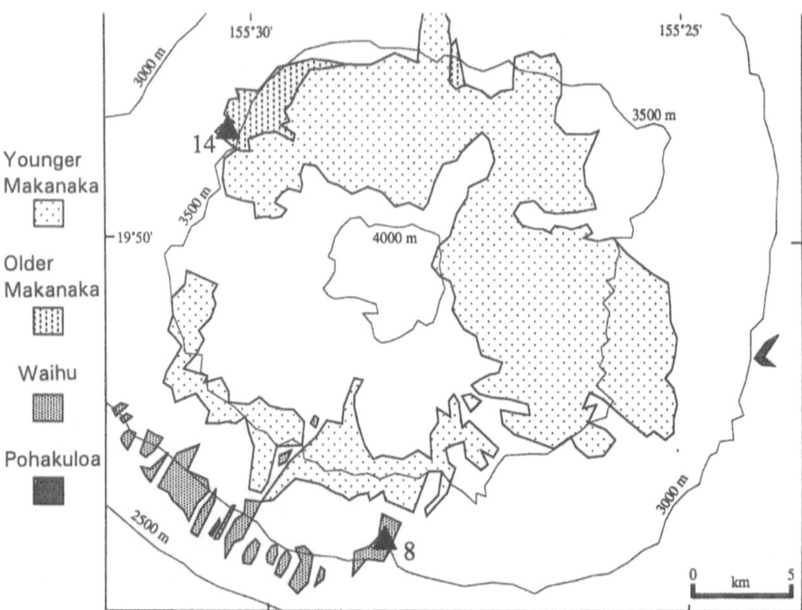

Figure 6. Ventifact sampling areas on Mauna Kea, Hawaii.

there is little development of rock varnish. What rock varnish occurs is interdigitated with other rock coatings. The pH of boulder surfaces was measured from 10 boulders on the Mono Basin moraine at Sawmill Canyon (Sharp and Birman 1963). A liter of deionized water (adjusted to a pH of 7 and kept at 25°C) was poured over 10 different boulder surfaces (with area ~10 cm^2), collected and measured with a pH meter. The pH ranged from 4 to 5, which is acidic enough to mobilize the manganese and wash away the varnish.

Two sites were sampled: ventifacts on the type Mono Basin moraine at Sawmill Canyon (Sharp and Birman 1963) that are coated with silica glaze (Figure 9), and ventifacts coated with oxalate on what Bursik and Gillespie (1993) identify as a "Mono Basin" till at Grant Lake (Figure 9). The Mono Basin moraine at Sawmill Canyon has been studied extensively to assess its age (e.g., Burke and Birkeland 1979, Gillespie 1982, Bursik and Gillespie 1993). According to Bursik and Gillespie (1993), the Grant Lake till has a "distinct subpopulation of case-hardened boulders." The boulders show evidence for aeolian abrasion and have calcium oxalate coatings.

In an active aeolian environment like the Mono Basin, where it is possible that ventifacts could be forming, rock coatings can answer a simple yes/no question as to whether ventifacts are fossilized. The silica glaze on the Mono Basin till indicates that ventifacts are not forming at the present. The oxalate-rich crusts at Grant Lake also suggest a fossil origin. There are clasts imbedded with the oxalate-rich crusts at Grant Lake (Figure 9) that have a geochemical

Figure 7. Ventifact used for cation-ratio dating at site 8 on Mauna Kea, Hawaii.

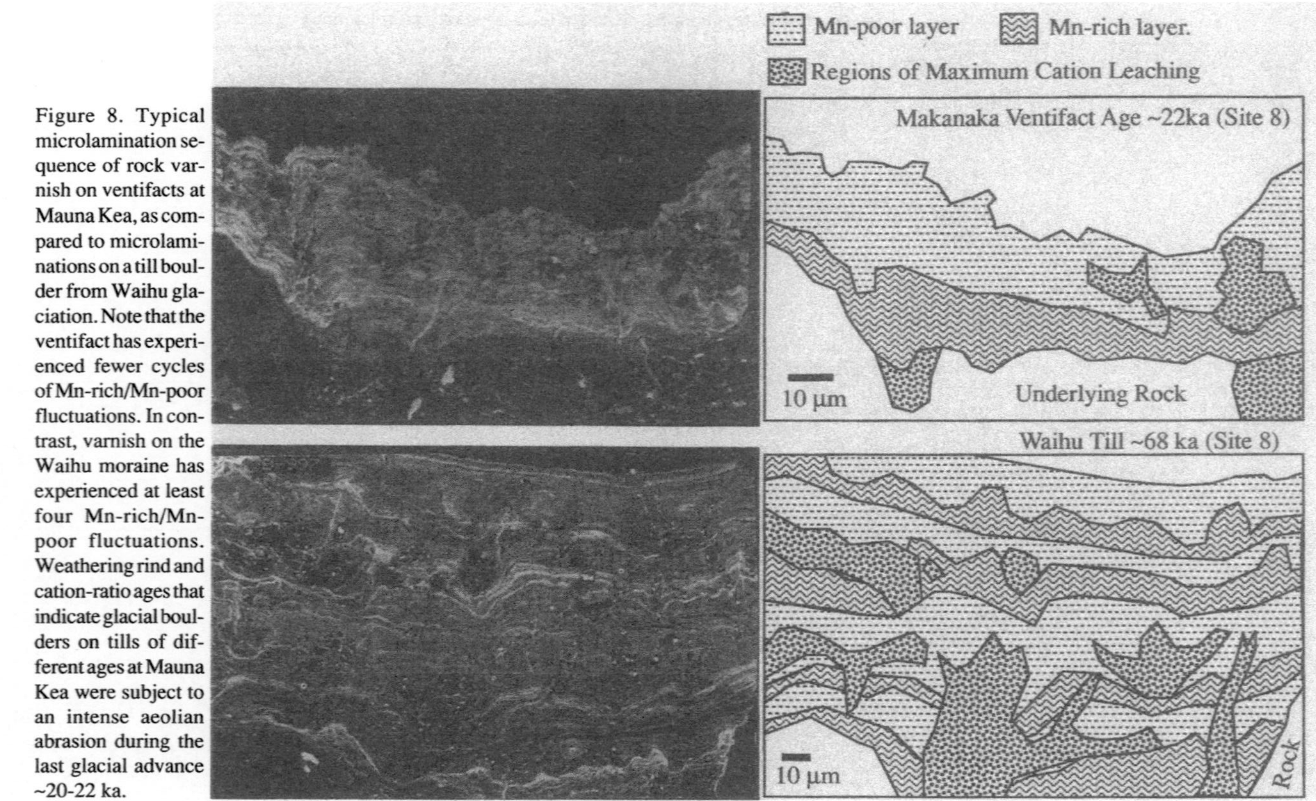

Figure 8. Typical microlamination sequence of rock varnish on ventifacts at Mauna Kea, as compared to microlaminations on a till boulder from Waihu glaciation. Note that the ventifact has experienced fewer cycles of Mn-rich/Mn-poor fluctuations. In contrast, varnish on the Waihu moraine has experienced at least four Mn-rich/Mn-poor fluctuations. Weathering rind and cation-ratio ages that indicate glacial boulders on tills of different ages at Mauna Kea were subject to an intense aeolian abrasion during the last glacial advance ~20-22 ka.

composition identical to Holocene tephra from the Mono Craters. It is likely that some of the oxalate-rich crust formed during the Holocene.

CONCLUSIONS

This study evaluated the potential for using different surficial alterations of ventifacts to study aeolian processes on alpine glacial moraines. The following conclusions summarize the results of this study:

(1) Many different types of surficial alterations take place on fossil ventifacts on glacial moraines. The growth of rock varnish and oxalate-rich crusts and the thickening of weathering rinds occur after aeolian abrasion has ceased. These coatings typically have a Moh abrasion hardness of less than 4-1/2, and are easily abraded by saltating quartz particles. The occurrence of these alterations indicates that ventifacts are not forming at present and are hence fossil.

(2) Aeolian activity is an integral part of the geomorphic system associated with alpine glaciers. The study of Pleistocene alpine glaciers is incomplete without analyzing the effects of aeolian processes. Investigators who use weathering and soils to date glacial moraines need to incorporate fossil ventifact data in their studies (cf. Burke and Birkeland 1979, Gillespie 1982, Bursik and Gillespie 1993). Studies of boulder weathering on glacial moraines have ignored the presence of ventifacts, despite their documentation and association with alpine glacier systems in the classic literature (e.g., Blackwelder 1929, Sharp 1949). Abrasion resets the weathering clock by reforming rock surfaces, and, where aeolian abrasion is sufficient to polish surfaces, impacts soil development (see Bach, this volume). Nor can investigators simply reevaluate their results by assuming that previously unidentified ventifacts correspond with the last glacial advance, because different ventifact populations can correspond to different glacial stages. This is illustrated in the Bishop Creek section of this paper.

(3) A multiple-technique approach to evaluate ages of ventifact populations reduces the chance of error associated with any one technique. By cross-checking results from the microstratigraphy of rock varnish, cation-ratio dating of varnish, analysis of weathering rinds and ^{36}Cl buildup, the chances of error in the data analysis is reduced.

(4) Utilizing different dating tools to study complex aeolian systems can produce synergistic results. Cosmogenic nuclide clocks (e.g., accumulation of ^{36}Cl) are not affected by aeolian abrasion of glacial boulders. Yet, the weathering phenomenon and rock coatings studied here are reset. Using these techniques in tandem, the geomorphic researcher can learn the age of the glaciation (through cosmogenic nuclides; cf. Zreda 1993), and the history of the last aeolian abrasion. Different dating techniques with different assumptions have the potential to yield unique insights into the history of aeolian processes.

(5) Glacial events play a critical role in controlling aeolian processes at

Silica Glaze
Rock Varnish
Ventifact on Moraine at Bloody Canyon
Underlying Ventifact
10 μm

Calcium Oxalate
Tephra or Silica Glaze
Ventifact on Moraine at Grant Lake
Underlying Ventifact
10 μm

Figure 9. BSE images of different types of rock coatings formed on ventifacts in the Mono Basin of eastern California. The upper image is silica glaze (with some interdigitation of rock varnish) that has formed on a fossilized ventifact on the Mono Basin moraine at Bloody Canyon (cf. Phillips et al. 1990). The lower image is a mixture of calcium oxalate and what may be tephra, formed on a ventifact from Grant Lake in the Mono Basin. The shard-like forms embedded calcium oxalate have been examined with the electron microprobe, and the chemistry is quite similar to Holocene volcanic ash in the Mono Basin. Another, less likely, possibility is that clasts are a type of silica glaze.

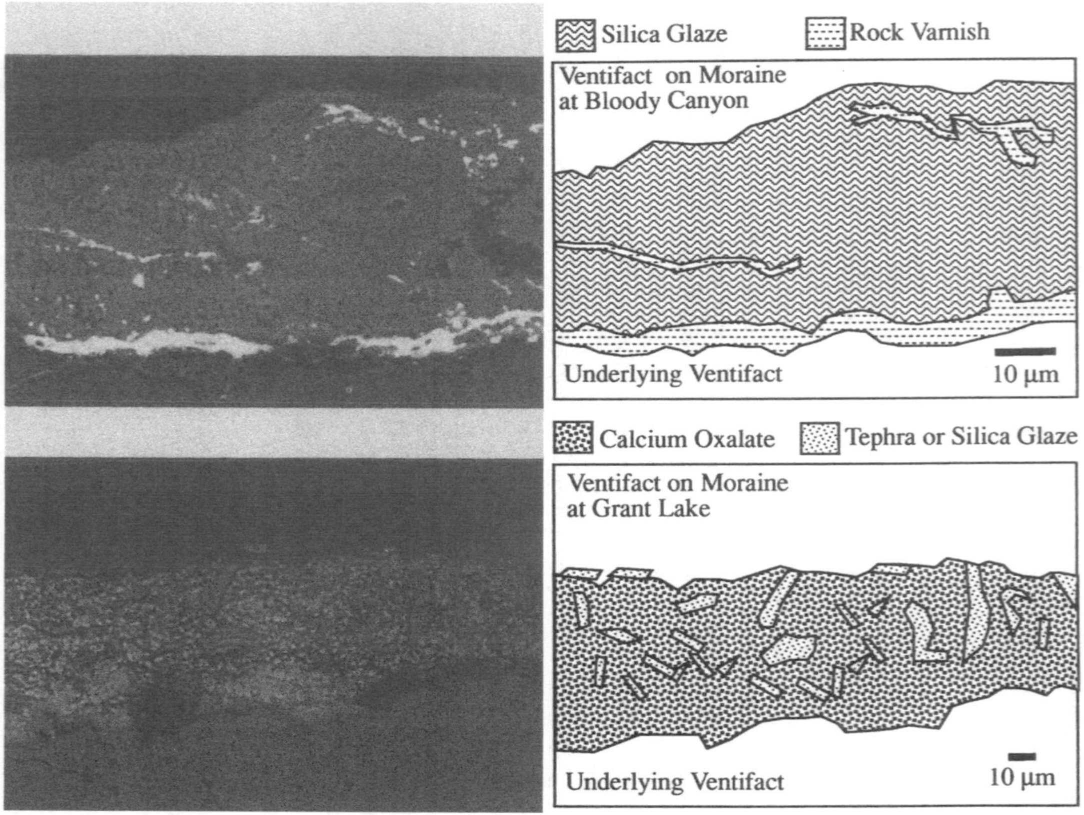

their margins. Ventifacts were abraded during glacial events at Bishop Creek in the eastern Sierra Nevada and Mauna Kea in Hawaii. At both sites, the age of the morainal boulders significantly pre-dated the cessation of aeolian abrasion. At Bishop Creek, the location and timing of aeolian abrasion moved along with the position of the glacier (see also Bach, this volume). The last major expansion of the ice cap on Mauna Kea corresponds with the last known period of aeolian abrasion.

ACKNOWLEDGMENTS

This research was funded by NSF grant 89-00403, NSF Presidential Young Investigator Award, and by a grant from the National Geographic Society. Thanks to D. Elliott-Fisk, F. Phillips, and M. Zreda for field expertise, T. Liu for sample preparation, J. Laity for comments on the manuscript, and A. Bach for Figure 2 and comments.

REFERENCES

Antevs, E. (1928) Wind deserts in Iceland. *Geographical Review*, v. 18, p. 675-676.

Beck, C. (1993) *Dating in Surface Context*. University of New Mexico Press, Albuquerque.

Bierman, P. , and A. Gillespie (1991) Accuracy of rock-varnish chemical analyses: implications for cation-ratio dating. *Geology*, v. 19, p. 196-199.

Blackwelder, E. (1929) Sandblast action in relation to the glaciers of the Sierra Nevada. *Journal of Geology*, v. 37, p. 256-260.

Bull, W. B. (1991) *Geomorphic Responses to Climatic Change*. Oxford University Press, Oxford.

Burke, R. M., and Birkeland, P. W. (1979) Reevaluation of multiparameter relative dating techniques and their application to the glacial sequence along the eastern escarpment of the Sierra Nevada, California. *Quaternary Research*, v. 11, p. 11-51.

Bursik, M. I., and Gillespie, A. R. (1993) Late Pleistocene glaciation of Mono Basin, California. *Quaternary Research*, v. 39, p. 24-35.

Cahill, T. A. (1992) Comment on "Accuracy of Rock-Varnish Chemical Analyses: Implications for Cation-Ratio Dating." *Geology*, v. 20, p. 469.

Chinn, T.J.H. (1981) Use of rock weathering-rind thickness for Holocene absolute age-dating in New Zealand. *Arctic and Alpine Research*, v. 13, p. 33-45.

Curtiss, B., Adams, J. B., and Ghiorso, M. S. (1985) Origin, development and chemistry of silica-alumina rock coatings from the semiarid regions of the island of Hawaii. *Geochemica et Cosmochimica Acta*, v. 49, p. 49-56.

Czajka, W. (1972) Windschliffe als Landschaftmerkmal. *Zeitschrift für Geomorphologie N.F.*, v. 16, p. 27-53.

Del Monte, M., and Sabbioni, C. (1987) A study of the patina called "scialbatura" on Imperial Roman marbles. *Studies in Conservation*, v. 32, p. 114-121.

Dorn, R. I. (1983) Cation-ratio dating: a new rock varnish age determination technique. *Quaternary Research*, v. 20, p. 49-73.

Dorn, R. I. (1989) Cation-ratio dating of rock varnish: A geographical perspective. *Progress in Physical Geography*, v. 13, p. 559-596.

Dorn, R. I. (1990) Quaternary alkalinity fluctuations recorded in rock varnish microlaminations on western U.S.A. volcanics. *Palaeogeography, Palaeoclimatology, Palaeoecology*, v. 76, p. 291-310.

Dorn, R. I., Cahill, T. A., Eldred, R. A., Gill, T. E., Kusko, B., Bach, A., and Elliott-Fisk, D. (1990) Dating rock varnishes by the cation ratio method with PIXE, ICP, and the electron microprobe. *International Journal of PIXE*, v. 1, p. 157-195.

Dorn, R. I., Jull, A.J.T., Donahue, D. J., Linick, T. W., Toolin, L. J., Moore, R. B., Rubin, M., Gill, T. E., and Cahill, T. A. (1992) Rock varnish on Hualalai and Mauna Kea Volcanoes, Hawaii.

Pacific Science, v. 46, p. 11-34.
Dorn, R. I., and Krinsley, D. H. (1991) Cation-leaching sites in rock varnish. *Geology*, v. 19, p. 1077-1080.
Dorn, R. I., Phillips, F. M., Zreda, M. G., Wolfe, E. W., Jull, A.J.T., Kubik, P. W., and Sharma, P. (1991) Glacial chronology of Mauna Kea, Hawaii, as constrained by surface-exposure dating. *National Geographic Research*, v. 7, p. 456-471.
Dorn, R. I., and Phillips, F. M. (1991) Surface exposure dating: review and critical evaluation. *Physical Geography*, v. 12, p. 303-333.
Gillespie, A. R. (1982) *Quaternary Glaciation and Tectonism in the Southeastern Sierra Nevada, Inyo County, CA*. Ph.D. dissertation, California Institute of Technology.
Glazovskiy, A. F. (1985) Rock varnish in the glacierized regions of the Pamirs (In Russian). *Data of the Glaciological Studies (Moscow)*, No. 54, p. 136-141.
Hall, K. (1989) Wind blown particles as weathering agents? An Antarctic example. *Geomorphology*, v. 2, p. 405-410.
Jennings, S. A., and Elliott-Fisk, D. L. (1993) Packrat midden evidence of late Quaternary vegetation change in the White Mountains, California-Nevada. *Quaternary Research*, v. 39, p. 214-221.
Jones, C. E. (1991) Characteristics and origin of rock varnish from the hyperarid coastal deserts of northern Peru. *Quaternary Research*, v. 35, p. 116-129.
Jones, D., Wilson, M. J., and McHardy, W. J. (1981) Lichen weathering of rock-forming minerals: applications of scanning electron microscopy and microprobe analysis. *Journal of Microscopy*, v. 124, p. 95-194.
Krinsley, D. H., and Manley, C. R. (1989) Backscattered electron microscopy as an advanced technique in petrography. *Journal of Geological Education*, v. 37, p. 202-209.
Kurz, M. (1986) In situ production of terrestrial cosmogenic helium and some applications to geochronology. *Geochimica et Cosmochimica Acta*, v. 50, p. 2855-2862.
Laity, J. E. (1991) Ventifact evidence for Holocene wind patterns in the east-central Mojave Desert. *Zeitschrift für Geomorphologie*, v. 84, p. 73-88.
Laity, J. E. (1994) Landforms of aeolian erosion. In A. D. Abrahams and A. J. Parsons (eds.) *Geomorphology of Desert Environments*. Chapman & Hall, London, p. 506-535.
McKenna-Neuman, C., and Gilbert, R. (1986) Aeolian processes and landforms in glaciofluvial environments of southeastern Baffin Island, N.W.T., Canada. In Nickling, W. G. (ed.) *Aeolian Geomorphology*. Allen & Unwin, London, p. 213-235.
Nishiizumu, K., Kohl, C. P., Arnold, J. R., Dorn, R., Klein, J., Fink, D., Middleton, R., and Lal, D. (1993) Role of in situ cosmogenic nuclides ^{10}Be and ^{26}Al in the study of diverse geomorphic processes. *Earth Surface Processes and Landforms*, v. 18, p. 407-425.
Nobbs, M., and Dorn, R. I. (1993) New surface exposure ages for petroglyphs from the Olary Province, South Australia. *Archaeology in Oceania*, v. 28, p. 18-39.
Perry, R. S., and Adam, J. (1978) Desert varnish: evidence of cyclic deposition of manganese. *Nature*, v. 276, p. 489-491.
Phillips, F. M., Zreda, M. G., Smith, S. S., Elmore, D., Kubik, P. W., and Sharma, P. (1990) A cosmogenic chlorine-36 chronology for glacial deposits at Bloody Canyon, eastern Sierra Nevada, California. *Science*, v. 248, p. 1529-1532.
Pineda, C. A., Jacobson, L., and Peisach, M. (1988) Ion beam analysis for the determination of cation-ratios as a means of dating southern African rock varnishes. *Nuclear Instruments and Methods in Physics Research*, v. B35, p. 463-466.
Pineda, C. A., Peisach, M., Jacobson, L., and Sampson, C. G. (1990) Cation-ratio differences in rock patina on hornfels and chalcedony using thick target PIXE. *Nuclear Instruments and Methods in Physics Research*, v. B49, p. 332-335.
Porter, S. C. (1979) *Geologic Map of Mauna Kea Volcano, Hawaii*. Geological Society of America Map and Chart Series, MC-30.
Powers, W. (1936) The evidence of wind abrasion. *Journal of Geology*, v. 44, p. 214-219.
Schlyter, P. (1991) Recent and periglacial wind action in Scania and adjacent areas of S Sweden. *Zeitschrift für Geomorphologie*, v. 90, p. 143-153.
Sharp, R. P. (1949) Pleistocene ventifacts east of the Big Horn Mountains, Wyoming. *Journal of Geology*, v. 57, p. 173-195.

Sharp, R. P., and Birman, J. H. (1963) Additions to the classical sequence of Pleistocene glaciations, Sierra Nevada, California. *Geological Society of America Bulletin*, v. 74, p. 1979-1086.

Traquair, J. A. (1986) Backscattered electron imaging as a tool for histochemically localizing calcium oxalate with the scanning electron microscope. *Canadian Journal of Botany*, v. 65, p. 888-892.

Watchman, A. (1991) Age and composition of oxalate-rich crusts in the Northern Territory, Australia. *Studies in Conservation*, v. 36, p. 24-32.

Whitney, J. W., and Harrington, C. D. (1993) Relict colluvial boulder deposits as paleoclimatic indicators in the Yucca Mountain region, southern Nevada. *Geological Society of America Bulletin*, v. 105, p. 1008-1018.

Zak, K., and Skala, R. (1993). Carbon isotopic composition of whewellite($CaC_2O_4.H_2O$) from different geological environments and its significance. *Chemical Geology*, v. 106, p. 123-131.

Zhang, Y., Liu, T., and Li, S. (1990) Establishment of a cation-leaching curve of rock varnish and its application to the boundary region of Gansu and Xinjiang, western China. *Seismology and Geology (Beijing)*, v. 12, p. 251-261.

Zreda, M. (1993) *Cosmogenic ^{36}Cl Chronology of Late Quaternary Glaciations: Glacial History, Correlations, and Paleoclimatic Implications*. Ph.D. dissertation, New Mexico Institute of Mining and Technology.

10 FINE MATERIAL IN ROCK FRACTURES: AEOLIAN DUST OR WEATHERING?

Niccole Villa,[1] Ronald I. Dorn,[1] and James Clark[2]
[1]*Department of Geography,* [2]*Department of Chemistry, Arizona State University*

ABSTRACT

Fine material in rock crevices from the deserts of southwestern North America and Hawaii, studied by light and electron microscopy, derive from both *in situ* weathering of the adjacent rock and the accumulation of aeolian dust. In some cases, such as quartz found in Hawaiian rock crevices, we see evidence for an aeolian origin. In other cases, the texture and chemistry of the fine material indicates a weathering origin. Fines in rock fractures are analogous to soils, and a general model for development of "fissuresols" is presented. Where rocks are friable and weathering is rapid, a residual fissuresol develops. Where dust storms are common and rocks are resistant to weathering, a cumulic fissuresol forms. A continuum likely exists between these two extremes in space today, from drier to wetter climates. Fissuresols can be tens of thousands of years old and experience drastically different climates. Therefore, the relative importance of weathering and aeolian input can shift over time.

INTRODUCTION

A topic in aeolian geomorphology that has received little attention is the ubiquitous fine material in rock fractures. Although desert dust has been examined in detail (e.g., Goudie 1978, Pye 1987), only a few geomorphologists have explored fine material in rock crevices. Coudé-Gaussen et al. (1984) present evidence for an aeolian orgin for the fines found in crevices in granitic rocks in the Sinai Peninsula. In every fissure we have ever forced open with a rock hammer in arid lands, material has been found both adsorbed to crevice sides and resting loosely in the joint. We have observed fines in rock crevices from arid lands in Africa, Asia, Australia, North America, and South America, and in dozens of lithologies ranging from basalt to granodiorite to limestone. If Coudé-Gaussen et al. (1984) are correct in assuming that most fines in rock crevices are aeolian dust, then rock fissures represent an aerially extensive terrestrial dust trap, second only to plants. The question we address in this study is whether fines are aeolian in origin, as Coudé-Gaussen et al. (1984) contend, or *in situ* from rock weathering on crevice walls.

Dust in rock crevices has both theoretical and applied significance. Fines absorb water and support plant life. Fines expand and contract with wetting and drying, aiding in the weathering of rocks. Fines also contain salts that are remobilized and reprecipitated to shatter cobbles (Amit et al. 1993). The material in rock crevices can influence the nature of rock coatings on crevice walls. Such coatings may include rock varnish, calcrete, or amorphous silica.

Desert Aeolian Processes. Edited by Vatche P. Tchakerian. Published in 1995 by Chapman & Hall, London. ISBN 978-94-010-6519-1

Rock varnish in fractures is believed to cause the orange color of Ayers Rock in Australia (Dorn and Dragovich 1990).

The issue of whether fines in rock fissures are from *in situ* weathering or aeolian fallout has applied relevance. One of the difficulties in interpreting the geochemistry of dust in experiments is the role of humans. Even in "remote areas" dirt roads that cut across calcic soil horizons can be prominent sources of calcium carbonate. If fines in rock crevices are cumulic, they could serve as natural "background" for studies in aeolian geochemistry. This could prove useful in a comparison with contemporary pollution studies. Because heavy metals such as copper and zinc preferentially adsorb onto fines, dust in rock fractures can be used for identifying areas with higher concentrations of heavy metals.

The purpose of this paper is to (a) assess the genesis of fines in rock fissures by examining samples from the Sonoran Desert, Death Valley, California, and Hawaii at millimeter and micrometer scales; and (b) present a general model for the development of fines in rock crevices.

METHODS

Mesic and xeric samples were selected from the Sonoran Desert of Arizona, Death Valley National Monument in California, and Hawaii. The Sonoran Desert was emphasized because of the abundance of cumulic loess deposits as a result of frequent dust storms (Brazel 1989). In contradistinction, we anticipated that fines from weathering would most likely be in more moist regions where dust storms are less common.

All samples were collected from the tops of rock outcrops to avoid input by slope wash. There is the possibility for transport of fines to rock fractures by ants and other organic agents. While we cannot completely rule out this possibility, we did not observe any evidence of "organic transport." Also, the fissures were all "tight"—with openings less than 2 mm at the top—not allowing the movement of likely transport organisms. Still, we cannot truly test this hypothesis until monitoring studies are conducted.

After removal with a rock hammer, samples were placed in tissue paper for gentle transport to the lab, mounted in epoxy, and polished for cross-sectional analysis of the fine material/rock interface. Samples were examined with light and backscatter electron microscope (BSE). In BSE, both the chemistry and texture are imaged simultaneously; brightness is a function of average atomic number (Krinsley and Manley 1989). The contact between the fine material and the adjacent rock was analyzed chemically with a JEOL wavelength dispersive electron microprobe with ZAF corrections and a 30-second counting time. We used both a 2- and a 10-micrometer spot size in order to study this fine/rock interface. The larger spot size was used to average the geochemistry of the smallest particles, whereas the smaller spot size was used to obtain quantitative data on specific grains.

In Death Valley, we compared the chemistry of quartzite and the adjacent fines in rock fissures. About 10 grams of loose fine material was collected from each fissure, and 10 grams was powdered from the adjacent quartzite (from a position in the center of the rock). These powders were then homogenized in a flux of lithium metaborate. The resultant homogenized beads were mounted in epoxy, polished, and carbon coated. Their composition was analyzed by an electron microprobe with a 30-micrometer spot size. Five separate measurements were made on each bead. Probe totals were normalized to 100% (Table 1) to account for the lithium metaborate flux.

RESULTS

Kitt Peak, Arizona

The Kitt Peak granodiorite samples were collected along an environmental gradient from about 900 m, where creosote bush (*Larrea divaricata*) dominates, to about 2000 m, where chaparral vegetation is the dominant species. Contrary to our expectations that the desert site would have the clearest loess signal and the high elevation sites mostly weathering, both sites showed evidence of weathering and dust.

Figure 1 shows the presence of fine particles next to the unweathered rock at the 2000 m elevation site. The microprobe transect shows a similarity in chemistry across the fines/rock boundary, suggesting in-situ weathering. Figure 2 presents another sample from this high elevation site, but the texture consists of finer particles. The microprobe transect reveals a very noisy chemical signature. Certain elements (Na, Al, K, Ca, Ti) are present in the fine particles, but not in the rock which is composed of quartz. Trace elements in the quartz (Mn, Fe) are also found in much greater concentrations in the adjacent fines.

Figures 1 and 2 represent the range of observed chemistries and textures found in the Kitt Peak study. Textures and chemistries characteristic of aeolian dust appear to occur more frequently at lower elevations. The aeolian materials are most likely from dust storms associated with summer convective thunderstorms (Brazel 1989). However, these and similar observations indicate that both weathering and aeolian deposition occurs at all elevations at Kitt Peak.

Sedona, Arizona

Samples were collected from joint fractures in a fluvial sandstone member of the Supai formation in the Schnebly Hill area of Sedona. Despite the high friability of the sandstone, there was evidence of both internal and external origin of the fracture constituents. While we scanned the cross section, there were quartz grains scattered throughout the fines that were similar in size and shape to the grains in the rock. It seems plausible that weathering dislodged these grains from the crevice wall.

Table 1
Electron microprobe measurements of bulk chemistry of fine material in rock fissures in quartzite, Death Valley, Califomia.

Collection site	Quartzite		Rock fissure			
Sea level, Hanaupah Canyon Fan, desert scrub vegetation near playa margin;	Al_2O_3	0.57	Na_2O	3.52%	MgO	3.41%
	SiO_2	98.70	Al_2O_3	10.44%	SiO_2	66.28%
	Fe_2O_3	0.73	P_2O_5	2.47%	SO_3	0.27%
pH of fissure fine material 9.8 ± 0.7			CaO	4.20%	K_2O	2.93%
			TiO_2	0.72%	MnO	0.10%
			Fe_2O_3	5.37%	BaO	0.29%
~1000 m, Panamint Range, desert scrub, on crest of hill, several kilometers from playa	SiO_2	98.56	Na_2O	2.08%	MgO	4.27%
	CaO	0.24	Al_2O_3	25.49%	SiO_2	46.30%
	Fe_2O_3	1.20	P_2O_5	3.80%	SO_3	0.38%
pH of fissure fine material 8.5 ± 0.3			CaO	5.18%	K_2O	3.55%
			TiO_2	0.90%	MnO	0.11%
			Fe_2O_3	7.77%	BaO	0.17%
~2000 m, Panamint Range, juniper dwarf woodland, east-facing slope	SiO_2	99.03	Na_2O	1.80%	MgO	3.87%
	Fe_2O_3	0.97	$A1_2O_3$	33.10%	SiO_2	39.3%
			P_2O_5	5.27%	SO_3	0.29%
pH of fissure fine material 8.2 ± 0.7			CaO	4.18%	K_2O	2.67%
			TiO_2	0.97%	MnO	0.21%
			Fe_2O_3	8.17%	BaO	0.17%
~3000 m, Panamint Range, limber bristle-cone pine woodland, east-facing slope	Al_2O_3	0.37	Na_2O	1.22%	MgO	3.52%
	SiO_2	99.13	Al_2O_3	29.11%	SiO_2	44.89%
	Fe_2O_3	0.50	P_2O_5	4.72%	SO_3	0.31%
pH of fissure fine material 8.0 ± 0.5			CaO	3.94%	K_2O	3.17%
			TiO_2	0.85%	MnO	0.25%
			Fe_2O_3	7.90%	BaO	0.12%

Evidence for an aeolian origin for the fines is seen in Figure 3, owing to the fibrous texture for the dust material. Quantitative electron microprobe measurements (10 mm spot size), on the inorganic fraction of the interface between fines and rock, show a relative enrichment in Ca, Mg, Na, and P as compared to the underlying sandstone, indicating a contribution from an outside source. The wavelength dispersive mode on the electron microprobe was used qualitatively to determine the carbon signal of the fibrous material. Carbon typically appears as black in BSE images, because of the low atomic number (Z = 6). The BSE image in Figure 3 was taken with very low contrast, in order to bring out the filamentous structure of the organic material in the fines.

Tempe Butte, Arizona

Figure 4 illustrates a typical fracture in the andesite at Tempe Butte, Arizona,

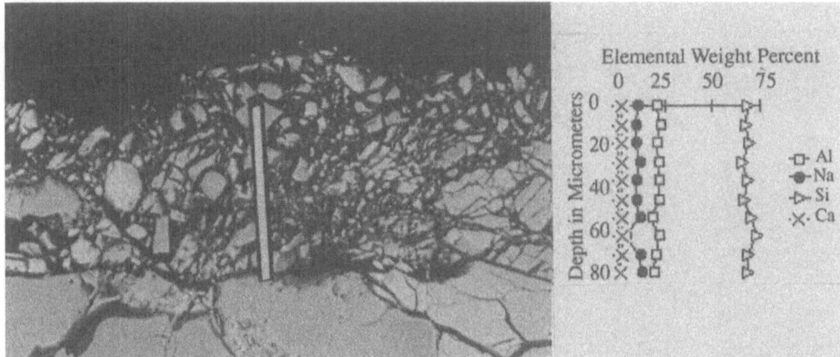

Figure 1. Backscatter (BSE) electron microscope image of a joint face in granodiorite sampled from ~2000 m at Kitt Peak. The line indicates the transect where electron microprobe measurements were made. The adjacent chart presents electron microprobe measurements (2 mm spot size) that are consistent with a plagioclase mineralogy for both the debris and the adjacent unweathered rock face.

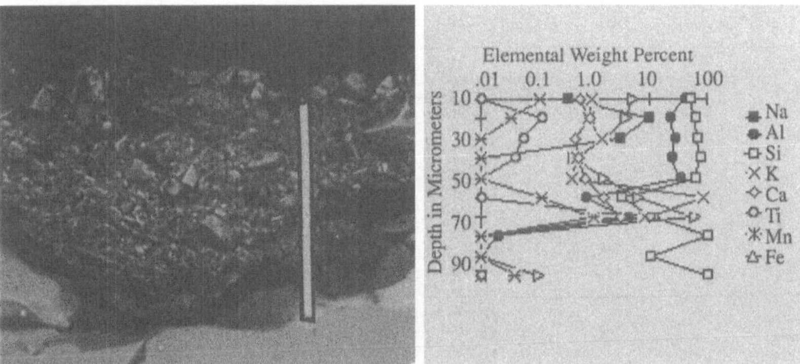

Figure 2. BSE image of a joint face in granodiorite sampled from 2000 m at Kitt Peak, within a meter of Figure 1. The corresponding electron microprobe measurements (2 mm spot size) along the transect show elements in the fine material that are not present along this fracture in quartz.

where the texture and geochemistry of the fine material is distinct from the underlying rock. In this sample, Mg, P, Fe, and Ca are more abundant in the dust than the underlying rock. This sample was collected at the very top of a fissure to avoid contamination of materials from above.

At Tempe Butte, we conducted a separate test for the relative contribution of aeolian dust and weathering products. Ten 2 x 2 cm samples were collected from rock fractures on a prominant south-facing knob of andesite. The upper part of these fractures were all touching the surface of the rock, so there was no source of weathered material from higher up in the rock fracture. First, the distribution of fine material was mapped. Then, this material was washed away and scrubbed gently with tap water and a toothbrush, revealing rock coatings

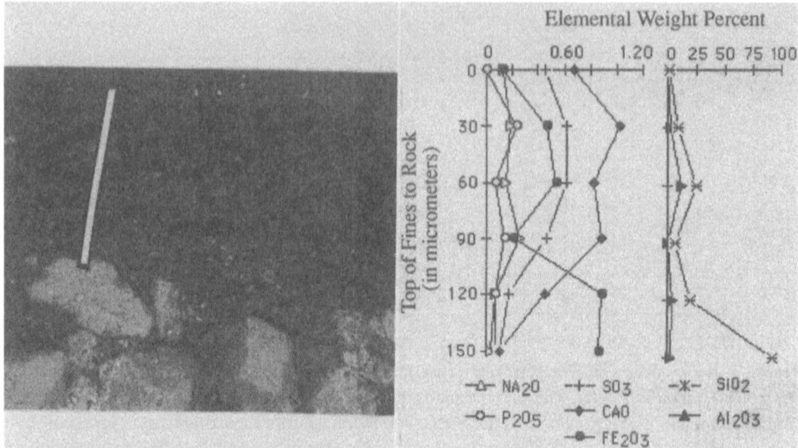

Figure 3. BSE image of a joint face in Supai sandstone sampled from the Sedona region of Arizona, showing the fibrous texture of the dust material.

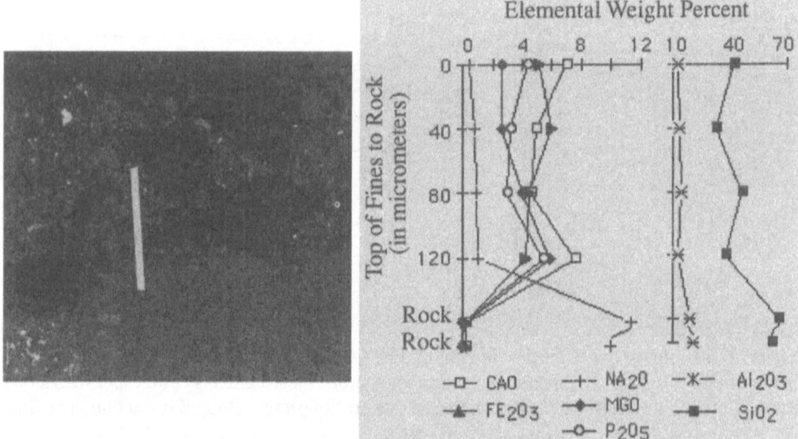

Figure 4. BSE image of fracture in andesite at Tempe Butte in Arizona. In this case, there was no coating between the fine material and the adjacent rock. The corresponding probe profiles (10 mm spot size) show a distinct change in chemistry from the fine material in the crevice to the adjacent rock.

of crack varnish and calcium carbonate. Lastly, the distribution of these coatings was mapped (Figure 5).

The only potential source of weathered fines appears to be from places where a coating was not present. Light microscope examination of these fissure sides reveals a thin weathering rind (<2 mm). Evidence for *in situ* clays or spalling of silt- or clay-sized particles in the weathering rind was not found. Similar results were also obtained by Colman and Pierce (1981).

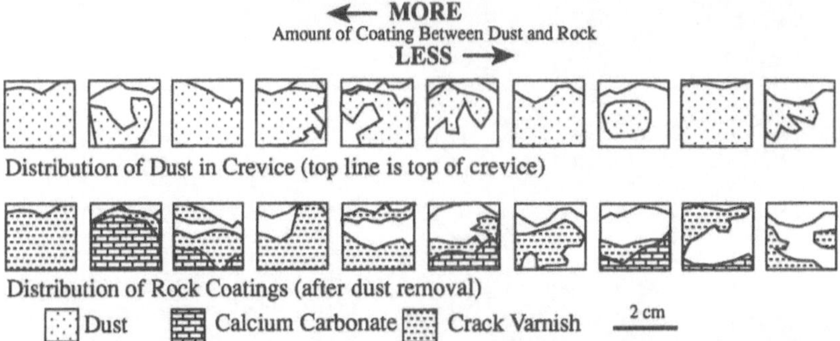

Figure 5. The distribution of dust and the underlying rock coatings in andesite fissures at Tempe Butte, Arizona. When rock coatings of orange varnish or carbonate rest directly under the dust, the source of the fine material cannot be from the underlying andesite rock, especially when the coating is at the top of the fracture. The lack of a coating could allow any weathering of the andesite to contribute to the fissuresol.

Figure 6 illustrates a lightly weathered rock, where rock varnish separates the rock from the dust. The fine material could not have weathered from the immediately adjacent andesite because of the presence of the rock coating. The combination of rock coatings separating dust from rock, and the lack of silt and clay weathering products, indicates that the fines have an aeolian origin.

We caution the possibility for an "optical illusion" effect in studies of dust in rock fissures. Our initial qualitative field observations revealed that basalt and andesite joint fractures in the Tempe Butte, Arizona, area contain a plethora of fine material, more so than granodiorite at Kitt Peak or sandstones at Sedona. This may simply reflect the texture of rock weathering products. Granodiorite weathers to grus, sandstone to sand. On the other hand, the weathering of basalt and andesite produces weathering rinds and cobble-sized angular fragments (Colman and Pierce 1981). We suspect, therefore, that the fines in basalt/ andesite fractures look like "pure" loess, and the lack of visual contamination by sand-sized material creates the illusion that extrusive rocks are more efficient dust traps. A test of this hypothesis requires controlled monitoring studies.

Death Valley, California

Death Valley is a graben located in one of the most arid regions of North America. Yet, adjacent to this desiccated lowland are the semi-arid slopes of the Panamint Range. The latter rise to over 3000 m. The higher elevations are mantled with coniferous vegetation. We collected quartzite samples from the floor of Death Valley to the 3000-m-elevation environmental gradient. Quartzites contain only minor amounts of trace elements, and thus are ideal for assessing the amount of external components to rock crevice fines.

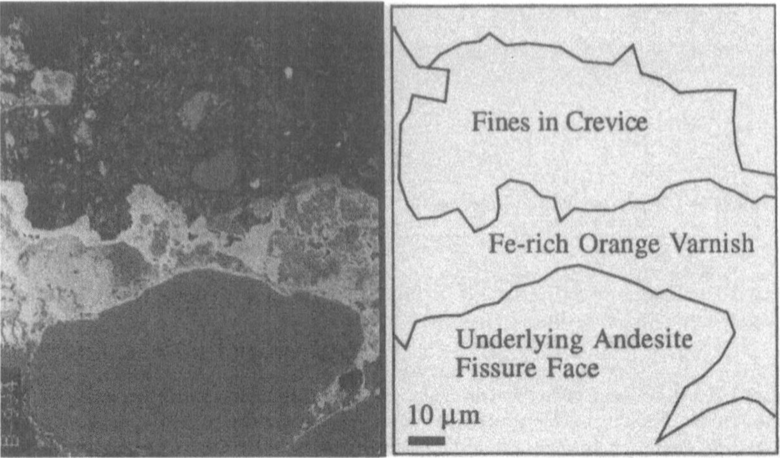

Figure 6. BSE image of fracture in andesite at Tempe Butte in Arizona. Taken from same outcrop less than a meter from the sample in Figure 4, this fracture displays a coating of orange (Mn-poor, Fe-rich) rock varnish that separates the fines from the rock. It is quite likely that the fines are incorporated into the orange crack varnish as it grows; this is indicated by the detrital grains within the orange varnish.

The bulk analysis of the fine material in quartzite crevices in Death Valley National Monument reveals distinct geochemistries (Table 1). There are abundant elements in the rock fissure fines that are not present in measurable quantitites in the quartzite. These elements are found in both arid environments, such as adjacent to the salt playa, as well as in the subalpine environment, among the coniferous vegetation.

There is also evidence that quartzite weathering is contributing silica to fissures. Observations of cross sections with BSE provided textural evidence for etching on crevice walls (Figure 7). This etching may be similar to the solution features observed on quartz grains with the secondary electron microscope (e.g., Tchakerian 1991). It is possible that the higher concentration of SiO_2 in the fissure dust at the playa margin is the result of higher pH, gypsum, and halite aiding in the dissolution of quartzite (Table 1). In contrast, frost weathering of quartzite could have contributed to the higher SiO_2 concentrations in fissures atop the Panamint Range (Table 1).

Hawaii

The higher elevations of the volcanoes in Hawaii are above the trade-wind inversion, and are extremely arid. Plant cover is extremely sparse. Figure 8 shows fines in rock fractures from within a few hundred meters of the summits of three of the larger volcanoes in the Hawaiian chain: Mauna Kea and Hualalai on Hawaii, and Haleakala on Maui. The presence of quartz is especially significant, given the fact that free quartz is rare in basalts. The quartz probably originates from Asia and is transported by upper-level winds (Beget et al. 1993).

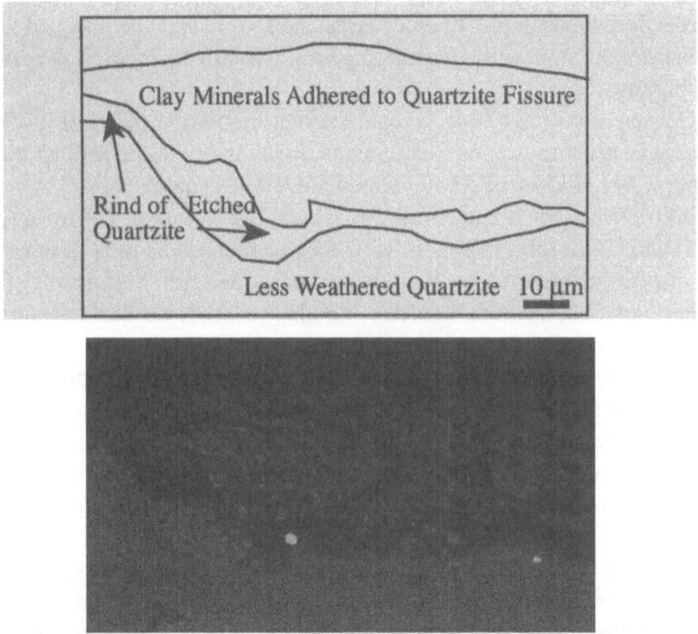

Figure 7. BSE image of clay minerals next to quartzite, from 2000-m site in Death Valley (see Table 1). The bright particle is a clast of barium sulfate. Note the enhanced etching of the quartzite along the fissure margin.

The presence of quartz in the Hawaiian rock fractures suggests an aeolian origin for some of the fracture constituents. The top BSE image in Figure 8 is of a rock fracture just beneath glacially polished basalt near the summit of Mauna Kea. It is probable that these fractures have been receiving aeolian dust input since Mauna Kea was last deglaciated about 15,000 yr B.P. (Dorn et al. 1991). Individual quartz grains identified by electron microprobe analysis are indicated in Figure 8. The middle BSE image is of a rock fracture near the summit of Haleakala Volcano, Maui. The bottom image is an SEM micrograph of a sample of tephra from near the summit of Hualalai Volcano, Hawaii, and shows the surface topography. The quartz in Figure 8 was identified by EDS as pure Si.

The upper two images in Figure 8 show evidence of silica glaze (Curtiss et al. 1985) between the quartz and the basalt. The enrichment of silica in the glaze, sometimes 40% SiO_2 more than the underlying basalt, has been an uncertainty in silica glaze research in Hawaii (Curtiss et al. 1985). We suggest that some of the silica is derived from the weathering of quartz loess.

DISCUSSION

Our analysis indicates that the fine materials in arid-land rock fractures derive from both aeolian dust and weathering. In some instances, aeolian dust appears to be the dominant source. In other cases, the texture and chemistry of the fine material is consistent with a weathering origin. In still other cases, the evidence is ambiguous.

Geoscientists have long recognized that the parent material of dryland soils can be a composite of weathered bedrock and cumulic aeolian material (Jenny 1941, Nikiforoff 1949, Marchand 1970, Yaalon and Ganor 1973, Mabbutt 1979, Gerson and Amit 1987, Bach, this volume). Following Jenny (1941) and Nikiforoff (1949), we view the fine material in rock fissures as the start of pedogenesis. Accordingly we propose the term "fissuresols" for the evolution of fines in rock crevices. We also recommend that fissuresols be further subdivided into residual soil or a cumulic soil types.

Figure 9 presents a general model of fissuresol development that is consistent with the evidence presented in this study. On one end of the spectrum, fines in rock fissures are completely external and form a cumulic fissuresol. On the other end, fines are entirely derived from weathering. Polygenetic fissuresols form when both weathering and aeolian dust contribute to the development of fines.

Lithology certainly plays a key role in determining whether the fissuresol is residual or cumulic. Friable rocks favor the development of residual fissuresols that are characterized in the field by a lack of rock coatings and a flaked texture on the crevice sides. Lithologies resistant to weathering would favor the development of cumulic fissuresols, which are characterized in the field by rock coatings on crevice sides. We are presently testing this model with artificial fissures that are placed in different natural settings and monitored over time. We are controlling most of the important variables involved in fissuresol development, such as vegetation, microclimate, topography, lithology, and dust storm frequency.

Climatic change may also be an important variable. Arid climates with abundant dust storms (Brazel 1989) would favor cumulic fissuresols. Moister climates tend to enhance weathering and the development of residual fissuresols. We suspect that a polygenetic fissuresol, however, might not truly represent a penecontemporaneous combination of dust and weathered crevice walls. Inputs from aeolian dust or weathering could be periodic. A rock fracture forced open during sample collection at Ayers Rock, Australia, yielded a varnish radiocarbon age of ~27 ka (Dorn and Dragovich 1990). This is best interpreted as a minimum age for the fracture. This example indicates that rock crevices could be potentially old enough to have experienced drastically different climates. It is quite possible that a fissuresol, now found in an arid climate and receiving mostly dust, could have fragments left by rock weathering during a more humid climate.

ACKNOWLEDGMENTS

This research was funded by an REU supplement to a NSF Presidential Young Investigator Award (to Dorn). Thanks to Tanzhuo Liu for assistance in sample

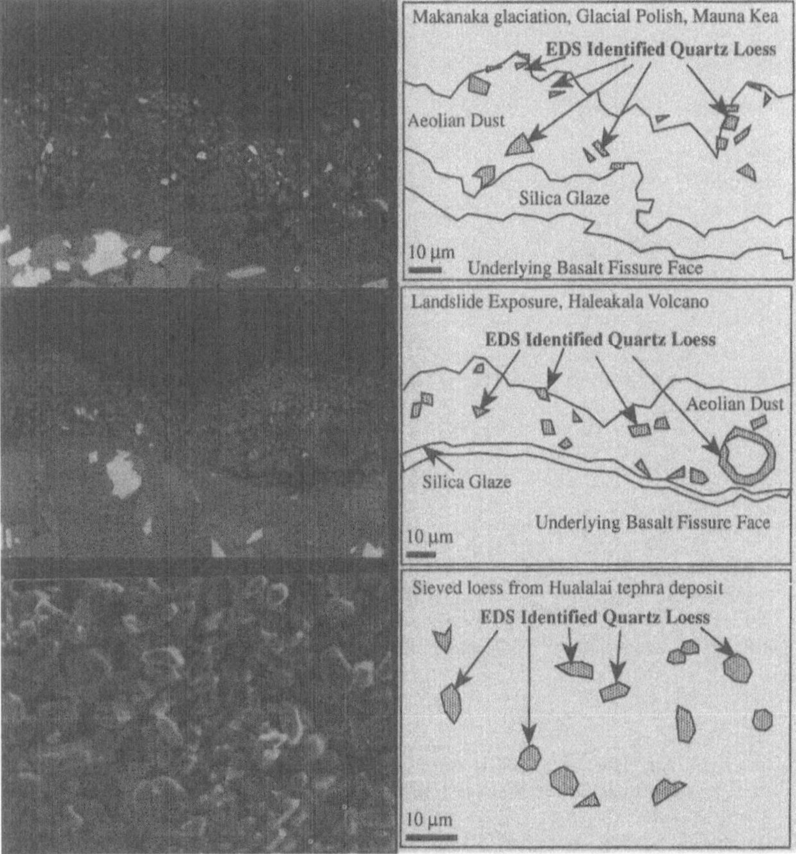

Figure 8. Electron micrographs of quartz in Hawaii. Top: BSE image of fracture near Mauna Kea summit. Middle: BSE image of fracture from Haleakala summit. Bottom: secondary electron image (showing topography) of tephra deposit near summit of Hualalai.

preparation, Tom Paradise and Greg Pope for discussions, and Kevin White (University of Reading), Andrew Goudie (University of Oxford), and Vatche Tchakerian (Texas A&M University) for valuable comments on the manuscript. However, the conclusions remain our own.

REFERENCES

Amit, R., Gerson, R., and Yaalon, D. H. (1993) Stages and rate of the gravel shattering process by salts in desert Reg soils. *Geoderma*, v. 57, p. 295-324.

Beget, J. E., Keskinen, M., and Severin, K. (1993) Mineral particles from Asia found in volcanic loess on the island of Hawaii. *Sedimentary Geology*. v. 84, pp. 189-197.

Brazel, A. J.- (1989) Dust and climate in the American southwest. *Paleoclimatology and Paleometeorology: Modern and Past Patterns of Global Atmospheric Transport*. Kluwer Academic Publishers, New York, p. 65-96.

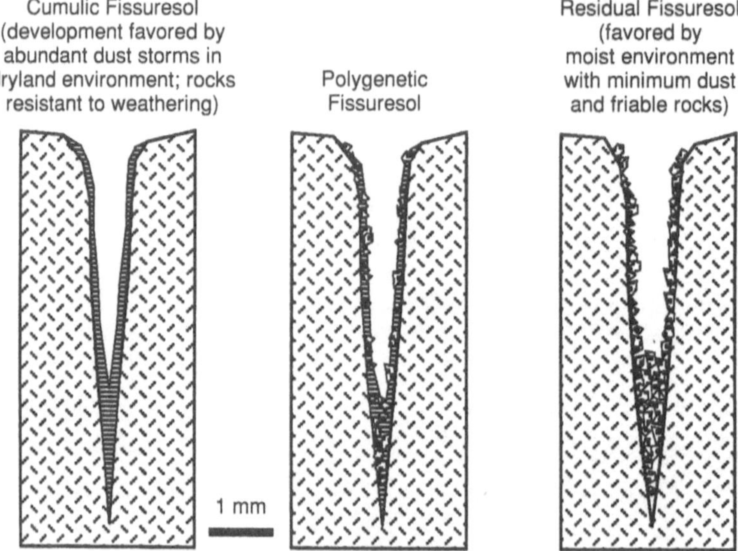

Figure 9. Generalized model of fissuresol development in rock fractures.

Colman, S. M., and Pierce, K. L. (1981) *Weathering Rinds on Andesite and Basaltic Stones as a Quaternary Age Indicator, Western United States.* U.S. Geological Survey Professional Paper, v. 1210.

Coudé-Gaussen, G., Rognon, P., and Federoff, N. (1984) Piegeage de poussières éoliennes dans des fissures de granitoides due Sinai oriental. *Compte Rendus de l'Academie des Sciences de Paris II*, v. 298, p. 369-74

Curtiss, B., Adams, J. B., and Ghiorso, M. S. (1985) Origin, development, and chemistry of silica-alumina rock coatings from the semiarid regions of the island of Hawaii. *Geochimica et Cosmochimica Acta*, v. 49, p. 49-56.

Dorn, R. I., and Dragovich, D. (1990) Interpretation of rock varnish in Australia: Case studies from the Arid Zone. *Australian Geographer,* v. 21, p. 18-32.

Dorn, R. I., Phillips, F. M., Zreda, M. G., Wolfe, E. W., Jull, A. J. T., Kubik, P. W., and Sharma, P. (1991) Glacial chronology of Mauna Kea, Hawaii, as constrained by surface-exposure dating. *National Geographic Research,* v. 7, p. 456-471.

Gerson, R., and Amit, R. (1987) Rates and modes of dust accretion and deposition in an arid region, the Negev, Israel. *Journal Geologic Society London Special Publication,* v. 35, p. 157-169.

Goudie, A. S. (1978) Dust storms and their geomorphological implications. *Journal of Arid Environments*, v. 1, p. 291-310.

Jenny, H. (1941) *Factors of Soil Formation.* McGraw-Hill, New York.

Krinsley, D. H., and Manley, C. R. (1989) Backscattered electron microscopy as an advanced technique in petrography. *Journal of Geological Eduction,* v. 37, p. 202-209.

Mabbutt, J. A. (1979) Pavements and patterned ground in the Australian stony deserts. *Stuttgarter Geographische Studien,* v. 93, p. 107-123.

Marchand, D. E. (1970) Soil contamination in the White Mountains, eastern California. *Geological Society of America Bulletin,* v. 81, p. 2497-2505.

Nikiforoff, C. C. (1949) Weathering and soil evolution. *Soil Science*, v. 68, p. 219-230.

Pye, K. (1987). *Aeolian Dust and Dust Deposits.* Academic Press, London.

Tchakerian, V. P. (1991). Late Quaternary aeolian geomorphology of the Dale Lake sand sheet, southern Mojave Desert, California. *Physical Geography*, v. 12, p. 347-369.

Yaalon, D. H., and Ganor, E. (1973) The influence of dust on the soils during the Quaternary. *Soil Science*, v. 116, p. 146-155.

11 MODELING SEASONAL PATTERNS OF BLOWING DUST ON THE SOUTHERN HIGH PLAINS

James M. Gregory,[1] Jeffrey A. Lee,[2] Gregory R. Wilson,[3] and Udai B. Singh[1]

[1]*Department of Civil Engineering,*
[2]*Department of Economics and Geography, and*
[3]*Department of Plant and Soil Science, Texas Tech University*

ABSTRACT

The Southern High Plains of the United States are notorious for blowing dust, especially during the "dust bowl" of the 1930s. Dust is an environmental and health hazard that can be mitigated with understanding and management. To aid in the understanding of wind erosion, a process-based mathematical simulation model has been developed at Texas Tech University. This paper overviews the components of the model, and illustrates its use with a long-term simulation of expected dust hours for each month of the year. Predicted values were compared to average monthly dust hours reported for Lubbock, Texas, from 1947 to 1989. Predictions matched measured values well, especially considering that the measured data were not used to calibrate or obtain regression coefficients. It is concluded that the Texas Tech model is a reasonable simulator of soil movement, dust concentration, and visibility.

INTRODUCTION

Soil erosion by wind, suspension of dust, and the associated reduction in visibility are serious environmental hazards in arid and semi-arid regions. High winds, low relative humidity, and bare soil created conditions of near zero visibility and triggered a 164-car accident that killed 17 people in 1991 in California on Interstate 5 (Wilson et al. 1993). A similar combination of high winds, low relative humidity, and bare soil resulted in a dust storm that swept across Western China on May 5, 1993, leading to the death of over 60 children who became disoriented because of low visibilibity and were blown or fell into the Yangtze River (Xia Xun Cheng, Lanzhou Institute of Desert Research, personal communication). Other problems associated with wind erosion are sediment deposition on highways, railways, and irrigation and drainage ditches, crop damage, loss of topsoil and soil nutrients, and health problems associated with fine particles. Obviously, wind erosion and dust emission is a problem that needs to be understood and controlled, especially as human impact on arid and semi-arid lands increase. Wind erosion modeling was initially advanced with the work of Bagnold (1941), who included several empirical relations and insights to the processes that affect threshold friction velocity and maximum transport rate for dune sand. Chepil (1957) advanced wind erosion

Desert Aeolian Processes. Edited by Vatche P. Tchakerian. Published in 1995 by Chapman & Hall, London. ISBN 978-94-010-6519-1

research by studying wind erosion on soil, instead of sand, a much more complex material.

In 1965, the first complete wind erosion equation (WEQ) for soil with various degrees of crop and residue cover was published by United States Department of Agriculture researchers (Woodruff and Siddoway 1965). While this equation was very useful for considering management alternatives, it was not process based and had major flaws in both the field length effect (Gregory 1984a, Stout 1990) and erodibility factor (Fuller 1988, Hance 1988). Recent efforts by USDA researchers include the development of a more physically based Wind Erosion Prediction System (WEPS) and the collection of more accurate data (Hagen 1991). While a better model to predict soil loss or movement is a critical first step for understanding wind erosion, it is also essential that off-site damage associated with sedimentation, suspension, and visibility be considered because it is estimated to be 45 times more costly than on-site damage to crop production (Dregne 1988). Recent efforts by researchers at Texas Tech University have focused on a wind erosion model that predicts soil movement, dust suspension, visibility, and PM_{10} (dust particles smaller than 10 micrometers in diameter, which can cause respiratory ailments). The objective of this paper is to overview the major elements of the Texas Tech Erosion Analysis Model (TEAM) and to test the model's ability to simulate seasonal patterns of blowing dust at Lubbock, Texas, located on the Southern High Plains.

MODEL OBJECTIVES

The objective of TEAM is to provide accurate predictions of soil movement, dust suspension, length of visibility, and concentrations of PM_{10} using a process-based, mathematical simulation. The model is responsive to soil variations, including soil type and manipulations because of tillage, cover variations, including plants, plant residue, stable aggregates, or rocks, and variations in surface soil moisture associated with rainfall and relative humidity. The model is designed to operate with a reasonable amount of input data to be acceptable from the user point of view. The primary focus has been single-event simulation; however, long-term probabilistic evaluations are also possible.

MODEL SYNOPSIS

Various processes act and interact to govern the rate of soil movement by wind. Wind erosion can be viewed as the interaction of three major energy components: wind (energy), soil (energy resistance), and plant cover and surface roughness (energy reduction) (Figure 1). The magnitude of the wind velocity at a specified height and the shape of the wind profile determine the rate at which energy is transferred to the field surface. Soil particles can move only if

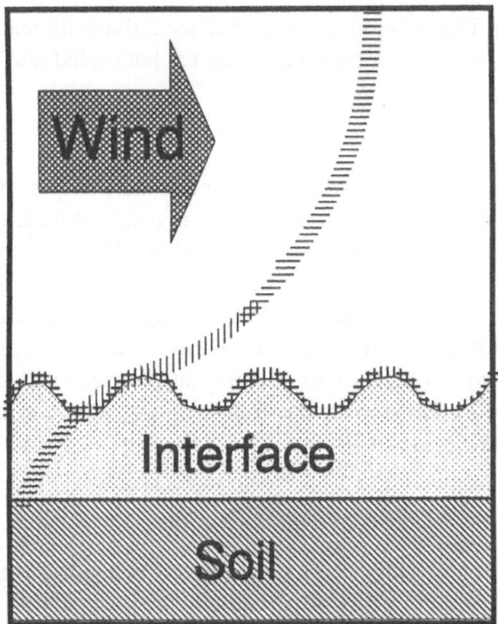

Figure 1. The interaction of the three energy components of the wind erosion process.

wind energy is sufficient to overcome both gravity and any inter-particle cohesion. Soil aggregates too large to move behave as a solid mass and act as cover in the interface zone to protect small loose particles from the energy of the wind. Cover from aggregates and plants protect soil from erosion by reducing the energy intensity at the bottom of the wind profile (top of the interface zone). Air trapped in the interface zone reduces energy intensity. Energy reduction is proportional to the fraction of cover and with the square of the interface zone thickness (Gregory 1984b). The roughness at the field surface causes drag on the wind and affects the shape of the wind profile. Changes at the plant and soil surface affect both the wind profile and the energy transferred to the loose soil at the field surface.

Dimensional-Based Detachment Function

Soil particles must be small and in a loose condition to move. This condition can be achieved through tillage, weathering, or by the erosion process through abrasion of soil aggregates. The detachment process (removal of particles from the surface) appears to be universal for both wind and water erosion and can be expressed with the following equation (Wilson and Gregory 1992, Gregory 1992):

$$m = [(\rho_{bs}/\tau_s)\ f(\theta)]K_E \tag{1}$$

where m = mass of soil detached (kg), ρ_{bs} = soil bulk density (kg/m^3), τ_S = soil shear strength (N/m^2), θ = soil shear angle (dimensionless), and K_E = kinetic energy input (J). This relationship was obtained through dimensional analysis and was shown to match measured results for both wind and water erosion (Wilson and Gregory 1992).

Energy Source

Wind is the energy source for the detachment and transport processes. The force component of the work to detach soil is obtained from the fluid shear stress associated with the shape of the wind profile. The most convenient term to use is friction velocity, defined as the square root of the ratio of fluid shear stress over fluid density. If winds are strong enough to exceed the threshold for movement of particles and dust, mechanical mixing dominates thermal mixing, and a neutral boundary layer exists (Greeley and Iversen 1985, p. 47). Friction velocity can be obtained with the following equation for neutral profile conditions (Abtew et al. 1989):

$$U_* = (0.4\, U_Z) \,/\, \ln[(Z\text{-}D)/z_o] \qquad (2)$$

where U_* = friction velocity (m/s), U_Z = velocity at height Z (m/s), Z = height above the soil surface at which wind velocity is measured (m), D = displacement height (average height of the roughness elements, m), and z_o = aerodynamic roughness (m).

The aerodynamic roughness is the elevation difference between the average surface roughness and the height of the extrapolated intercept of the log relationship where wind velocity goes to zero. The displacement height is the elevation difference between the lower surface base, such as the bottom of a ridge or the soil base from which aggregates protrude, and the average height of the upper surface of the roughness elements. The total distance of z_o and D is the amount that the wind profile is displaced above the soil base (Abtew et al. 1988).

Displacement height can be predicted with the following function (Abtew et al. 1988):

$$D = F_c H_a \qquad (3)$$

where F_c = the fraction of cover by the roughness elements, and H_a = the average height of the upper surface of the roughness elements.

The aerodynamic roughness can be estimated with the following two equations (Gregory 1991):

$$z_{ors} = 0.13\,(H_m\text{-}D) \qquad (4)$$

and

$$z_o = z_{ors} + [0.13\,(H_{ms} - D_s) - z_{ors}]\,[1 - \exp^{-J}] \qquad (5)$$

where $J = (H_{ms}\, W_e)/S_e^{\,2}$, z_{ors} = the aerodynamic roughness for a surface with 0.3 or more fraction of cover by elements (z_{ors} is predicted with Equation 4 using

H_m and D for the elements that provide the major cover), H_m = maximum height of element providing the major cover, H_{ms} = the maximum height of sparse cover elements (tallest elements in system), D_s = the displacement height of all elements including the sparse elements, W_e = the width of sparse elements, and S_e = spacing of sparse elements (centerline to centerline).

Energy Reduction (Cover)

Gregory (1984b) and Gregory and McCarty (1986) used the concept of relative soil loss as a function of canopy and residue cover for soil erosion by rain drop splash, overland flow, and wind. The concepts used to develop the equation were based on three principles. The first is that relative soil loss is proportional to the ratio of kinetic energy that reaches the soil to the maximum kinetic energy at the top of the cover. The second concept is that standing canopies and surface residue intercept turbulent pulses of wind and thus reduce the kinetic energy that reaches the soil surface. The final concept is that layers of standing air absorb the impact of wind energy and reduce the kinetic energy that is transferred to the soil. The following equation was developed to predict relative energy as a function of fraction and height of cover:

$$S = (1 - F_r) / \{1 + [(h_r/h_s) - 1]F_r\}^2 \qquad (6)$$

where S = cover factor, which expresses the fraction of wind energy that reaches the eroding soil surface, F_r = fraction of cover, h_r = height of cover (m), and h_s = micro-relief (typically 0.01 m). Equation 6 predicted the reduction from potential soil loss for wind erosion for variations in ridge height and clod cover with an $R^2 = 0.99$ (Gregory 1984b).

Maximum Transport Rate

Equation 1 was expanded by Gregory et al. (1993) through the evaluation of kinetic energy from the wind profile, cover, and particle size effects. The development is detailed; however, the resulting equation relates detachment to soil strength, particle size, particle size distribution, cover, friction velocity, and threshold friction velocity. The resulting detachment equation is:

$$m / (WLt) = C_1 \rho_f (\rho_{bs}/\tau_s) (D_{50}/D_r) \{[1 + C_2 [(\sqrt{D_{75}} - \sqrt{0.08}) / \sqrt{D_r}]\} [SU_*^2 - (0.8U_{*t}/G_f)^2]U_* \qquad (7)$$

where W = width of eroding surface (m), L = length of eroding surface (m), t = time (hr), C_1 = coefficient obtained from calibration (0.004 sec/hr), C_2 = dimensionless coefficient obtained from calibration (125), ρ_f = density of fluid (1.23 kg/m^3 for air), D_{50} = median particle diameter of primary (none aggregated) soil particle size distribution (mm), D_{75} = 75th percentile of primary particle size, D_r = reference particle diameter (1.0 mm), U_{*t} = threshold friction velocity (m/s), and G_f = a gust factor (1.5 for field conditions and 1.0 for wind tunnel conditions).

The maximum soil transport rate by wind occurs when the soil surface is completely covered with loose erodible particles. Wilson and Gregory (1992) showed that loose soil shear strength can be determined by equating it to wind shear at threshold conditions. The following equation was developed for loose soil detachment and maximum transport rate with the length of detachment set to unity:

$$m / Wt = 0.004\, \rho_{bs}\, [G_f / (0.8U_{*t})]^2\, (D_{50} / D_r) \quad (8)$$
$$\{1 + 125\, [(\sqrt{D_{75}} - \sqrt{0.08}) / \sqrt{D_r}]\}\, [SU_*^2 - (0.8U_{*t}/G_f)^2]\, U_*$$

The maximum transport equation was calibrated with published data reported from wind tunnel studies and fit the experimental data with an R^2 of 0.95, which was significant ($\alpha = 0.001$) (Gregory et al. 1993). The equation was also tested with field data from Svasek and Terwindt (1974) shown in Figure 2 and data from Nickling (1978) shown in Figure 3. The upper and lower curves were generated with Equation 8 by considering the effect of relative humidity on threshold friction velocity.

Threshold Friction Velocity

Bagnold (1941) derived an equation for threshold friction velocity which is a square-root function of particle diameter (for particle diameters larger than about 0.1 mm). For arid regions, this relation is adequate for most predictions, but in semi-arid or sub-humid regions, the effect of soil moisture is a major factor. Gregory and Darwish (1990) expanded Bagnold's relationship to include the effect of electrostatic bonding (for small diameter particles) and soil moisture. The following equation was developed:

$$U_{*t} = 0.118\, \{21.2\, D_{50}\, [1 + 0.01W_a + (0.0045 / D_{50}^2) +$$
$$(0.75 / D_{50})\, [\exp -0.1^{(W_a/W_w)}]\, (W_a - W_c)]\}^{0.5} \quad (9)$$

where 0.118 = dimensionless constant, 21.2 = a constant ($m^2/(s^2\,mm)$), W_a = soil water content expressed as a percentage, W_w = wilting point of soil expressed as a percentage, and W_c = water attached to clay, in scratches on the particle surface, or internal to the surface of the particle expressed as a percentage. Soil water and clay water were related to relative humidity (Puri et al. 1925) and an equation was empirically developed to describe the relationship by Gregory (1991). The term $W_a - W_c$ is set to zero when W_c exceeds W_a.

Length Effect

Chepil's measurements on the length effect (Chepil 1957) provided a classic data set that related soil movement to field length and soil type. Bagnold (1936) had investigated the effect of length on sand movement in a wind tunnel; however, the length to maximum movement for sand was only 2 to 4 m compared to hundreds of meters for soil (Figure 4). A physical explanation for the length effect was first presented by Gregory (1984a). Two concepts were

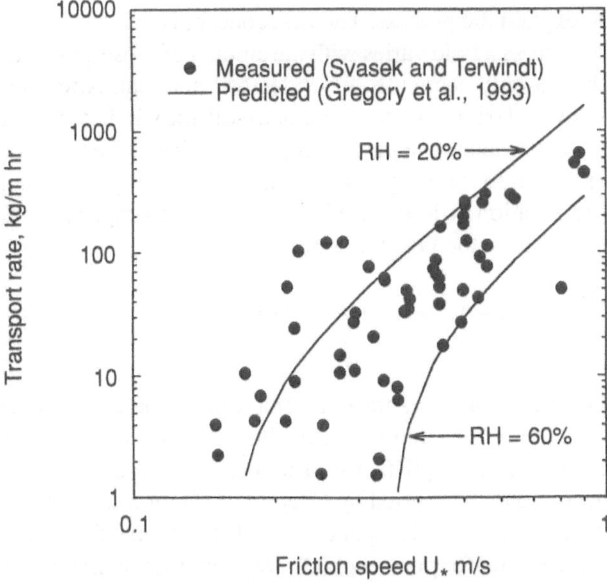

Figure 2. TEAM predicted and measured field transport data from Svasek and Terwindt (1974) (D_{50} = 0.250 mm and D_{75} = 0.281 mm) for relative humidities of 20 and 60 percent (reproduced for Gregory et al. 1993).

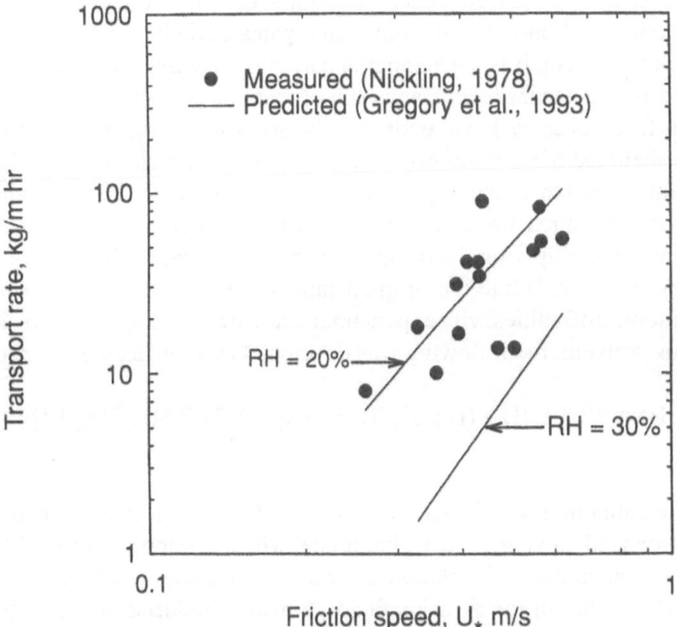

Figure 3. TEAM predicted and measured field transport data from Nickling (1978) (D_{50} = 0.063 mm and D_{75} = 0.125 mm) for relative humidities of 20 and 60 percent (reproduced for Gregory et al. 1993).

identified to explain the process. The first concept is that the change in actual transport rate across a field varies with the amount of loose soil on the surface. New detachment only occurs from bonded soil material. After the surface is completely covered with loose (non-bonded) soil, the wind energy for transport is used to re-detach and move the existing loose soil and the transport rate is at a maximum condition for the given friction velocity. Gregory (1984a) derived the following equation to describe this process for the simplest condition of no incoming sediment at the beginning of the field:

$$L_f = 1 - \exp^{-A_f L\,(D_t/D_{tl})} \tag{10}$$

where L_f = length factor, A_f = abrasion factor (dimensionless), D_t = detachment of bonded soil (kg/m^2/hr), and D_{tl} = detachment of loose soil (kg/m^2/hr).

The dimensionless length factor varies between 0.0 and 1.0. This equation matched measured data reported by Chepil (1957) for a sandy loam soil with an R^2 of 0.81 and was significant at $\alpha = 0.001$. It also matched measurements made in a wind tunnel for pure sand ($D_t = D_{tl}$) reported by Bagnold (1936) for a friction velocity of 0.92 m/s with an R^2 of 0.95, which was significant at α = 0.001. The abrasion factor, A_f, was set to 1.0 for these analysis.

Equation 10 is valid, however, only for sandy soils or pure sand at high winds because it only considers one of the two processes that affect the length relationship. The second concept of describing the length effect is that the total kinetic energy of particles abrading aggregates initially increases exponentially with field length as the number of particles increase. The collision of one loose particle detaches others, and each of them detaches more, etc. At about 35% surface cover by loose particles, the solid soil surface becomes sufficiently sheltered from direct impact from saltating particles to the point, that exponential growth in the number of particles is curtailed, and the abrasion adjustment factor reaches an upper limited of 1.0. Gregory (1984a) considered this effect by multiplying the detachment ratio in Equation 10 by an abrasion adjustment factor. While the original function produced accurate results, it caused some difficulties with exponential overflow on computers. To circumvent this problem, the following was developed as a replacement function:

$$A_f = 0.1 \exp \{2.3\, L/[L + (D_{tl}/(D_t S U_*^2)) \exp [-1.7\, (D_t S U_*^2/D_{tl})L]\} \tag{11}$$

The calibration for Equation 11 was obtained with data from Bagnold (1936) for sand movement in a wind tunnel with a friction velocity of 0.36 m/s. The R^2 obtained for this relationship was 0.97 and was highly significant (α = 0.001). The use of the abrasion factor, however, reduced the fit of the high friction velocity data from Bagnold to an R^2 of 0.80. Several uncertainties exist with the data collected by Chepil (1957); however, with reasonable assumptions of friction velocity and clod size distribution, Equations 10 and 11 yielded

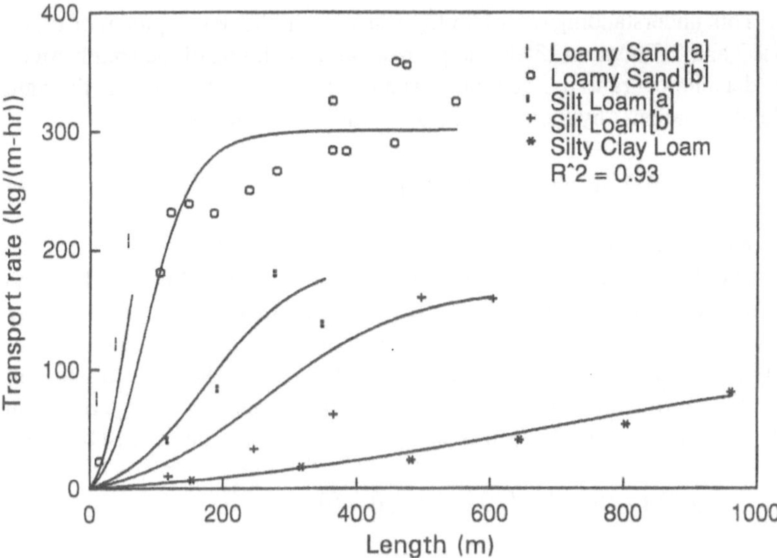

Figure 4. Length effect from data reported by Chepil (1957).

an R^2 of 0.93 ($\alpha = 0.001$) for five High Plains soils. The textural class of these soils ranges from a loamy sand to a silty clay loam (Figure 4).

Erodibility Effect

In the literature, many different definitions for soil erodibility are found. In this paper soil erodibility will be defined as the mass detached per unit of kinetic energy. One key to understanding wind erosion, especially the length effect, is the recognition that at least two soil erodibilities exist: one for cohesive soil and one for loose soil. To detach soil particles from a solid soil, input energy must break the bonding force between particles. To detach loose soil, the input energy only has to overcome the force of gravity and any liquid bonds, such as surface tension.

Wilson and Gregory (1992) defined soil erodibility, E, for the solid (or aggregated) part of the soil as:

$$E = (\rho_{bs}/\tau_s)\, f(\theta) \tag{12}$$

where E = erodibility (kg/J). Loose soil erodibility, E_l, was defined (Wilson and Gregory, 1992) as:

$$E_l = N[\rho_{bs}/(\rho_f U_{*t}^2)] \tag{13}$$

where E_l = loose soil erodibility (kg/J), and N = a number obtained from calibration.

This understanding of erodibility was used in the development of Equations 7 and 8. Equation 13 also helps in the understanding of the length effect. The detachment ratio in Equations 10 and 11 is obtained by dividing the right side of Equation 7 by the right side of Equation 8, which gives:

$$D_t/D_{tl} = (C_3\,\rho_f\,\rho_{bs}\,/\tau_s)\;/\;\{0.004\,\rho_{bs}\,[G_f/(0.8\,U_{*t})^2]\} \tag{14}$$

where C_3 = calibration coefficient.

The detachment ratio is essentially the ratio of erodibilities. Wilson and Gregory (1992) noted that the $\tau_s/\,\rho_{bs}$ term in Equation 12 had the same dimensions as crushing energy (the energy required to crush a defined mass of cohesive soil) as defined by Skidmore and Powers (1982). A convenient method to estimate soil erodibility for solid soil is crushing energy. Based on data form Chepil (1957), when crushing energy is used, Equation 14 becomes

$$D_t/D_{tl} = (5\rho_f\,/E_c)/\;\{0.004\,\rho_{bs}\,[G_f/(0.8\,U_{*t})^2]\} \tag{15}$$

where E_c = crushing energy (J/kg). The advantage of Equation 15 over Equation 14 is that crushing energy can be measured relatively easily as compared to shear strength (Skidmore and Powers 1982).

Reference Zone Concentration

The rate of soil movement, X, for a given field length and soil condition is obtained by multiplying the results from Equations 8 and 10. The height of saltation is determined by Owen (1964):

$$H_s = U_*^2/(2g) \tag{16}$$

where H_s = height of saltation (m), and g = gravity (m/s^2). The flow rate of air in the saltation layer is calculated with the following equation:

$$Q_a = 5.1\,(H_s)\,(U_*)\,(W)\,(3600) \tag{17}$$

where Q_a = air flow rate (m^3/hr). The value of $5.1U_*$ defines the wind velocity at the top of the roughness elements (Gregory et al. 1993). Concentration of sediment in the saltation layer is obtained by dividing the mass per time times width divided by the air flow rate from Equation 17:

$$C_s = X\,W\,/\,Q_a \tag{18}$$

where C_s = concentration in the saltation zone (kg/m^3), and X = rate of soil movement (kg/m/hr).

Suspension Concentration

The average reference height for the reference dust concentration is the saltation height divided by 2.0. Distribution of dust concentration with height is then predicted with the following equation modified from Anderson and Hallet (1986) (Gregory et al. 1991):

$$C_H = Cs\ (2H/Hs)^{-[Us/(0.4U_*)]} \tag{19}$$

where C_H = concentration at desired height H (kg/m^3), H = height at which concentration is predicted (m), and U_s = particle settling velocity (m/s).

Visibility Prediction

The length of visibility depends on light penetration. Gregory (1987) derived an equation to predict visibility as a function of dust concentration and particle size. The total light penetration through a system of various particle sizes can be predicted by evaluating the probability of penetration through all concentrations of various particle sizes in the system.

$$P_T = \exp\ [-1500\ (1/p)\ \mathbf{Isu}(_{i=1},{}^{n},\ (C_i/D_i))] \tag{20}$$

where P_T = fraction of light that directly penetrates a distance of 1.0 m, C_i = concentration of particles for size class i (kg/m^3), D_i = average diameter of size class i (mm), ρ = particle density (2650 kg/m^3), i = size class, and n = number of size classes.

The effective average particle diameter at a given height can be determined by rearranging Equation 20:

$$D_{eff} = -1500\ (1/\rho)\ (C_T/\ln P_T) \tag{21}$$

where D_{eff} = the effective average particle diameter (mm), and C_T = the total concentration of all particle size classes at height H.

Length of visibility at a given height can then be predicted using the following equation (Gregory 1987):

$$L_V = [(-\rho\ \ln P)/1500]\ (D_{eff}\ /\ C_T) \tag{22}$$

where L_v = length of visibility (m). The factor P is often set to 0.02 (2.0% of light coming from the target), the lower limit of detection with the human eye (Robinson 1968).

LONG-TERM APPLICATION

Wind erosion on a typical dryland cotton field near Lubbock, Texas, was simulated using TEAM. The purpose of the simulation was to estimate amount of erosion and the number of hours of blowing dust occurring on the field for each month. Lee et al. (1992) performed a similar test with an earlier version of TEAM. However, the simulation presented here is made with a significantly improved model which directly predicts dust hours. The following information was determined for the field, to be used as input to the model.

The soil is a sandy loam, typical of the region, with 80% sand, 10% silt and 10% clay. Soil erodibility, based on clay content, is 0.11 kg/J (Wilson and Gregory 1992).

Wind speed distributions for each month were determined from hourly meteorological data for Lubbock. Wind speeds were adjusted from anemometer height to a standardized height of 10 m following the procedure outlined in Lee et al. (1993). Wind speeds were simulated with a cyclic function derived to match wind speeds for each hour of the day for each month, based on hourly wind speed recorded by the National Weather Service. From measurements, the 40-year average wind speed values for each hour of the day for each month were determined. It was observed that values at night were essentially constant. During the day, the change in wind speed followed a parabolic function. Deviations from average hourly values were simulated with a probabilistic function developed by Gregory (1989). Similar cyclic functions were used to estimate hourly relative humidity values.

Field length is 400 m. Visibility as observed at the weather station was assumed to be equal to the visibility at the end of the eroding field.

Cotton is planted in late May, with a 1.0 m spaced row laid out perpendicular to the prevailing wind direction. The crop begins growing in early June, with 5-cm-high-plants for the first week of June, 7.5 cm for the second and third weeks, and 10 cm for the last week of June and first week of July. From the second week of July to harvest in November, the plants are higher than 15 cm, which is sufficient to completely protect the surface from erosion. After harvest, 40-cm standing plant stalks are left on the field, affording a 3% effective cover in the field from 15 November to 31 December, when the field is tilled and the stalks are incorporated into the soil. The field has no vegetative cover from 31 December to 31 May, and erosion control is done with 3.0-cm-diameter soil aggregates covering 50% of the surface. The plant growth information is consistent with dryland cotton growth experiments done in the region by Barker et al. (1985, 1989) and Bilbro (1991). Rain occurs one out of six days, leaving the soil too wet for erosion to occur that day.

These input data are reasonable for the region, though spatial and temporal variations exist (Lee et al. 1993). The model output used in this study is total erosion (kg/m width at the end of the field) and hours of blowing dust from the field for each month. More specifically, predicted dust hours are the number of hours that visibility is reduced below 11 km because of blowing dust at a height of 10 m. The results are shown in Figure 5.

Data Used to Test the Model Output

The lack of long-term erosion data precludes a direct test of the predicted erosion amounts in the region; however, predicted and measured dust hours can be compared. Comparisons were made to the average monthly values for the years 1947 through 1989 obtained from the United States National Weather Service at Lubbock, Texas. A relative index of the amount of blowing dust in an area can be derived from weather records, as described in greater detail in Lee et al. (1993). At First Order Weather Stations, a blowing dust event begins when horizontal visibility is reduced below 11 km and ends when visibility exceeds 11 km. The time, visibility and wind speed are recorded at hourly intervals during an event, and also when significant changes in visibility occur. Visibility can be used to predict the concentration of dust in the air with the equation presented by Patterson and Gillette (1977), where:

$$M = c / v^{\gamma} \tag{23}$$

where M = concentration of dust in the air (kg/m^3), c = empirical coefficient (2 x 10^{-5} kg/m^3/km), v = visibility (km), and γ = empirical coefficient (1.07). Patterson and Gillette (1977) empirically determined the values of c and γ for the Southern High Plains. Visibility is recorded for the weather observer's eye height, approximately 1.7 m. The wind speed at anemometer height is converted to speed at observer eye height, $u_{1.7}$, using the procedure of Lee et al. (1993).

An estimate of dust transport at observer height, $E_{1.7}$, for one visibility reading is made with (Lee et al., 1993):

$$E_{1.7} = M\, u_{1.7}\, t_s \tag{24}$$

where $E_{1.7}$ = dust transport at 1.7 m (kg/m^2), $u_{1.7}$ = wind velocity at 1.7 m (m/s), and t_s = duration of reading (s). $E_{1.7}$ is the amount of dust passing through a square meter centered at 1.7 m. $E_{1.7}$ is summed for all visibility readings during a blowing dust event to create an index of the amount of blowing dust during the event. $E_{1.7}$ is a relative measure and both the horizontal and vertical extent of the dust during the event are not included. It is, however, the most detailed measure of sediment transport obtainable from weather records.

$E_{1.7}$ was determined for all blowing dust events recorded at Lubbock for the years 1947 to 1989. The sum of $E_{1.7}$ for all events of a given month were averaged, and the results are shown in Figure 5. Also in Figure 5 are the erosion amounts for each month predicted with TEAM. The reader should note that these are two different measures of erosion with different units. To allow comparison between the two measures of erosion, they are plotted in Figure 6 as the percentage of total annual erosion occurring each month.

Another measure of erosion is the number of hours of blowing dust occurring in a given time period. The same data set used to determine values of $E_{1.7}$ can be used to determine dust hours for Lubbock (Lee et al. 1993). The average dust hours per month for Lubbock from 1947 to 1989 are shown in Figure 5 along with the predicted dust hours. The empirical dust hours derived

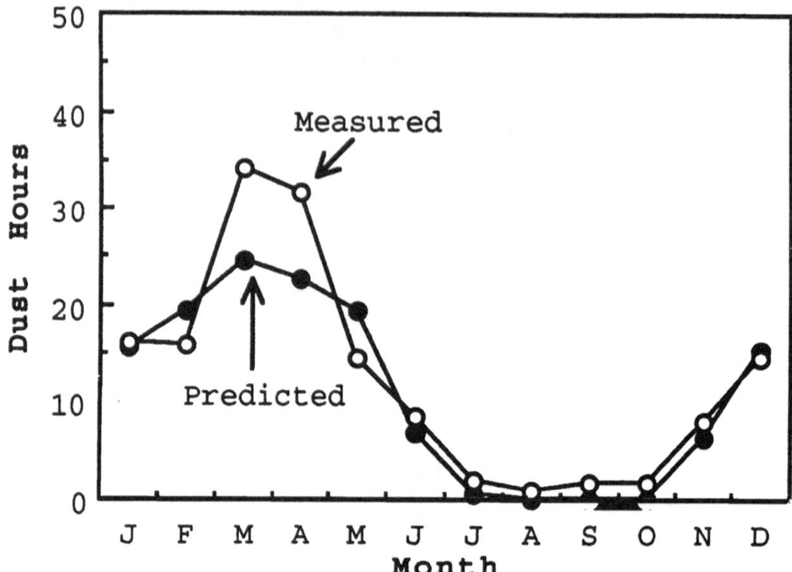

Figure 5. Average number of hours of blowing dust predicted with TEAM and measured by the National Weather Service.

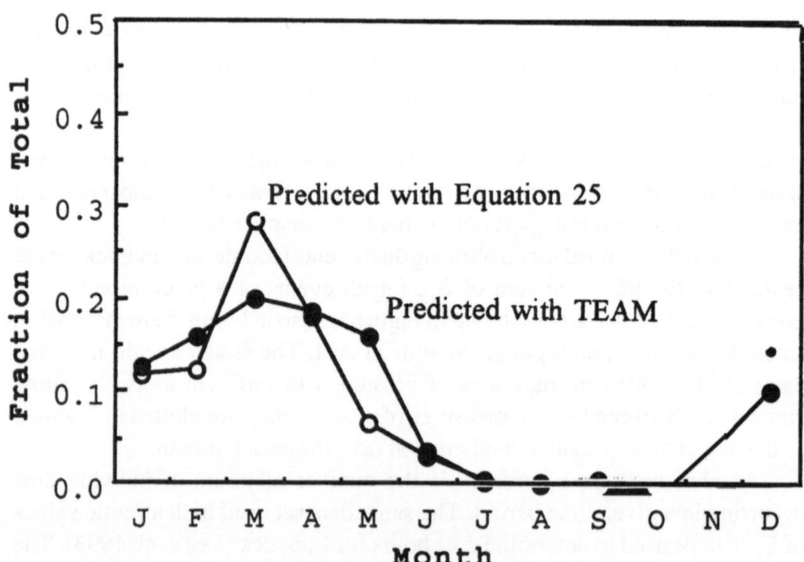

Figure 6. Fraction of total annual dust erosion for each month predicted with TEAM and empirically determined from National Weather Service records.

from meteorological data and the predicted dust hours from TEAM are both for horizontal visibility less than 11 km, but for different heights above the surface. The empirical data are for eye height (about 1.7 m) and the TEAM predictions are for 10 m. Dust concentration, and therefore visibility, varies with height, so the two measures are not directly comparable. The comparison is reasonable, though, because the weather observations are made at Lubbock International Airport, which is not a source area for dust. On a non-erodible surface downwind from the site of erosion, the dust has diffused to where the visibility at eye height is less than it was at the eroding field.

RESULTS AND DISCUSSION

The amount of long-term erosion occurring on fields in the region is not known, owing to a lack of field data. As mentioned earlier, an indirect test of the model output is to use the calculated values of $E_{1.7}$ averaged for each month. The two data sets are measures of amounts of erosion, but with different units. There also is considerable error likely in both sets of data, so the following comparison should be viewed with healthy skepticism.

Figure 6 shows the seasonal pattern of erosion simulated with TEAM and derived from the empirical visibility record plotted as a fraction of average yearly total erosion occurring each month. The seasonal pattern of erosion is similar in both data sets, with higher erosion rates in late fall, winter, and early spring when wind speeds are higher and the soil is unvegetated, and lowest in late spring, summer, and early fall, when crop cover protects the surface. The measured erosion is higher than the simulated erosion in March and the opposite pattern exists in May. For other months, the agreement is relatively close. It is likely that land management techniques not incorporated in the model simulation are largely responsible for the discrepancies in March and May. In May, farmers are more likely to protect the soil from erosion by creating aggregates because erosion after planting can kill young plants from abrasion by saltating sand grains. In March, the incentive to minimize wind erosion is not as critical.

The empirical and TEAM predicted dust hours are shown in Figure 7 and show a similar seasonal pattern. The predicted values are lower than those measured in March and April, and are close in all other months. Variations in land management may account for the March and April discrepancies, though additional research is required to find the actual causes.

CONCLUSION

The Texas Tech Erosion Analysis Model is a process-based mathematical simulation which provides estimates of wind erosion and dust generation on farm fields. A test of the ability of the model to simulate the seasonal pattern

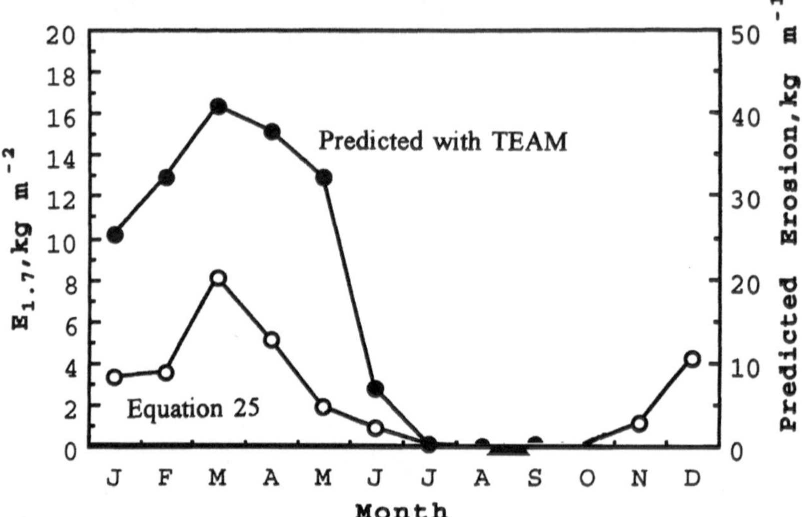

Figure 7. Seasonal pattern of sediment transport predicted with TEAM and empirically derived from National Weather Service data ($E_{1.7}$) for Lubbock, Texas.

of wind erosion at Lubbock, Texas, shows that TEAM is reasonably accurate for such applications, though additional work is needed to further test and verify the model.

ACKNOWLEDGMENTS

We would like to thank Doug Sherman for critically evaluating an earlier version of the manuscript.

REFERENCES

Abtew, W., Gregory, J. M., and Borrelli, J. (1989) Wind profile: estimation of displacement height and aerodynamic roughness. *Transactions ASAE*, v. 32, p. 521-527.

Anderson, R. S., and Hallet, B. (1986) Sediment transport by wind: toward a general model. *Geological Society of America Bulletin*, v. 97, p. 523-535.

Bagnold, R. A. (1936) The movement of desert sand. *Proceedings of the Royal Society of London*, v. 157, p.594-620.

Bagnold, R. A. (1941) *The Physics of Blown Sand and Desert Dunes*. Methuen, London.

Barker, G. L., Hatfield, J. L., and Wanjura, D. F. (1985) Cotton plant response to wind and water stress. *Crop Plant Research*, v. 12, p. 33-47.

Barker, G. L., Hatfield, J. L., and Wanjura, D. F. (1989) Influence of wind on cotton growth. *Transactions ASAE*, v. 32, p. 94-104.

Bilbro, J. D. (1991) Relationship of cotton dry matter production and plant structural characteristics for wind erosion modeling. *Journal of Soil and Water Conservation*, v. 46, p. 381-384.

Chepil, W. S. (1957) Width of field strip to control wind erosion. *Kansas Agricultural Experiment Station Technical Bulletin 92*, Kansas State University. Manhattan, Kansas.

Dregne, H. E. (1988) Wind erosion: an international perspective. *Proceedings of the 1988 Wind Erosion Conference*, Texas Tech University, Lubbock, Texas.

Fuller, W. W. (1988) Problems in the application of the wind erosion equation to field sites. *Proceedings of the 1988 Wind Erosion Conference*, Texas Tech University, Lubbock, Texas.

Greeley, R., and Iversen, J. D. (1985) *Wind as a Geological Process on Earth, Mars, Venus and Titan*. Cambridge University Press, Cambridge.

Gregory, J. M. (1984a) Analysis of the length effect for soil erosion by wind. Paper presented at the winter Meeting of American Society of Agricultural Engineers, New Orleans, LA. Paper no. 842540.

Gregory, J. M. (1984b) Prediction of soil erosion by water and wind for various fractions of cover. *Transactions ASAE*, v. 27, p. 1345-1350, 1354.

Gregory, J. M. (1987) Visibility prediction from dust concentration and particle size. Paper presented at the Summer Meeting of the American Society of Agricultural Engineers, Baltimore, MD. Paper no. 872032.

Gregory, J. M. (1989) Wind data generation for Great Plains locations. Presented at the International Winter Meeting of the American Society of Agricultural Engineers, New Orleans, LA. Paper No. 892664.

Gregory, J. M. (1991) *Wind Erosion: Prediction and Control*. Report prepared for the U.S. Army Corps of Engineers, Waterways Experiment Station, Vicksburg, Mississippi.

Gregory, J. M. (1992) Erosion equation derivation similar to the USLE. Paper presented at the Summer Meeting of American Society of Agricultural Engineers, Charlotte, NC. Paper no. 922051.

Gregory, J. M., and McCarty, T. R. (1986) Maximum allowable velocity prediction for vegetated waterways. *Transactions ASAE*, v. 29, p. 748-755.

Gregory, J. M., and Darwish, M. M. (1990) Threshold friction velocity prediction considering water content. Paper presented at the Winter meet of American Society of Agricultural Engineers, Chicago, IL. Paper No. 902562.

Gregory, J. M., Singh, U. B., Lee, J. A., Fedler, C. B. (1991) Dust hours, visibility, and wind erosion prediction. Paper presented at the Summer Meeting of the American Society of Agricultural Engineers, Albuquerque. Paper No. 914007.

Gregory, J. M., Wilson, G. R., and Singh, U. B. (1993) Wind erosion: detachment and maximum transport rate. Paper presented at the Summer Meeting of the American Society of Agricultural Engineers, Spokane. Paper No. 932050.

Hagen, L. J. (1991) A wind erosion prediction system to meet user needs. *Journal of Soil and Water Conservation*, v. 46, p. 106-111.

Hance, W. G. (1988) Application and problems of the USDA wind erosion equation. *Proceedings of the 1988 Wind Erosion Conference*, Texas Tech University, Lubbock, Texas.

Lee, J. A., Gregory, J. M., and Singh, U. B. (1992) A comparison of model predictions for wind erosion and visibility records at Lubbock, Texas. *Papers and Proceedings of the Applied Geography Conference*, v. 15, p. 10-14.

Lee, J. A., Wigner, K. A., and Gregory, J. M. (1993) Drought, wind and blowing dust on the Southern High Plains of the United States. *Physical Geography*, v. 14, p. 56-67.

Nickling, W. G. (1978) Aeolian sediment transport during dust storms: Slims River Valley, Yukon territory. *Canadian Journal of Earth Science*, v. 15, p. 1069-1084.

Owen, P. R. 1964. Saltation of uniform grains in air. *Journal of Fluid Mechanics*, v. 20, p. 225-242.

Patterson E. M., and Gillette, D. A. (1977) Measurements of visibility vs mass-concentration for airborne soil particles. *Atmospheric Environment*, v. 11, p. 193-196.

Puri, A. N, Crowther, E. M., and Keen, B. A. (1925) The relation between the vapor pressure and water content of soils. *Journal of Agricultural Sciences*, v. 15, p. 68-88.

Robinson, E. (1968) Effects of air pollution on visibility. In A. C. Stern (ed.) *Air Pollution*, v. 1. Academic Press, New York.

Skidmore, E. L., and Powers, D. H. (1982) Dry soil-aggregate stability: energy-based index. *Soil Science Society of America Journal*, v. 46, p. 1274-1279.

Stout, J. E. (1990) Wind erosion within a simple field. *Transactions ASAE*, v. 33, p. 1597-1600.

Svasek, J. N., and Terwindt, J.H.J. (1974) Measurements of sand transport by wind on a natural beach. *Sedimentology*, v. 21, p. 311-322.

Wilson, G. R. and Gregory, J. M. (1992) Soil erodibility: understanding and prediction. Paper presented at the Summer Meeting of American Society of Agricultural Engineers, Charlotte, NC. Paper no. 922049.

Wilson, G. R., Gregory, J. M., and Brownell, J. R. (1993) Analysis of the 1991 California wind erosion disaster. *Proceedings of the 7th U.S. National Conference on Wind Engineering*, University of California, Los Angeles, v. 2, p. 859-868.

Woodruff, N. P., and Siddoway, F. H. (1965) A wind erosion equation. *Soil Science Society of America Proceedings*, v. 29, p. 602-608.

12 REVERSED DESERTIFICATION ON SAND DUNES ALONG THE SINAI/NEGEV BORDER

Haim Tsoar,[1] Victor Goldsmith,[2] Steve Schoenhaus,[2] Keith Clarke,[2] and Arnon Karnieli[3]

[1]Department of Geography and Environmental Development, Ben-Gurion University of the Negev; [2]Department of Geology and Geography, Hunter College of The City University of New York; [3]Remote Sensing Laboratory, Ben-Gurion University of the Negev

ABSTRACT

Changes in the Israeli/Egyptian border since 1948 have affected the land use patterns of the Bedouin tribes inhabiting these areas. These changing patterns are clearly discerned from aerial photographs and satellite images in this area of extensive linear dunes. The most recent change occurred after Israel's withdrawal from the Sinai in 1982. Field observations show that the vegetation in the Negev recovered from overgrazing in a very short time, with a 180% increase in the number of Negev shrubs, coincident with a continued decrease in the number of shrubs on the Sinai side of the border. Analysis of the reflectance of the Landsat-MSS for the recovery years (1984, 1987, and 1989) shows a statistically significant difference between the two sides of the border for the four MSS bands. However, the spectral brightness curves of the four bands failed to reveal any specific "signature" typical for vegetation. Field measurements with a spectrometer indicate that the curve shape of the reflectance of the Landsat-MSS images is swayed by the biogenic crust in the Negev that makes the well known contrast in albedo with Sinai.

INTRODUCTION

The process of desertification and land degradation is of great concern in drylands that once supported agriculture and animal grazing. Anthropogenic factors accelerate desertification processes (Helldén 1991), and, contrary to the accepted notion that desertification processes on sand dunes becomes irreversible unless there is direct human interference (Reichelt 1989), vegetation can be established on sandy areas that had been fenced off from grazing (Otterman 1981, Warren and Harrison 1984). This process, although rare, is known as reversed desertification. Sand fixation protects the available sand moisture, decreases sand transport and erosion, and increases the amount of fines, thus promoting more vegetation growth (Tsoar 1990).

Remote sensing is a viable and appropriate tool for studying desertification and climate change in arid and semi-arid environments (e.g., Prince et al. 1990). Remote sensing often provides the least expensive data that can be acquired and frequently is the only type of information available in remote arid zones.

Arid environments afford excellent field sites because of their considerable geographic location, cloud-free weather, bright spectral properties, high

Desert Aeolian Processes. Edited by Vatche P. Tchakerian. Published in 1995 by Chapman & Hall, London. ISBN 978-94-010-6519-1

air and surface temperatures, high evapotranspiration rates, and sparse vegetation. Otterman (1981) found that the vegetated Negev sand dunes do not exhibit characteristic reflectance curves common to green vegetation. This was interpreted by Otterman et al. (1990) to result from plant debris littering the surface and the shadowing effect of desert shrubs.

The purpose of this study is to document and analyze the reversed desertification in the Negev and to compare it with adjacent Sinai sections using multitemporal Landsat-MSS imagery, aerial photographs, and field spectral measurements and observations.

STUDY SITE AND METHODS

Physical Setting

An area that has undergone reversed desertification is the Israeli/Egyptian border in the northwestern Negev and northeastern Sinai deserts (Figure 1). Although created by an arbitrary political border in 1906, the Negev and Sinai are considered one geographical desert with many linear sand dunes. The study area is located in a climatic transition zone, with an annual average rainfall between 100 mm north of Nizzana and 104 mm in Revivim, to 175 mm in the northern Haluza sands (Figure 1). The rainy season begins in October and ends in May. From 1944 to1989, the highest recorded precipitation in Revivim was about 202 mm and the lowest, about 32 mm (Figure 2). The area has been inhabited by Bedouin nomads, who use the sand surface for growing castor, beans, barley, wheat, and watermelons. Goats, camels, and donkeys graze the dunes and follow the seasonal moves of the Bedouin (Tsoar and Møller 1986).

The Contrast in Albedo Between the Negev and Sinai

Since the advent of satellite images in the 1960s, the sandy area of the Sinai/Negev border has attracted the attention of many geoscientists and ecologists because of the high contrast in albedo between the bright Sinai and the darker Negev (Lowman 1966, Otterman 1974, Otterman et al. 1975, Muehlberger and Wilmarth 1977, Adams et al. 1978, Noy-Meir and Seligman 1979, Otterman 1981, Danin 1983, Warren and Harrison 1984, Otterman and Tucker 1985, Waisel 1986, Tsoar and Møller 1986, Otterman et al. 1990, Danin 1991, Meir et al. 1992).

Qualitative analysis of aerial photographs from 1944 and 1945 showed sparse vegetation and no contrast in albedo among the linear dunes of the Sinai and Negev desert border areas (Noy-Meir and Seligman 1979). After the creation of Israel in 1948, some of the Bedouin tribes migrated from the Negev (Israeli side) into the Sinai (Egyptian side) because of the Israeli military occupation. Owing to security reasons, the tribes that remained in the Negev were relocated to areas surrounding Beer-Sheva. However, between 1949 and 1957, there were sporadic infiltrations of Bedouin from the Sinai across the border to the sandy areas in the Negev (Meir et al. 1992). Post-migration

Figure 1a. Location map of the study area in the Southeastern Mediterranean region: enlarged map.

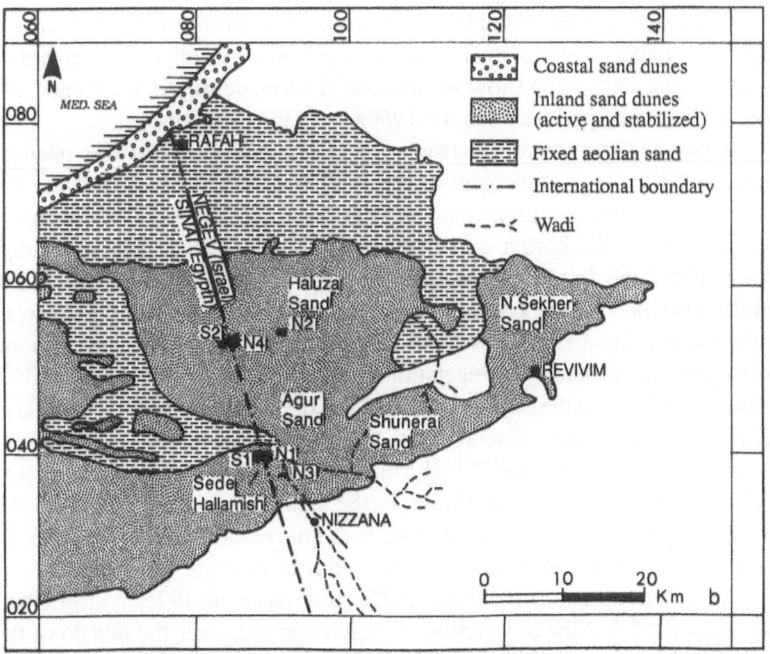

Figure 1b. Location map of the study area in the Southeastern Mediterranean region: location of the six sample sites and sources of rainfall data.

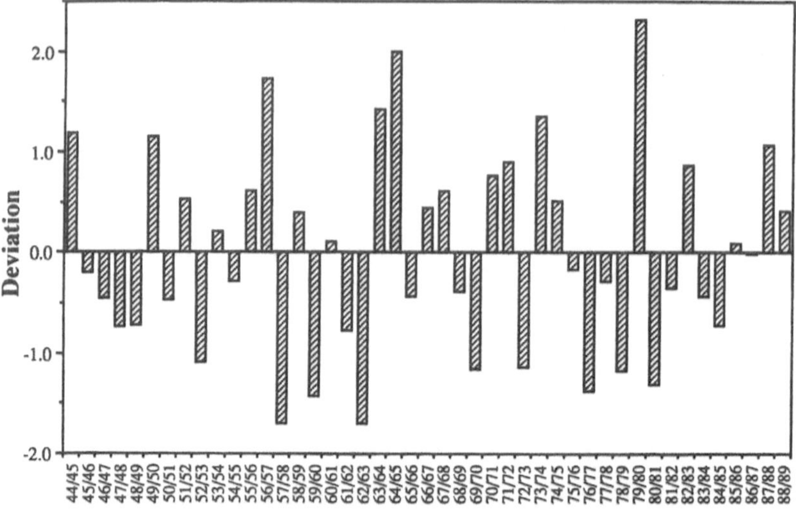

Figure 2. The yearly deviation of rainfall in Revivim for the period 1944/45-1988/89. The deviation values are the number of standard deviation units each year's rainfall value is from the annual average rainfall which is 104 mm.

pressures in the northern Sinai led to the removal of shrubs by the Bedouin for firewood, shelter, cultivation and livestock grazing. Quantitative analyses of shrub volume since 1945 show an increase in shrub density in the Negev and a decrease in the Sinai between the 1950s and 1968 (Figure 3).

In late 1967, the border between the Negev and the Sinai was opened, allowing for the free movement of the Bedouin. The Negev side along the border was severely overgrazed and the vegetation destroyed. As seen in Figure 3, both sides of the border experienced continuous decrease in vegetation density after 1968. In 1982, the political border was reestablished and Bedouin activity curtailed on the Negev side. Shortly after, the vegetation began to recover in the degraded Negev areas (Figure 4). This land use change in the Negev brought the density of vegetation in 1989 to its highest value since 1945. On the other hand, in the Sinai, vegetation density has continued to decrease at an alarming rate, and is presently at its lowest value since 1945 (Figure 3).

As seen in Figures 2 and 3, high variation in rainfall is not well correlated to changes in the number of shrubs in the same years, thus leading to the conclusion that the amount of moisture is not necessarily a limiting factor in vegetation growth on sand dunes (Tsoar and Møller 1986, Tsoar 1990).

Several investigators (Charney 1975, Otterman et al. 1975, Waisel 1986) have suggested that the destruction of vegetation increases the albedo of the sand, thus lowering the soil temperature and creating a stable atmosphere with lower possibility of rainfall. This has been refuted by field measurements by

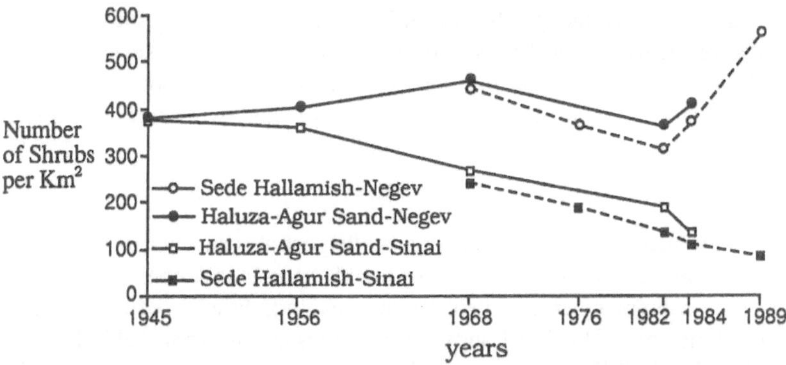

Figure 3. Changes in shrub density in two locations on both sides of the border between 1945 and 1989, as determined from aerial photography analysis (after Meir et al. 1992).

Figure 4. A photograph of a recovered linear dune on the Negev side of the border taken on October 2, 1992. The darker appearance of the dune surface in the background is a result of the formation of the biogenic crust.

Balling (1988). Thus, it appears that vegetation can grow back naturally as human and animal pressures are removed.

In addition, the presence or absence of vegetation influences dune geomorphology. During the 1968-1982 period, straight linear vegetated dunes in the Negev changed into linear-braided and sinuous seif (sharp-crested) dunes with no vegetation (Tsoar and Møller 1986, Tsoar 1989).

An important change that occurred in the area during the period of vegetation recovery after 1982 is the establishment of filamentous cyanobacteria on the sand surface (Figure 4). This organism (which embodies glutinous gossamer) forms a crust that is composed mostly of dust-sized particles trapped by its gelatinous surface (Danin et al. 1989).

Methods

Data were analyzed using the Landsat multispectral scanner (MSS) taken on three different dates: July 20, 1984; July 13, 1987; and June 8, 1989. Only images taken in the summer were selected in order to exclude the effect of annual plants and soil moisture. The four Landsat-MSS bands are shown in Table 1.

The ERDAS-PC image processing system was used to perform all the image analysis for this study. The three Landsat-MSS scenes were rectified to the Cassini-Soldner (Israeli grid) map projection. Ground control point locations were digitized from several available maps on an Altek ACT 34-1 digitizing tablet using MICRODIJ software.

From each of the three whole Landsat-MSS scenes, 601-rows by 601 column images were rectified. These included the Sede Hallamish study area. From these, three smaller (437 rows by 371 columns) co-registered images were cut to include the most pertinent areas while excluding those with no data (this was done because the study area was at the extreme northwest of all three scenes).

Photochromatic aerial photographs of the Sede Hallamish study area were obtained for 1968, 1982, and 1989. All photos were video digitized using a Sony SSC-D5 Video Camera equipped with a Nikon Nikkor 28 mm lens. The 1982 and 1989 photo sets were rectified to the Soldner-Casini projection with 8-m and 79-m pixel size, respectively, and 1.5 RMS tolerance.

From visual inspection of the rectified 1989 photo images and field observations, sample sites were located that were representative of the vegetated Negev side in Sede Hallamish and the unvegetated Sinai in Egypt. These were labeled samples N1 and S1, respectively. Polygons enclosing the samples were digitized on the screen and the map coordinates of the vertices recorded. These coordinates were matched as closely as possible to the coarser resolution Landsat images to produce polygons containing N1 and S1 samples on each working Landsat-MSS image.

Further visual examination of the 1989 image resulted in four more samples being digitized on screen (see Figure 1):

N2—an area in the middle Haluza dune field that appears darker than the southern Sede Hallamish dune fields (N1).

N3—a Negev dune area composed of three small polygons just west of the wadi Nizzana that we suspect is vegetated because of its darker appearance than the surrounding areas (high MSS band 7 reflectance versus lower reflectance in MSS 4 and 5, in the 1989 image). The aerial photo also shows this area to be darker than adjacent dune areas.

Table 1
Landsat-MSS bands

MSS band	Wavelength μm	Color
4	0.5-0.6	Green
5	0.6-0.7	Red
6	0.7-0.8	NIR
7	0.8-1.1	NIR

S2—a bare dune area in the northern Sinai that represents an area that differs visually from the first Sinai sample south of it by being less yellow.

N4—a Negev dune area opposite S2.

In situ upwelling radiance of sand and vegetation in the N1, N2, N3, S1, and S2 sites were measured using the LICOR-1800 field spectrometer. This instrument had been set up for 2-μm wavelength increments between 0.4 μm and 1.1 μm. Readings from the hand-held spectrometer were performed from 1 m high on a clear day on October 2, 1992. The reflectance values were calculated by relating the target radiances to the downwelling irradiation as measured by a cosine-corrected receptor.

Atmospheric correction was not applied to the MSS scenes. Such a correction is usually done by subtending the lowest brightness value in an area that does not normally reflect much radiation from each pixel for each band, using a different low value for each band. The dark area in the image is usually in deep shadow or an open body of water. These were not found in the images, since the study area is in an arid region with high sun angles.

ANALYSIS AND RESULTS

Statistical Analysis of the Brightness Values of the Sample Sites

Descriptive statistics, including minimum, maximum, mean, standard deviation, and covariance matrix, were extracted for each of the six sample sites on each of the three Landsat-MSS images. These statistics were used for two types of comparisons:

(1) Comparison of means between sample pairs (all combinations) on the same satellite image using the test described by Silk (1979, p. 169). One-tailed tests of significance were used when comparing Negev samples with Sinai samples (the null hypothesis in each case being that there is no difference between the brightness values of the samples in the comparison, and the alternate hypothesis, that the Negev side brightness values are less than the Sinai side). Two-tailed tests were used in comparing two Sinai samples or two

Table 2
Difference in Z-scores of 1984 image samples. Band numbers are in parentheses. Critical values of Z: $\alpha = 0.05$ (One-tailed: 1.65, two-tailed: 1.96) $\alpha = 0.01$ (One-tailed: 2.33, two-tailed: 2.58).

Sample	N2	N3	N4	S1	S2
N1	3.72 (4)	1.10 (4)	9.47 (4)	-5.19 (4)	-47.18 (4)
	19.49 (5)	7.99 (5)	-7.27 (5)	-6.01 (5)	-17.38 (5)
	22.68 (6)	9.27 (6)	-11.43 (6)	-4.44 (6)	-16.68 (6)
	22.92 (7)	10.47 (7)	-11.46 (7)	-3.62 (7)	13.26 (7)
N2		-0.09 (4)	11.42 (4)	7.81 (4)	42.32 (4)
		1.38 (5)	9.33 (5)	22.50 (5)	29.61 (5)
		0.72 (6)	7.86 (6)	23.07 (6)	33.89 (6)
		0.62 (7)	6.61 (7)	22.41 (7)	30.84 (7)
N3			4.43 (4)	2.68 (4)	13.06 (4)
			5.26 (5)	9.30 (5)	10.57 (5)
			4.39 (6)	10.47 (6)	12.72 (6)
			3.98 (7)	11.45 (7)	14.33 (7)
N4				4.57 (4)	-23.23 (4)
				-10.72 (5)	-15.79 (5)
				-13.25 (6)	-20.82 (6)
				-12.61 (7)	-18.67 (7)
S1					-33.52 (4)
					-6.32 (5)
					-6.52 (6)
					-6.69 (7)

Negev samples because there was no a priori knowledge. A BASIC program was written to perform the calculations. Results are summarized in Tables 2, 3, and 4.

(2) Comparison of means of the same samples over time according to the test suggested by Wonnacott and Wonnacott (1977, p. 217). A SAS program was written to compute the Z-scores. Results are summarized in Table 5.

Because of the lack of multispectral satellite data from the transitional 1982 period, video-digitized panchromatic aerial photographs were used as an estimate for the existing conditions during that time. The 1982 and 1989 digital photographs, which included samples N1 and S1, were rectified as before, but were resampled to match the 79-m spatial resolution of the MSS images. Sample statistics for N1 and S1, based solely on the panchromatic waveband (0.40-0.70 μm), were extracted because these were the most representative sites and the available aerial photographs did not include all six of the sample sites. This was done with the understanding that statistics from the panchro-

Table 3
Difference in Z-scores of 1987 image samples. Band numbers are in parentheses. Critical values of Z are the same as in Table 2.

Sample	N2	N3	N4	S1	S2
N1	12.28 (4)	-2.12 (4)	6.55 (4)	-10.32 (4)	-42.38 (4)
	22.21 (5)	5.04 (5)	12.40 (5)	-17.98 (5)	-25.53 (5)
	23.34 (6)	5.58 (6)	-13.88 (6)	-8.02 (6)	-8.01 (6)
	21.11 (7)	5.73 (7)	-16.21 (7)	-12.59 (7)	-23.70 (7)
N2		-6.14 (4)	15.91 (4)	18.42 (4)	44.79 (4)
		-6.40 (5)	8.17 (5)	35.70 (5)	40.70 (5)
		-8.92 (6)	6.02 (6)	26.63 (6)	26.63 (6)
		-7.84 (7)	3.07 (7)	28.92 (7)	39.60 (7)
N3			-0.05 (4)	1.53 (4)	11.18 (4)
			-1.44 (5)	11.19 (5)	12.40 (5)
			-3.87 (6)	7.27 (6)	7.27 (6)
			-5.34 (7)	12.93 (7)	17.82 (7)
N4				-4.00 (4)	-30.56 (4)
				-25.34 (5)	-29.56 (5)
				-16.54 (6)	-16.54 (6)
				-24.33 (7)	-33.37 (7)
S1					-23.84 (4)
					-5.02 (5)
					0.00 (6)
					-6.11 (7)

matic imagery could not be compared directly with the multispectral imagery because of the differences in spectral range, but that they might be of value in comparing sample sites on the same image. Results show that the albedo of S1 and N1 samples were not statistically different in 1982 but were very different in 1989.

The N3 sites showed low reflectance, and along with the 1989 data, the lowest values (Figure 5). Field observations indicate that all N3 sites were in interdune areas, covered with wilting annuals and a high surficial silt and clay content. Examination of the N3 sites from nine aerial photographs during different years, revealed the appearance of shrubs only in years with above average rainfall. The volume of annuals in 1989 was higher than in 1987 and 1984 because of the higher amounts of rainfall during the three months of the rainy season. The N3 annuals are found exclusively where soils have large percentages of fines. Owing to the textural characteristics of dune sands, annuals cannot even get established during above average precipitation years (Tsoar 1990).

Table 4
Difference in Z scores of 1989 image samples. Band numbers are in parentheses. Critical values of Z are the same as in Table 2.

Sample	N2	N3	N4	S1	S2
N1	19.29 (4)	31.67 (4)	-3.91 (4)	-7.47 (4)	-48.77 (4)
	26.11 (5)	38.21 (5)	-20.83 (5)	-20.08 (5)	-20.20 (5)
	21.26 (6)	24.79 (6)	-14.09 (6)	-4.09 (6)	-8.14 (6)
	25.18 (7)	23.04 (7)	-18.04 (7)	-19.01 (7)	-29.13 (7)
N2		17.66 (4)	11.96 (4)	20.72 (4)	60.46 (4)
		13.26 (5)	7.06 (5)	41.15 (5)	41.23 (5)
		7.73 (6)	4.93 (6)	20.70 (6)	28.61 (6)
		1.18 (7)	4.44 (7)	39.24 (7)	48.80 (7)
N3			25.72 (4)	31.93 (4)	59.58 (4)
			20.50 (5)	51.51 (5)	51.57 (5)
			11.37 (6)	24.47 (6)	30.63 (6)
			5.13 (7)	35.67 (7)	43.01 (7)
N4				-9.42 (4)	-41.01 (4)
				-38.77 (5)	-38.88 (5)
				-15.20 (6)	-20.86 (6)
				-31.60 (7)	-39.35 (7)
S1					-25.24 (4)
					-0.74 (5)
					-2.23 (6)
					-6.25 (7)

Values of Brightness Contrast Between the Negev and Sinai

To make comparisons between the Negev and Sinai samples on satellite imagery and aerial photography, adjacent Sinai and Negev areas, S1-N1 and S2-N4, were selected for further study. Figures 6 and 7 show the satellite image raw sample and average brightness values for these locations. The brightness curves are all similar and indicate very low values in MSS 4, very high in MSS 5 and 6, and somewhat less in MSS 7. As the prior visual analyses from the 1960s and 1970s showed, the Sinai side has higher brightness than the Negev side. The figures show that this is true in all three images in this study in each MSS band. The differences seem to be greater in the N4-S2 comparison than in the N1-S1.

Comparisons of Samples from the Same Image

1984 Satellite Image: The results of the difference of the Z-scores of the means are shown in Table 2. There were statistically significant differences between

Table 5
Difference of means Z-scores for the same samples over time.

Period	1984-1987				1987-1989				1984-1989			
Band	4	5	6	7	4	5	6	7	4	5	6	7
Sample												
N1	-5.79	-10.52	16.82	6.42	-0.85	0.71	-14.23	7.11	-5.66	-9.33	-4.16	15.32
N2	-13.41	-9.74	7.80	5.56	-9.48	-8.15	-12.75	-2.10	-22.87	-20.13	-6.42	-0.80
N3	1.90	2.75	10.40	10.49	-16.58	-17.71	-23.00	-10.57	-16.72	-13.35	-6.42	0.13
N4	-7.66	-9.93	5.52	-0.42	-8.19	-8.68	-12.86	-1.10	-14.97	-27.30	-10.79	-2.14
S1	1.98	3.23	8.92	17.55	-0.12	4.81	-5.86	12.30	1.00	6.10	-0.72	31.73
S2	-0.07	0.00	5.43	21.19	9.80	0.00	-4.82	24.58	12.43	0.00	-3.08	42.10

almost all the samples at an $\alpha = 0.01$, using a one-tailed test for comparisons of a Sinai and Negev sample or a two-tailed test for two Sinai or two Negev samples. Qualitative analysis of the Z-scores shows that MSS 5, 6, and 7 best differentiate two Negev samples except for N2 and N3, which were not significantly different in any bands. N3 and N4 showed smaller, but statistically significant Z-scores. The two Sinai samples were significantly different in all four bands, especially MSS 4, with S1 being less reflective than S2.

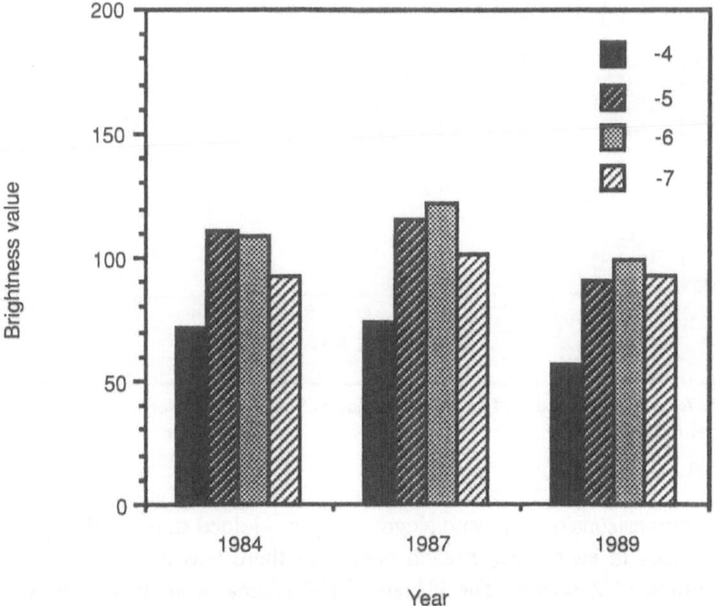

Figure 5. Changes in brightness values in N3 in the four bands (Table 1) for the three years of the investigation.

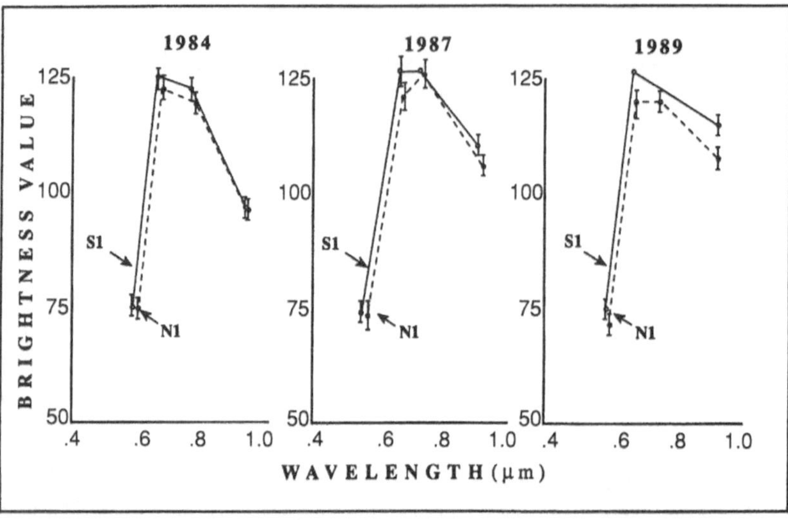

Figure 6. Average and standard deviation of adjacent Sinai (S1) and Negev (N1) locations from Landsat-MSS bands 4-7.

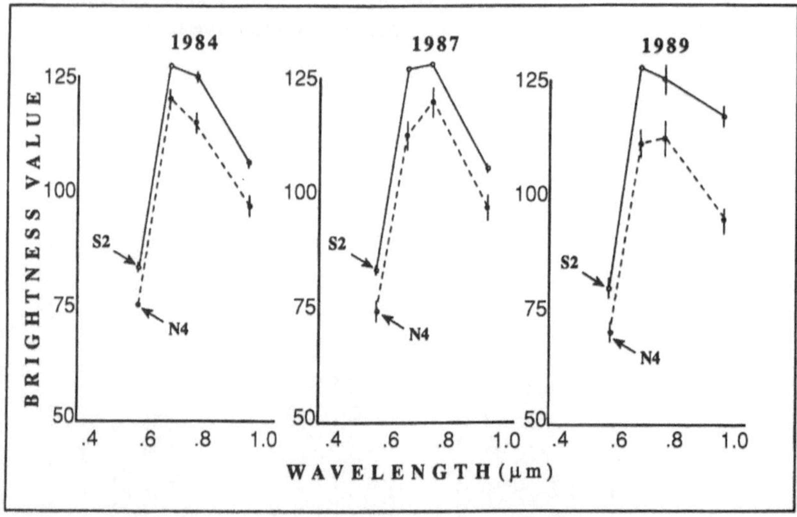

Figure 7. Average and standard deviation of adjacent Sinai (S2) and Negev (N4) locations from Landsat-MSS bands 4-7.

Comparisons of Sinai and Negev samples yielded statistically significant differences in each case in each band, yet there was a great range in the magnitude of Z-scores. The N1 and S1 Z-scores in all bands showed the smallest values. The largest values were exhibited by N1 and S2 and N2 and S2, both especially in MSS 4, and N4 and S2.

1987 Satellite Image: There were statistically significant differences between all sample combinations in all bands (Table 3). Comparison of the Negev samples shows the largest differences between N1 and N2 in all bands except MSS 4, between N4 and N1, except MSS 4, and N4 and N2, mostly in MSS 4. The combinations of N4 and N3, N3 and N2, and N1 and N3 showed the smallest magnitude of Z-scores. The comparison of Sinai samples was most significant in MSS 4, with no difference in MSS 6. The greatest differences between the Negev and Sinai samples were in comparing N1 and S2 (especially MSS 4, but not MSS 6), N2 and S1, N2 and S2, N4 and S2, and N4 and S1 (except MSS 4). The smallest differences were found in S1 and N3 (especially MSS 4), and N1 and S1 (except for MSS 5).

1989 Satellite Image: Differences between samples in almost all situations seemed to be most pronounced in the 1989 image (Table 4). Negev samples were well differentiated in all bands except for N2 and N3 (MSS 7). The Sinai samples were again most distinct in MSS 4. The differences between the Negev and Sinai samples were most pronounced in combinations N1-S1 (MSS 5 and MSS 7), N1-S2 (especially MSS 4, 5, and 7), and N2-S2, N2-S1, N3-S1, N3-S2, N4-S1, N4-S2.

Comparisons of the Same Samples Over Time

MSS Band 7: Almost all the band-by-band comparisons between the same samples over time showed statistically significant Z-scores at $\alpha = 0.01$ (Table 5). Qualitative analyses of the Z-scores revealed that there were larger differences in MSS 7 in Sinai samples than in Negev samples over time. The Sinai samples were brighter in this band with increasing time. The Negev samples generally increased in brightness over time. However, N2, N3, and N4 showed some decreases in brightness in some of the time periods. The biggest increase in brightness magnitude for the Negev occurred during 1984-1987, and for N1, during 1984-1989.

MSS Band 6: In studying changes in MSS 6 for each sample over time, the magnitude of changes is greater in the Negev samples than in the Sinai samples. All samples showed higher reflectance in 1987 compared to 1984, and lower reflectance in 1989 than in 1987 or 1984. The Negev samples showed a decrease in brightness values between 1984 and 1989, while one of the Sinai samples showed no significant change, and the other, a smaller decrease in reflectance values than any of the Negev samples.

MSS Band 5: All the Negev samples except N3 showed highly significant decreases in brightness in MSS 5 from 1984 to 1987, while S1 showed an increase in brightness and S2 remained unchanged. Between 1987 and 1989, the results were similar to the above, except that N1 changed insignificantly and N3 showed much decreased brightness. Overall, between 1984 and 1989, the Negev samples showed a great decrease in brightness values, while S2 remained unchanged and S1 moderately increased in brightness.

MSS Band 4: S1, S2, and N3 showed no significant changes in brightness between 1984 and 1987. The Sinai samples and N1 showed no significant change between 1987 and 1989, while, for the entire period of 1984-1989, only S1 showed no change. The rest of the Negev samples showed a decrease in MSS 4 during the period of vegetation recovery.

Field Measurements

Observations and results of the spectral analysis indicate five spectral reflectance curves for each of the surface samples (Figure 8). Two spectra show bare sand in S1 and S2 and biogenic crusts in N1 and N2. The N3 site is dominated by dense wilting annuals over crust. The sand spectra have low reflectance in the blue band, a steep slope to about 0.6 µm, then a decreasing slope to about 0.8 µm, and a shallow dip from 0.8 to 1.1 µm. This wavy shape is somewhat similar to the Landsat-MSS spectra presented in Figures 6 and 7.

The biogenic crust spectra are generally characterized by the same convex shape in the visible band and concave shape in the NIR band. However, there is a slight dip of the crust spectra from 0.6 to 0.7 µm owing to the presence of biogenic microphytes such as cyanobacteria (blue-green algae), as well as lichens (Danin 1991). In each sand and crust pair, the spectra curves cross each other around 0.55 µm. The sand has a higher reflectance in the green, red, and NIR regions of the electromagnetic spectrum, and the biogenic crust has higher reflectance in the blue band.

The higher reflectivity of the crust in the blue band is probably present because of the existence of silt, clay, and cyanobacteria. Comparison between sites shows that in the red and NIR bands, the S1 and N1 spectra are higher than the respective S2 and N2 spectra. Field observations have confirmed the yellower appearance of the S1 sand relative to the S2 sand found to the north. This yellower semblance of the sand comes from a clayey envelope around the sand grains. It may indicate a greater age of sand or post-depositional modifications (Gardner and Pye 1981). On the other hand, S2 and N2 have high reflectance values in the blue band. The N3 crust has the lowest reflectance of all.

Figure 8 shows the reflectance of the *Artemisia monospérma*, which is the most widespread shrub in the study area. The dip in the red band (from chlorophyll absorption) is not pronounced and the spectra shape is similar to that of bare soil. Also, the overall vegetation spectra is much lower than the sand spectra, especially in the NIR, where it does not exceed 15%, while for bare sand it is more than double.

DISCUSSION AND CONCLUSIONS

The Negev side of the border recovered very quickly once protected from uncontrolled use by Bedouin nomads after April 1982. This quick recovery starts as the endemic vegetation begins to re-establish in the sandy desert. Biogenic crusts are formed when the vegetation cover reaches about 20%-30% of the area. Vegetation and the biogenic crust then act as a trap for aeolian dust particles. Both are thus responsible for the formation of the dark gray crust.

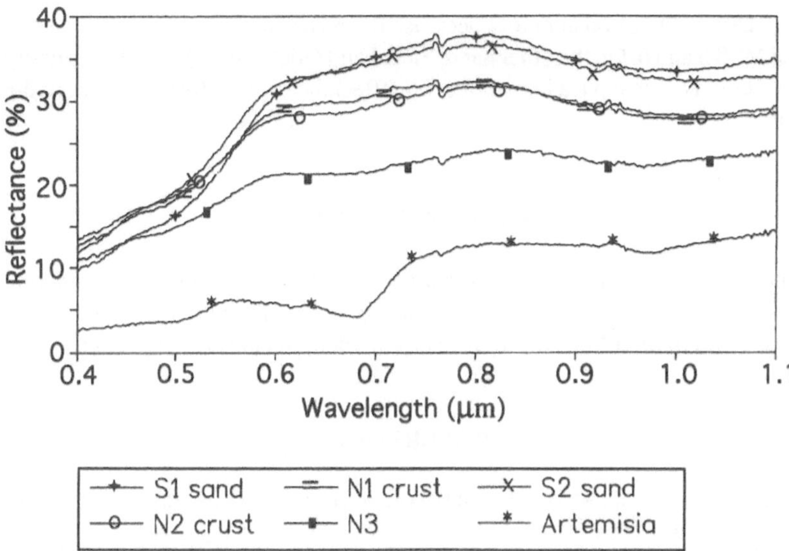

Figure 8. Reflectance of *Artemisia monospérma*, surface sand, and crust samples measured in the field by hand spectrometer.

There is a small difference between the shape of the spectra of the Negev vegetated sand dunes and the adjacent Sinai grazed sand dunes (Figures 6 and 7). Likewise, very little difference in spectra exists between the bare sand and crusted sand (Figure 8). The sandy desert vegetation, as exemplified by the *Artemisia monospérma*, has a relatively low reflectance, especially in the NIR band. Consequently, the effect of vegetation on the general reflectance of the Negev is not perceptible (compare Figures 6 and 7 with Figure 8). Hence, it is concluded that the well-known contrast in spectra between the Sinai and the Negev is not a direct result of vegetation cover but is primarily caused by the biogenic crust cover.

The difference of the Z-scores of the means (Tables 2, 3, and 4) shows an increasing difference in the red photosynthetic signal (MSS 5) between the Negev and Sinai from 1984 to 1989. This indicates that the biogenic crust with its trifling photosynthesis, is spreading. A similar increase in difference of the NIR (MSS 7) between the two sides of the border reflects the expansion of grazing in the Sinai (bare sand dunes are reflected in a high percentage in the MSS 7 band) and the increase in crust cover in the Negev (biogenic crust decreases its NIR reflectance).

The above conclusions are supported by the results of the difference of the Z-scores of the means over time (Table 5). The Sinai samples primarily show a significant increase in the NIR band (increasing grazing). For the Negev samples, there is a significant decrease in MSS 5 over time, indicating the increase of the biogenic crust over time.

Dry annuals considerably decrease the reflectance values as observed in the MSS 5 and 6. On the other hand, Siegal and Goetz (1977) found dead or dry vegetation not to have a large effect on reflectance values from the underlying soil.

ACKNOWLEDGMENTS

The authors are much obliged to Karl Szekielda for his time and stimulating discussion, to Dan Blumberg and Robert Balling for comments on the manuscript, and to Sara McLafferty for her kind help in analyzing the data.

REFERENCES

Adams, R., Adams, M., Willens, A., and Willens, A. (1978) *Dry Lands: Man and Plants*. The Architectural Press, London.

Balling, R. C. (1988) The climatic impact of a Sonoran vegetation discontinuity. *Climatic Change*, v. 13, p. 99-109.

Charney, J. G. (1975) Dynamics of deserts and drought in the Sahel. *Quaternary Journal of the Royal Meteorological Society*, v. 101, p. 193-202.

Danin, A. (1983) *Desert Vegetation of Israel and Sinai*. Cana, Jerusalem.

Danin, A. (1991) Plant adaptation in desert dunes. *Journal of Arid Environments*, v. 21, p. 193-212.

Danin, A., Bar-Or, Y., Dor, I. and Yisraeli, T. (1989) The role of cyanobacteria in stabilization of sand dunes in southern Israel. *Ecologica Mediterranea*, v. 15, p. 55-64.

Gardner, R., and Pye, K. (1981) Nature, origin and palaeoenvironmental significance of red coastal and desert dune sands. *Progress in Physical Geography*, v. 5, p. 514-534.

Helldén, U. (1991) Desertification—time for an assessment? *Ambio*, v. 20, p. 372-383.

Lowman, P. D. (1966) The Earth from orbit. *National Geographic*, v. 130, p. 645-671.

Meir, A., Tsoar, H., and Khawalde, O. (1992) *The Impact of Changes in Bedouin Land Use on the Physical Environment Along the Egyptian-Israeli Border Since the 1940s*. Research Report submitted to Jo Alon Regional Study Center, Department of Geography, Ben-Gurion University of the Negev, Beer Sheva. (in Hebrew)

Muehlberger, W. R., and Wilmarth, V. R. (1977) The shuttle era: A challenge to the Earth scientist. *American Scientist*, v. 65, p. 152-158.

Noy-Meir, I., and Seligman, N. G. (1979) Management of semi-arid ecosystems in Israel. In B.H. Walker (ed.) *Management of Semi-Arid Ecosystems*. Elsevier, Amsterdam, p. 113-160.

Otterman, J. (1974) Baring high-albedo soils by overgrazing: A hypothesized desertification mechanism. *Science*, v. 186, p. 531-533.

Otterman, J. (1981) Satellite and field studies of man's impact on the surface in arid regions. *Tellus*, v. 33, p. 68-77.

Otterman, J., Manes, A., Rubin, S., Alpert, P., and Starr, D. O. (1990) An increase of early rains in southern Israel following land-use change? *Boundary-Layer Meteorology*, v. 53, p. 333-351.

Otterman, J., and Tucker, C. J. (1985) Satellite measurements of surface albedo and temperatures in a semi-desert. *Journal of Climate and Applied Meteorology*, v. 24, p. 228-235.

Otterman, J., Waisel, Y., and Rosenberg, E. (1975) Western Negev and Sinai ecosystems: Comparative study of vegetation, albedo, and temperatures. *Agro-Ecosystem*, v. 2, p. 47-59.

Prince, S. D., Justice, C. O., and Los, S. O. (1990) *Remote Sensing of the Sahelian Environment*. Technical Centre for Agricultural and Rural Cooperation, Brussels.

Reichelt, R. (1989) Desertification in the Sahel: The exposing of the "old erg" of an earlier Sahara. *Natural Resources and Development*, v. 30, p. 104-113.

Siegal, B. S., and Goetz, A.F.H. (1977) Effect of vegetation on rock and soil type discrimination. *Photogrammetric Engineering and Remote Sensing*, v. 43, p. 191-196.

Silk, J. (1979) *Statistical Concepts in Geography*. George Allen & Unwin, London.

Tsoar, H. (1989) Linear dune-forms and formation. *Progress in Physical Geography*, v. 13, p. 507-528.

Tsoar, H. (1990) The ecological background, deterioration and reclamation of desert dune sand. *Agriculture, Ecosystems and Environment*, v. 33, p. 147-170.

Tsoar, H., and Møller, J. T. (1986) The role of vegetation in the formation of linear sand dunes. In W. G. Nickling (ed.) *Aeolian Geomorphology*. Allen & Unwin, Boston, p. 75-95.

Waisel, Y. (1986) Interactions among plants, man and climate: Historical evidence from Israel. *Proceedings of the Royal Society of Edinburgh*, v. 89B, p. 255-264.

Warren, A., and Harrison, C. M. (1984) People and the ecosystem: Biogeography as a study of ecology and culture. *Geoforum*, v. 15, p. 365-381.

Wonnacott, T. H., and Wonnacott, R. J. (1977) *Introductory Statistics for Business and Economics*. John Wiley & Sons, New York.

13 A REVIEW OF THE EFFECTS OF SURFACE MOISTURE CONTENT ON AEOLIAN SAND TRANSPORT

Steven L. Namikas and Douglas J. Sherman
Department of Geography
University of Southern California

ABSTRACT

Over the past several decades, a number of studies have shown that intergranular cohesion associated with the presence of moisture significantly increases the critical shear velocity required to initiate motion in sand grains, and decreases transport rates. This paper examines currently available models of moisture effects and compares model predictions for several hypothetical situations. Model predictions exhibit considerable disagreement regarding the magnitude of moisture effects. For 0.27-mm sands, predicted increases in threshold shear velocity associated with a 1% moisture content ranged from about 8% to 148% of the expected dry threshold velocity, and with 4% moisture increased to 47%-206% of the dry value. Based on the predicted threshold shear velocities, the expected transport rates at a 1% moisture content under a 0.50 m s^{-1} shear velocity range from no transport to more than 100 kg m^{-1} hr^{-1}.

INTRODUCTION

Attempts to understand and predict the behavior of aeolian sediment transport systems require specification of several key parameters describing wind and sediments. For ideal aeolian environments (i.e., uniform, clean, dry, non-cohesive sands), the required parameters are typically shear velocity and mean sediment diameter. Most modern aeolian sediment-transport models also require a threshold term for the initiation of sediment motion. Specification of this term is straightforward under ideal conditions, requiring only mean sediment diameter and sediment density. However, conditions in natural environments are seldom so simple. In particular, surficial moisture is often cited as having a potentially significant impact on the initiation of movement of sand by wind (Gares 1987, Sarre 1987, Kroon and Hoekstra 1990, Nickling and Davidson-Arnott 1990, Pye and Tsoar 1990, Sherman and Hotta 1990, Nordstrom and Jackson 1992), yet the current state of knowledge regarding this phenomena is poor. The pertinent literature is sparse, and includes aspects of studies dealing with soils (e.g., Bisel and Hsieh 1966, Azizov 1977) despite the fact that the moisture retention characteristics of such materials differ substantially from those of sand. Virtually all of the research on moisture effects has relied on experimentation, although some models have strongly theoretical roots. The purpose of this paper is to discuss the effects of moisture on the

Desert Aeolian Processes. Edited by Vatche P. Tchakerian. Published in 1995 by Chapman & Hall, London. ISBN 978-94-010-6519-1

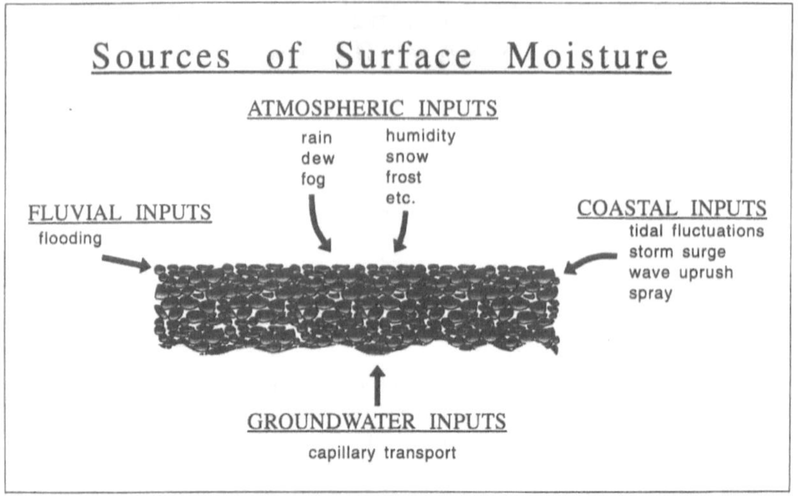

Figure 1. Sources of ground surface moisture.

initiation of sand movement by wind, and critically review past work and attempts to model these effects.

The presence of water in surface sediments is important because moisture increases the shear velocity required to initiate motion, and because it occurs frequently in most aeolian environments. There are several origins for moisture in surficial sands (Figure 1). The atmosphere is the primary source in arid regions, where moisture is added by precipitation, or by condensation of water vapor directly onto the surface. The atmosphere is also a source of moisture for coastal areas (marine or lacustrine), but substantial quantities of water are also added by wave uprush, overwash, or water-level fluctuations associated with tides or surges. In fluvial systems, including glacial outwash plains, surface sediments are left damp by receding flood waters or by atmospheric sources. Further, in any of these environments, moisture can migrate to surficial sediments via capillary flow from soil moisture or a near-surface water table.

Moisture content, w, is almost always expressed as a percent by weight of a sediment sample:

$$w = \frac{(w_s - w_d)}{w_d} \tag{1}$$

where w_s is the total sample weight, and w_d is the dry weight. Moisture content is usually determined by weighing an initially moist sample, followed by oven-drying and re-weighing of the sample to determine the weight of water evaporated (Hanks 1992). For this method it is desirable to restrict sediment sample depth to one or two mm in order to try to characterize conditions best representing the surface (e.g., Sarre 1988). Although there is no standard practice for obtaining moisture samples, it is imperative to preserve the sample in an airtight container until weighing is possible.

Moisture content can also be reported as a percent by volume (w_v). The only unambiguous method is:

$$w_v = \frac{(w_s - w_d)/\rho_w}{w_d/\rho_s} \tag{2}$$

where ρ_w is water density and ρ_s is sediment density. There are other methods of estimating moisture volumetrically. However, they involve complex procedures that require measurement or estimation of both the sediment porosity and volume of pore space occupied by water. As a consequence, this measure of moisture content is rarely used in studies of wind blown sand.

The upper limit on moisture content is controlled by sediment composition (density), and porosity associated with grain packing. If a porosity of 40% is assumed for quartz sands, the maximum possible moisture content is about 25%. Sediments that have been completely air-dried will have an apparent moisture content of 0%. However a molecular film of water, about 0.1 μm thick, will remain adhered to the grains (e.g., Terzaghi and Peck 1967). It is assumed that this water does not effect sediment transport processes.

It is presumed that surface tension, *T*, is the agent that imparts a cohesiveness between sediment grains when moisture is present in the pore spaces between grains. The surface of the water behaves as an elastic membrane as water molecules exert an attraction toward each other. If the edges of the membrane are stretched, as in capillary rise, its area is increased by an amount *A* as additional energy, *E*, is stored in the surface, $E = T \bullet A$. For water at 20°C, the surface tension is 0.075 N m^{-1}, and this value decreases with increasing temperature (Duncan et al. 1970). Capillarity results from the molecular attraction, adhesion, of the water to the surrounding solids (sand grains in this instance). Because this attraction exceeds the attraction between water molecules, cohesion, the edge of the water will be displaced away from the body of liquid until the adhesive forces and the cohesive forces are in equilibrium with gravity force. The adhesive attraction is greater for smaller contact radii:

$$h = \frac{2\sigma}{r\rho_w g} \tag{3}$$

where *h* is the height of water rise, *r* is the radius of the contact opening, and *g* is gravity (e.g., Buckman and Brady 1969). Therefore, in sediments, adhesion of the water molecules to the grains is a primary bonding process, and surface tension holds the water together.

Displacement of the water—solid contact above the still water level results in a lowered pressure within the liquid body—in effect, a suction which acts on the area of solid in contact with the liquid. Haines (1925) showed that the

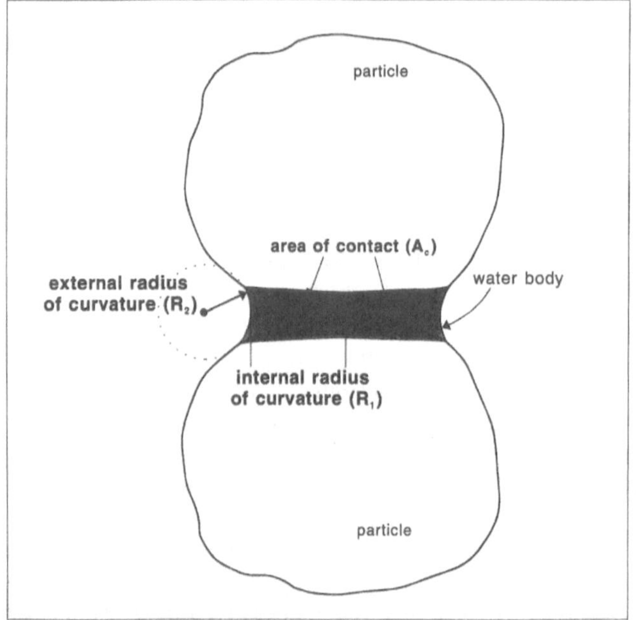

Figure 2. Definition sketch of parameters affecting capillary force generated by the presence of water in particle interstices.

magnitude of the capillary force (F_c) acting on spherical particles can be expressed as a function of particle-water contact area (A_c) and the difference (P) between the internal pressure of the water and the atmospheric pressure exerted on its free surface, such that:

$$F_c = A_c \cdot P = \pi R_1^{\,2} \cdot T \frac{(R_1 - R_2)}{R_1 R_2} \tag{4}$$

where R_1 and R_2 are the internal and external radii of curvature of the water body, respectively (Figure 2). Fisher (1926) expanded the formulation by incorporating an additional term to account for the tension because of the air-water interface:

$$F_c = A_c \cdot P = \pi R_1^{\,2} \cdot T \frac{(R_1 - R_2)}{R_1 R_2} \cdot 2\pi R_1 T \tag{5}$$

In order to initiate sediment transport, the additional motion-inhibiting force resulting from capillarity must be offset by greater shear stress.

We can categorize efforts to model moisture effects as being based primarily on empirically defined relation, or derived first from theory associated with surface tension. Although the theoretical treatments of capillary forces in spherical particle systems outlined above have been available for more than 60 years, most efforts have employed the former approach. Despite nearly 40 years of discussion and research concerning moisture effects on aeolian transport, considerable disagreement still exists regarding the nature of these phenomena. One point of agreement, however, is that all but the earliest of the models are based on a core threshold term first described by Bagnold (1936). By considering the balance of forces acting on a grain, he showed that the threshold shear velocity (U^*t) which must be exceeded to overcome the weight force acting to hold a particle in place and initiate motion could be expressed as:

$$U_{*t} = A\sqrt{\frac{\rho_s - \rho}{\rho} gd} \tag{6}$$

where A is the square root of the Shields threshold parameter, ρ and ρ_s are the density of air and sediment, respectively, g is acceleration due to gravity, and d is mean grain diameter. The value of A is usually assumed to be 0.1, although it varies over a range from 0.1 (Bagnold 1936) to 0.2 (Lyles and Woodruff 1972). Attempts to describe the effect of moisture on the threshold of motion have started with this expression and added empirical or theoretical correction terms representing functions of sediment moisture content.

EMPIRICAL MODELS: THRESHOLD EFFECTS

Chepil (1956)

Chepil (1956) conducted one of the earliest laboratory investigations into the effects of moisture on aeolian transport. Although the primary focus was on soils, one of the samples involved was a "typical" dune sand. Moisture contents considered were quite low, with maximum levels approximating the content which could be retained by a given soil at a pressure of 15 atmospheres (permanent wilting point). In the case of the sand sample this amounted to a moisture content of about 1.3% by weight. Chepil considered such low moisture levels to be present mainly in the form of hygroscopic water, that is, thin films weakly bound to the grains by adsorption, with little or no free water in the form of cuffs or wedges surrounding particle contacts which could induce cohesion by exerting capillary forces on adjacent grains. Starting with a basic transport expression essentially identical to that developed by Bagnold (1936), he found that observed moisture effects could be modeled as:

$$q_w = C \cdot \left(\sqrt{\frac{\tau - \lambda}{\rho}}\right)^3 \tag{7}$$

where q_w is the wet sand transport rate, C is an empirical constant accounting for the size, density, and shape of the grains, τ is the shear stress at the bed, and λ is a "resistance" parameter, which represents the additional bed stress necessary to overcome the effects of moisture, and is a function of the "equivalent moisture" content w_e (the ratio of actual moisture content to the moisture content retained at a pressure of 15 atmospheres, the permanent wilting point). From wind tunnel experiments, an empirical expression was developed for the resistance parameter:

$$\lambda = 6(w_e)^2 \tag{8}$$

It should be noted that this relationship was based on observations of several different soil types with a lone "typical dune sand" sample, and there was considerable scatter in the data on a log-log plot.

Although Chepil (1956) was primarily concerned with moisture effects on aeolian transport rates, his formulation of the moisture effect as an increase in bed resistance to shear stress allows the resistance parameter to be readily incorporated into a threshold stress expression:

$$U_{*t_w} = \sqrt{\frac{\tau_t + \lambda}{\rho}} \tag{9}$$

where U_{*tw} is the wet threshold shear velocity, and τ_t is the threshold bed stress under dry conditions. This modification enables us to directly compare the Chepil model with subsequent work.

The use of a moisture content measure indexed to soil water tension was unique to Chepil (1956), until the recent work of Gregory and Darwish (1990). Based on theoretical considerations it appears to be a superior approach to the gravimetric moisture content generally used by other workers; the smaller pore diameters present in a fine or poorly sorted sand should produce larger capillary forces (and hence stronger cohesion) for a given gravimetric moisture content, than the larger pores associated with increasing grain size and sorting (see discussions of soil moisture retention in Marshall and Holmes 1988 and Hillel 1971, for example). Thus, the primary advantage of this approach results from implicit recognition of the fact that for a given gravimetric moisture content the tension binding soil particles and water together will differ for soils with different mean grain sizes, sorting, clay contents, etc. Unfortunately, there appear to be problems with the experimental data used to calibrate the resistance parameter.

Chepil reported that no motion occurred in dune sands with an equivalent moisture content of about 1% when exposed to wind speeds of 14 m s^{-1} measured 0.15 m above the surface. From his original data, for dune sand with a 1% equivalent moisture content the resistance parameter value was about 0.7 N m^{-2}, so that the initiation of motion would not occur with shear velocity less than about 0.75 m s^{-1}. This value is quite high compared to the findings of subsequent research.

Although this perhaps explains why Chepil's findings have generally not been considered in subsequent studies, it is uncertain why such anomalous results were obtained. It is possible that some cementation of the surface layer occurred as a result of wetting and drying of the small clay fraction (reported as <0.3%). However, Chepil reported soil samples were "thoroughly" mixed before each run, which should have broken up any aggregates. He also reported a 15-atmosphere soil moisture of 1.28% for the dune sample, significantly larger than 0.05%-0.2% gravimetric contents at 15 atm measured by McKenna-Neuman and Nickling (1989) for several fine to medium sand samples. The discrepancy may have been due to higher clay contents or a relatively poor degree of sorting in Chepil's sample, either of which would increase the moisture tension associated with a particular gravimetric content. If such was the case, his results would not be expected to be comparable to studies focusing on clean, well-sorted sands.

Belly (1964)

Belly (1964) investigated moisture content effects on threshold shear velocity in a wind tunnel, using 0.44-mm sand grains and moisture contents ranging from about 0.15% to 4.0% by weight. Surface moisture contents up to about 0.3% were generated by varying the water vapor content of the air. Hotta et al. (1984) suggest that this approach is not strictly comparable with other studies since only the surface layer of sand was wetted and the underlying sediment remained dry. However, higher moisture levels (0.6%-4.0%) were generated by directly wetting the sediment, a procedure commonly employed in other studies. Therefore, his results should be generally comparable at all but the lowest moisture contents. His best fit relationship between moisture content and threshold shear velocity was:

$$U_{*_{t_w}} = U_{*_t}(1.8 + 0.6 \log_{10} w) \tag{10}$$

where w is moisture content in percent. Both Belly (1964) and Johnson (1965) speculate that this relationship may be applicable over a wide range of grain sizes if moisture effects scale constantly with grain size. In fact, size, sorting, and shape of particles exert significant influence on capillary forces at a given gravimetric moisture content (Childs 1969, Hillel 1971), so the applicability of the relationship to sediments with different size and sorting characteristics is uncertain.

The propriety of the relationship postulated by Belly is questionable when applied for moisture contents higher than those in his test cases. Sarre (1987) reports from field observations that transport rarely takes place at moisture levels approaching 24% (near or at saturation). Yet extrapolation of the Belly model suggests that motion will occur at these moisture levels under shear velocities about 2.5 times greater than the dry threshold. This suggests that at some point between 4% moisture content and saturation the Belly expression will begin to underestimate moisture effects, but that point and the magnitude of potential underestimation are unknown.

Logie (1982)

Logie (1982) studied the effects of moisture in the laboratory, subjecting a medium-sand surface with varying moisture contents (ranging from about 0.2% to 17%, by weight) to a constant wind (14 m s^{-1} measured at 0.05 m above the bed), and measuring the time required for motion to occur. Below moisture contents of about 1.2% motion was instantaneous, suggesting that the wet threshold shear velocity was exceeded under these conditions. No data were available for moisture contents between 1.2% and 4%, yet Logie suggested that a moisture content below 4% will have insignificant effects on transport. However, this is a substantial contrast to the findings of all other laboratory studies, and is probably only relevant under the circumstances of this study. Logie's reported shear velocity of 0.88 m s^{-1} is well in excess of the 0.23 m s^{-1} dry threshold expected for the 0.24-mm grains employed in her study, and thus the wind would be expected to be competent to overcome some degree of additional resistance to motion, such as that associated with the presence of moisture.

Logie's work does make apparent the importance of evaporation in the initiation of motion when sediments are wet, although the results are insufficient to allow a direct relationship to critical shear velocity. High moisture levels inhibited sediment motion until evaporation lowered the threshold shear velocity below the shear velocity of the incident wind field. At high moisture levels (8%-16%) the initiation of movement was delayed for substantial periods of time (7-10 hours in some cases), but no attempt was made to simulate realistic atmospheric moisture conditions so that it is uncertain as to how well these data reflect the importance of evaporation in natural systems.

Hotta et al. (1984)

Horikawa et al. (1982) and Hotta et al. (1984) have reviewed several investigations of the effect of moisture published in the Japanese literature. Empirical studies of threshold effects included those by Nakashima and Suematsu (1976), and Tanaka et al. (1954). Based on laboratory investigations with 0.45-mm sediment and moisture contents up to 4%, Nakashima and Suematsu (1976) developed an empirical expression:

$$U_{*t_{15}} = \sqrt{gd\frac{\rho_s}{\rho}}\sqrt{\frac{B'}{A'}}$$
$$where: A' = -2.0x10^{-7} + 22.0x10^{-7}e^{0.39w} \qquad (11)$$
$$B' = 1.0x10^{-5}e^{-0.34w}$$

where U_{t15} is the threshold wind speed measured at an elevation of 0.15 m. This is the only extant expression which predicts that the threshold of motion increases exponentially with moisture content. An apparent weakness is evident when the effect of a given moisture content is considered for different grain sizes, holding the densities constant. Since d is the only component of the model which varies, the *absolute* increase in predicted threshold wind speed will be larger for larger grain sizes, suggesting that the cohesive forces associated with the presence of a given moisture level will be enhanced by the larger pore spaces present within larger grains, in contradiction to theory.

On the basis of their review, Hotta et al. (1984) concluded that none of the empirical expressions presented thus far were generally applicable. Major problems included the lack of consistent methods for identifying the threshold of motion and measuring wind characteristics, and the extremely limited range of grain sizes and moisture contents studied. They attempted to develop a general model based upon data presented by Tanaka et al. (1954), deriving an empirical relationship between threshold shear velocity and moisture content as a grain-size-independent, linear function:

$$U_{*_{t_w}} = U_{*_t} + 7.5w \quad (12)$$

Use of the Tanaka et al. data is questionable, since other studies reviewed by Hotta et al. (1984), as well as the additional empirical work reviewed herein, all tend to indicate logarithmic or exponential threshold-moisture content relations, and a grain-size dependence.

It is apparent that a generally applicable empirical model of the effect of moisture on threshold shear velocity is not currently available. Additional laboratory work is needed over a broad-enough range of grain sizes, sorting, and moisture contents to be reasonably representative of naturally occurring conditions, and the role of evaporation must be clearly identified.

EMPIRICAL MODELS: TRANSPORT RATES

Several field studies considering moisture effects on aeolian transport rates have failed to shed substantial light on the problem. Svasek and Terwindt (1974) found that the transport rates measured in their field study were somewhat lower than predicted by theory at low shear velocities (<0.4 m s^{-1}) and suggested that moisture effects were responsible. However, the scatter in their data was such that they were unable to quantify a relationship. Borøwka (1980) found that transport rates tended to be slightly higher at a given wind velocity over a smooth "humid" surface than over a rippled dry surface, suggesting that surface roughness effects are more significant. However, no measurements of surface moisture were made so the general applicability of the observations are uncertain.

Sarre (1987)

Sarre (1987) reported on his incorporation of a moisture correction parameter into the Kadib (1965) transport equation, based on field observations of transport rates and moisture levels. Essentially the effect of moisture was represented as negligible for gravimetric contents up to 14%, followed by rapidly decreasing transport rates at higher moisture contents. Under a shear velocity of 0.64 m s^{-1} Sarre's approach predicts transport at 67% of the dry rate at 16% moisture, and 1.1% of dry rate at 22% moisture, with a near total cessation of movement at about 24%. This finding contradicts the results of controlled lab investigations reported by Hotta et al. (1984), which indicate that at a shear velocity of 0.60 m s^{-1}, transport rates at a moisture content of just 3% would be only 50% of the dry rate, at a 4% content only 3% of the dry rate, and at moisture contents above about 5% would be zero.

Hotta et al. (1984)
In their review, Hotta et al. (1984) were able to identify two empirical expressions relating transport rate to moisture content; those proposed by Iwagaki (1950) and Nakashima and Suematsu (1976) (Equations 11 and 12, respectively):

$$q_w = 0.3(U_{100} - 6.0) \tag{13}$$

$$q_w = A' \frac{\rho_s}{g} U_{15}{}^2 (U_{15} - U_{15t}) \tag{14}$$

where U_{100} and U_{15} are wind speeds at elevations of 1.00 m and 0.15 m. The former expression (Equationa 11) is overly simplistic as it assumes a fixed threshold wind speed of 6 ms^{-1} and does not allow for variable moisture contents. Hotta et al. found the transport rates predicted by the latter (Equation 12) to be too small, and the expression is unable to account for the observation that at higher shear velocities the transport rate over wet sand approaches that for dry sediment (Horikawa et al. 1982).

Hotta et al. (1984) proposed the relationship:

$$\begin{aligned} q_w &= K \frac{\rho}{g} (U_* + U_{*_{t_w}})^2 (U_* - U_{*_{t_w}}) \\ where:\ U_{*_{t_w}} &= U_{*_t} + 7.5 w I_w \end{aligned} \tag{15}$$

which is composed of the transport model developed by Kawamura (1951) for a dry surface, modified by replacement of the dry threshold shear velocity terms with wet surface equivalents. The wet surface threshold is Equation 12 in this paper, with the inclusion of an evaporation index (I_w), ranging from 0 to 1, to account for time-varying surface moisture conditions. A series of laboratory wind tunnel investigations were conducted, employing moisture levels ranging from about 2%-8% (the widest range of moisture content considered in controlled lab investigations to date), which achieved reasonable agreement with the model.

The evaporation index is only of use when the shear velocity is below the wet threshold level and above the dry threshold. At shear velocities greater than the wet threshold it is assigned a value of 1.0. However, the authors suggest that the relationship between I_w and surface moisture content in situations where the shear velocity is between the dry and wet thresholds is too complex to be amenable to an analytic solution. Since I_w must be determined empirically, the predictive utility of the proposed model is questionable.

In general, the empirical work on moisture effects reviewed above provides a less than conclusive explanation of surface moisture effects on aeolian transport. Differences in experimental method, including measurement of wind/shear velocity, identification of initiation of motion, and measurement of moisture content, make intercomparison of studies difficult and uncertain. Variations in the characteristics of the sediments studied and the moisture contents considered add to this problem, and introduce the additional complication that the range of sediment-moisture conditions studied to date certainly does not provide a comprehensive or even representative picture of what could be expected to occur in nature.

The results of these studies are sufficient to suggest that at some set of combinations of low moisture contents and high shear velocities, transport rates will approximate the dry rate. It appears that the particular combinations may vary for sediments with different size and sorting characteristics. Quantitative identification of the appropriate combinations must be considered tentative at best. Hotta et al. (1984) indicate that at shear velocities above about 1.50 ms^{-1} transport above surfaces with moisture contents up to 8% will not differ from the dry rate. Sarre (1989) indicates that moisture contents below about 14% will have negligible effects on transport rates, even at shear velocities as low as about 0.60 ms^{-1}. Neither study identifies effects associated with sediment characteristics.

THEORETICAL MODELS: THRESHOLD EFFECTS

Kawata and Tsuchiya (1976):

Kawata and Tsuchiya (1976; reported in Horikawa et al. 1982 and Hotta et al. 1984) developed a surface-tension based model to explain moisture effects:

$$U_{*_{t_w}} = U_{*_t}\sqrt{1+D}$$

$$where:\ \mathrm{D} = \frac{2\sqrt{6}}{5}\sqrt{\alpha_1\alpha_2}\sqrt{\mathrm{n}_o}\sqrt{\frac{\rho_s}{\rho_w}\frac{T\sqrt{w}\cos\xi}{(\rho_s-\rho)gd}} \tag{16}$$

where α_1 and α_2 are constants, n_0 is the number of contact points for a grain, ρ_w is water density, and ξ is the angle of contact between the grain and water. This model predicts that for a given moisture content, the magnitude of the relative increase in threshold shear velocity will decrease with increasing grain size, as expected given that surface tension and adhesive forces are more effective in the smaller pores associated with finer grain sizes. Horikawa et al. (1982) found the model to provide a fairly good fit to data derived from controlled lab studies. Although establishing a basic theoretical background, the model has a weakness in its reliance on gravimetric moisture content. Given a constant moisture content it will predict the same threshold effect for sediments of the same mean grain size regardless of the degree of sorting. Yet in the field the degree of sorting can vary considerably, and thus the associated pore configurations and the cohesion associated with a given moisture content can vary as well.

McKenna-Neuman and Nickling (1989)

In a recent study, McKenna-Neuman and Nickling (1989) attempted to provide a more complete basis for assessing the effects of moisture on aeolian transport. By considering capillary forces associated with the presence of moisture they developed the expression:

$$U_{*_{t_w}} = U_{*_t} \cdot \left[1 + \frac{6}{\pi d^3 g(\rho_s - \rho)} \cdot 2\cos\beta \cdot F_c\right]^{1/2} \tag{17}$$

$$F_c = \frac{\pi T^2}{P} G \tag{18}$$

where β is the particle resting angle, F_c is the force resisting motion associated with moisture, T is surface tension, P is soil moisture tension, and G is a nondimensional coefficient representing particle contact geometry.

The basic concept underlying the derivation of F_c is that of capillary force. Surface tension (T), a constant for a given liquid at a given temperature, acts to hold the surface of a liquid contained within a solid flat. When the solid-liquid attraction exceeds the liquid-liquid, the liquid is pulled up the solid. To maintain the curved liquid-gas interface, pressure on the liquid side must be less than the pressure on the gas side of the interface, and this lower pressure results in an attractive force or suction binding the solid and liquid. The magnitude of the suction acting on a unit area of particle surface is equivalent to the pressure difference (P) between the two sides of the liquid-gas interface:

$$A_p - W_p = P \tag{19}$$

where A_p is the pressure on the gas side of the interface (atmospheric pressure in the case of soil moisture) and W_p is the pressure on the liquid side of the interface. Given a system of two spherical particles, the magnitude of P can be shown geometrically to be dependent on the internal (R_1) and external (R_2) radii of curvature of the water wedge (see Haines 1925, Childs 1969), such that:

$$P = -T\left(\frac{1}{R_1} - \frac{1}{R_2}\right) \tag{20}$$

As moisture content increases, R_1 will increase and the moisture tension or pressure deficiency per unit area of particle surface will decrease. However, at the same time R_2 must also increase resulting in a greater increase in affected surface area and a net increase in the total moisture tension acting on the particles. It is worth noting that P is denoted by a variety of names in the literature, including moisture tension, matric tension, and matric suction.

The manner in which the magnitude of R_1 and R_2 vary with changing water content is a function of the shape of particle contacts. The G parameter included in F_c accounts for the geometric properties of the grain-grain contact within the water wedge. The relation between F_c and G as represented in McKenna-Neuman and Nickling's model is derived from consideration of idealized grain to grain contacts in the form of cones of different sizes. Thus, they are representative of angular to subangular contacts although not necessarily of rounded particle contacts. While the range of theoretically possible values for G is quite large, (about 0 to 500) the actual range of G values found in their experiments on fine to medium size sands was fairly small (about 0.2 to 3.0).

McKenna-Neuman and Nickling (1989) evaluated their model in a laboratory wind tunnel experiment, using grain sizes from 0.19 to 0.51 mm and moisture contents up to about 2.4%, by weight. The model provided a reasonable fit to the data, particularly above permanent wilting point (PWP) moisture levels. The authors suggest that significant departures from the model at moisture levels below PWP might indicate a change in the physical character of the water wedges at very low moisture contents. They argue that relating shear velocity threshold to gravimetric moisture content is inappropriate, given that sediments with different mean grain sizes and degrees of sorting have differing ability to retain moisture. As mean grain size and degree of sorting decrease, a constant gravimetric moisture content will generate larger capillary forces.

The proposed model has several limitations which should be noted. It considers a relatively small range of moisture. The authors justify this by stating that the range of water contents studied "brackets the critical range of gravimetric moisture contents that appear to significantly affect the susceptibility of sand-sized material to wind erosion." However, other studies have found that transport rates continue to decrease as moisture content increases from 2% up to at least 8% (e.g., Hotta et al. 1984), suggesting that the effect of moisture continues to increase as content exceeds 2%, and that wind can and does move sands under such conditions. The upper limit of model applicability occurs in a physical sense when the moisture content becomes large enough that the wedges at individual contacts begin to coalesce, changing the geometry of the individual water bodies. It is uncertain as to what happens at such moisture levels, although Sherman (1990) has suggested that the filling of pore spaces with water will reduce surface roughness and, therefore, shear velocity.

From a practical perspective a key limitation lies in the requirement for moisture tension data as a model input rather than moisture content. Standard techniques for measuring moisture tension (e.g., tensiometers, pressure chamber testing; see Hanks 1992) require samples of greater thickness than the uppermost few millimeters of sand that are exposed to wind action and are of primary concern here. Yet the moisture characteristics of underlying sediments may differ significantly from those at the surface, making it difficult to get data representative of actual surface conditions. Hysteresis further complicates the problem. A given water content will result in greater moisture tension when a sediment is drying out than when it is being wetted, thus, a range of moisture tensions is possible for a given sediment at a given water content (Childs 1969).

Gregory and Darwish (1990)

Gregory and Darwish (1990) have attempted to work around the need for moisture tension data by relating it empirically to w_e, thereby allowing moisture content to be used directly once the permanent wilting point moisture content is established for a given sediment. Considering a system of two spherical particles with one resting directly on top of the other, they proposed the expression:

$$U_{*tw} = A\sqrt{\frac{\rho_s}{\rho} gd}.\left[1 + \frac{6}{\pi d^3 g \rho_s} + 0.01w + F_b\right] \tag{21}$$

$$\begin{aligned} F_b &= F_{dc} + F_{wc} \\ F_{wc} &= A_c P \\ F_{wc} &= a_1 e^{-a_2 w_e}\left(w - w_c\right) \end{aligned} \tag{22}$$

where $A = 0.118$, F_b is bonding force, F_{dc} is dry cohesive force (assumed to be zero for sand-sized grains), F_{wc} is cohesive force due to moisture, A_c is the area of the wet contact, a_1 and a_2 are empirically determined constants (with suggested values of 0.00075 and 0.1, respectively), and w_c is the proportion of moisture (in percent) absorbed by clay particles.

Aside from a slight variation in the dry threshold shear velocity, the first two terms are identical to McKenna-Neuman and Nickling (1989). Three main differences distinguish the two models. First, since Equation 21 envisions one particle resting directly on top of a second, a trigonometric correction for the direction of the cohesive force is not needed in the Gregory and Darwish approach. Second, Gregory and Darwish include a term to account for the effect of the additional weight force which must be overcome as a result of water clinging to the grain. Finally, Gregory and Darwish take a substantively different approach to representing the cohesive force due to moisture.

The wet cohesive force is modeled as the product of moisture tension and the area acted upon by that force. Conceptually, the use of a two-sphere model to account for moisture contact area is simpler and appears less realistic than the variable cone shapes employed by McKenna-Neuman and Nickling (1989). However, in practice the A_c contribution to F_{wc} is subsumed into the a_1 coefficient, the value of which was determined empirically through a best-fit to several data sets. Therefore, incorporation of a more sophisticated geometric model would not be expected to alter model predictions. The key assumption underlying the formulation proposed by Gregory and Darwish is that moisture tension can be adequately represented as a function of w_e, and thus related directly to moisture content once the permanent wilting point content is established for a given sediment. In best-fitting the model to several published data sets, the range of a_2 values obtained by Gregory and Darwish was relatively small (0.00062 to 0.00078, although their own lab data gave a value of 0.0012) which argues favorably for their parameterization. However, as outlined above, hysteresis can produce different moisture tensions in a given sediment at a given moisture content. The data sets employed by Gregory and Darwish for calibration were taken from studies in which moisture was added to dry samples—if the procedure were reversed and saturated samples allowed to dry to the same water contents, different moisture tensions would be expected.

The w_c variable represents an attempt to account for moisture which is absorbed by clay particles or lost in the interstices of aggregate grains and not available to bind grains. The value of w_c must be determined empirically for a given sediment, again through best-fit procedures. A difficulty with the form of w_c becomes evident upon examination of Equation 22. It is apparent that for moisture levels below w_c the wet cohesive force assumes a negative value and acts to reduce the predicted threshold shear velocity. However, for relatively clean sands the problem is negligible (Gregory and Darwish obtained best-fit w_c values of 0.00-0.06 for the data of Belly (1964) and McKenna-Neuman and Nickling (1989), for example).

THEORETICAL MODELS: TRANSPORT

Ismailov et al. (1991)

Ismailov et al. (1991) have proposed what they indicate to be a theoretically derived formula for transport effects: (23)

$$q_w = c_a \, w^{c_b/c_c}\left(e^{c_b U} - 1\right)$$

where boundary conditions:

$$U = U_1;\ w = w_1;\ q_w = q_{w1}$$
$$U = U_1;\ w = w_2;\ q_w = q_{w2}$$
$$U = U_2;\ w = w_1;\ q_w = q_{w3}$$

are used to determine constants as:

$$c_a = q_{w1}\left(w_1^{-c_b/c_c} \,/\, e^{c_b U_1}\right)$$
$$c_b = \left(\ln\left(q_{w1}/q_{w3}\right)\right) / \left(U_1 - U_2\right)$$
$$c_c = c_b\left(\ln\left(w_1/w_2\right)\right) / \left(\ln\left(q_{w1}/q_{w2}\right)\right)$$

Essentially, transport rates (q_{w1}, q_{w2}, q_{w3}) measured at combinations of two different wind speeds (U_1, U_2) and moisture contents (w_1, w_2) are used to calibrate the model. However, no physical interpretation of the relationship is evident upon examination, nor do the authors provide one in their text. Further, although they state that transport rates should be proportional to shear velocity, they elect to use wind velocity (measured at an unstated elevation) in their transport expression. Additionally, since grain parameters are not included in the relationship it requires calibration for each sediment type of concern, thus rendering it of little practical utility. Finally, since there is no consideration of evaporation it cannot be expected to represent natural conditions adequately.

COMPARISON OF THRESHOLD MODELS

The models of moisture effects on threshold shear velocity were plotted against a range of moisture contents to allow comparison of the relationships (Figure

3). A constant grain size of 0.27 mm was used for all models. Where possible a full 0%-24% range of moisture contents was considered, representative of conditions from dry to saturation assuming a porosity of about 40%, but the actual data limits are indicated in each case. Two scenarios were used for the Gregory and Darwish (1990) model. The "clean" prediction employed a PWP moisture content of 0.2% (reported by McKenna-Neuman and Nickling [1989] for their 0.27-mm sample) and a w_c value of 0.00. The "clay" prediction used a w_c value of 1.04 as reported by Gregory and Darwish (1990) for their samples and a PWP content of 3.0% (a weighted average of values they reported for 0.22- and 0.36-mm sediments).

In two cases it was not possible to simply evaluate the formulae at each moisture content. McKenna-Neuman and Nickling's (1989) expression requires moisture tension values, which cannot be simply related to gravimetric content. However, they graphically present their model in relation to gravimetric content for the 0.27-mm grain size. This curve was digitized to obtain the data for Figure 2, but the moisture contents evaluated ranged only up to about 1.4%. As the relation between gravimetric moisture and moisture tension is not known, the model could not be extended to higher moisture levels. The model of Nakashima and Suematsu appears to have been specified incorrectly in Hotta et al. (1984), as attempts to solve for critical threshold velocities directly resulted in obviously erroneous values. A graph of the relation presented in Hotta et al. (1984) was digitized for incorporation into Figure 2, and again the relation could not be extended to higher moisture contents. The 0.45-mm grain size used in Hotta et al. (1984) was corrected to 0.27 mm by scaling down predicted threshold shear velocity values using the law of the wall.

It is apparent from Figure 3 that considerable variation exists among model predictions. Given that most of models are built around the Bagnold (1936) dry threshold model, it would be expected that the models will tend to converge on the dry threshold of 0.24 m s^{-1} as zero moisture contents are approached. The dry threshold used by McKenna-Neuman and Nickling (1989) was determined experimentally, and is slightly higher. In the cases of Belly (1964) and Kawata and Tsuchiya (1976), the model formulations do not provide for a smooth transition from dry to wet conditions. The effect of the w_c component is apparent in the Gregory and Darwish (1990) clay prediction, which does not begin to provide reasonable values until the w_c value (1.04) is exceeded by the moisture content.

A trend toward increasing divergence among predictions with increasing moisture content is readily apparent. At a moisture content of only 1% by weight, substantial differences in predicted threshold shear velocity are evident, with increases ranging from about 8% to 148% of the expected dry value. Ignoring the high and low extremes, predicted increases still range from about 31% to 128%. It should be noted that, with the exception of Kawata and Tsuchiya (1976), all of the models are calibrated to empirical observations at this low moisture level. At a moisture content of 4%, the four models which can reasonably be expected to still be applicable (the three whose data limits have

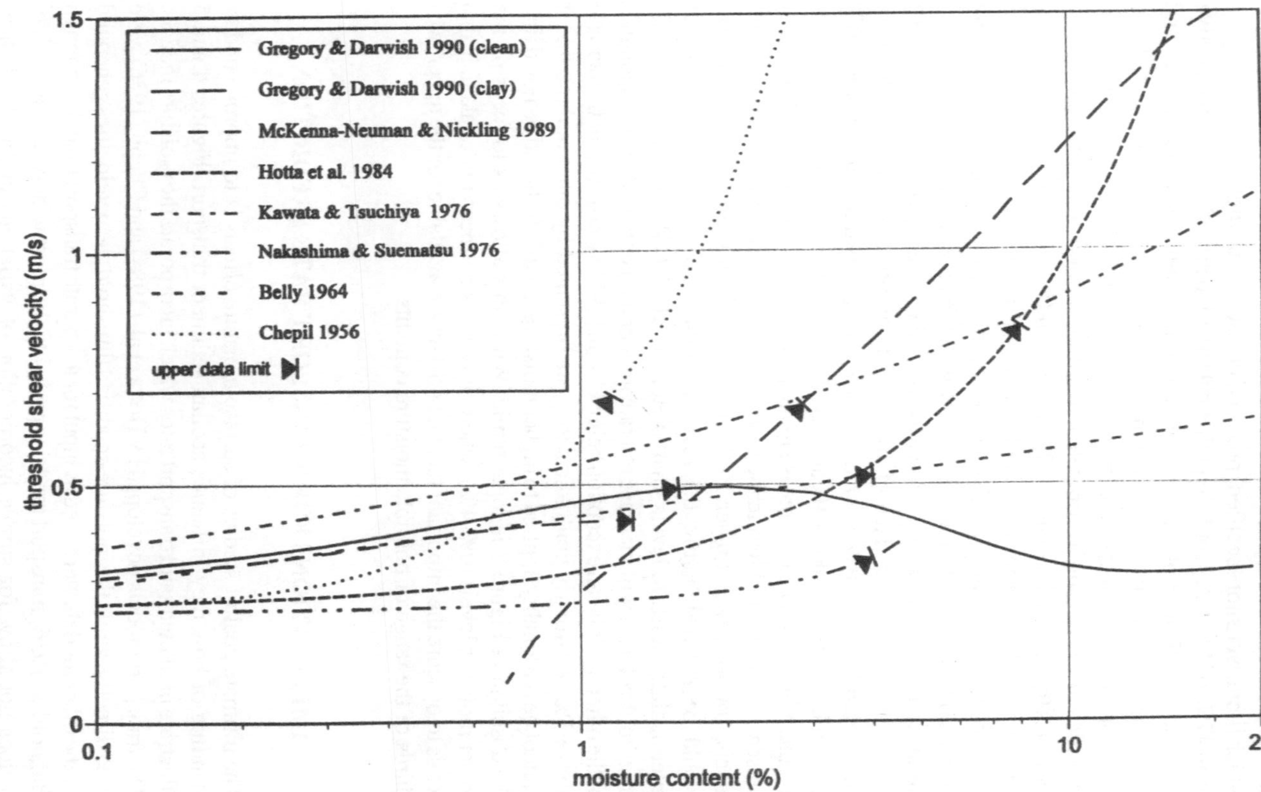

Figure 3. Model predictions of increase in threshold shear velocity with increasing moisture content, for 0.27-mm grain size.

not been exceeded and the theoretical model of Kawata and Tsuchiya, for which limits were not specified) predict increases in threshold velocity ranging from 47% to 206%. A simple hypothetical situation provides some perspective as to the significance of this level of variation. Using the law of the wall and assuming a zero displacement height and a roughness length of $2_{d50}/30$, the lower prediction indicates a wind velocity of about 34 km/hr (measured at an elevation of 1 m) would be competent to initiate motion, while the higher value indicates that a velocity of nearly 72 km/hr would be required. The potential for error in modeling the movement of sediment is apparent.

Above 4% moisture the divergence becomes acute, and application of any of the models to moisture levels greater than 8% appears unreasonable given the spread in possible results. A weakness in the Gregory and Darwish (1990) model is apparent in its lack of applicability at moisture contents which are large relative to the PWP content. Once actual moisture exceeds the PWP content by a factor of about 10, any subsequent increase in w results in a larger decrease in F_{wc} and a net decrease in the predicted threshold value. While this presents a serious problem in the case of clean beach sands at higher moisture levels, for soils with higher PWP contents (>2-3%) the saturation content would be reached before the reversal point. The only available data at intermediate moisture levels, that of Hotta et al. (1984) tends to suggest that threshold velocity will increase at a rate greater than obtained by extrapolation of the other models (an exception being Chepil [1956] whose dramatic increase above 1% so strongly contradicts the data presented by other authors that it cannot be reasonably applied to higher moisture levels). It should be noted that the abscissa in Figure 3 is logarithmic. While this allows us to focus on the lower moisture levels (0%-4%), where most work has been undertaken, it also tends to obscure the important fact that no data are available for the upper two-thirds of the range of possible moisture contents.

IMPLICATIONS FOR TRANSPORT RATE PREDICTION

The ultimate goal of studies of surface moisture effects is to improve understanding of how they influence aeolian sediment transport. Figures 4 and 5 illustrate predicted sand transport rates which incorporate the moisture models. The transport models developed by Hotta et al. (1984) and Chepil (1956) were used directly, while the moisture-modified threshold shear velocities generated by the other models were used as input for a separate transport model. Although Bagnold's (1936) transport model is probably the most widely used approach, it does not allow for simple incorporation of variations in threshold shear velocity. However, Lettau and Lettau (1977) have proposed a similar model that incorporates an explicit threshold velocity term, which was adopted for use here:

$$q = C'\sqrt{\frac{d}{D}}\frac{\rho}{g}(U_* - U_{*_t})U_*^2 \tag{24}$$

where C is equal to 4.2.

Under a shear velocity of 0.50 m s^{-1}, transport predictions derived from all models go to zero at relatively low moisture levels, with the exception of Gregory and Darwish (1990) "clean," which exceeds its practical limit at a content of about 2%. However, there is considerable disagreement as to what content this will occur at (Figure 4). Given constant moisture contents above 5%-6% (i.e., assuming evaporation is negligible) it would perhaps be safe to predict zero transport, but with a content of 1% predictions from 0.0 to 0.03 kg m^{-1} s^{-1} are available. When applied to larger spatial and temporal scales, this amounts to the difference between no transport and moving about 270 m^3 of material across a 1-km-long strip over a four-hour period. At the higher shear velocity used in Figure 5 the disparity is clearer, and it becomes difficult to place much confidence even in the identification of a limiting moisture content above which no transport would be expected.

CONCLUDING REMARKS

Considerable evidence is available to support the argument that surface moisture may significantly affect the transport of sand by wind. However, the exact nature of the effect is still not well understood or quantified. In particular, very little work has been done at intermediate and high moisture contents, and there is a critical need for investigations at moisture levels above about 4%. None of the proposed models provide a theoretical consideration of the role of evaporation, suggested by the studies of Hotta et al. (1984) and Logie (1982) to be important to transport rates. Given that the presence of moisture in surface sediments affects aeolian transport, it becomes necessary to develop an understanding of both the moisture levels that can be expected in particular environments and how those moisture levels vary over time. If, as several investigators have suggested, moisture levels greater than a few percent may effectively eliminate transport, the role of evaporation in reducing moisture levels to the point where transport can occur needs to be examined.

It is important to emphasize that the presence of even relatively small amounts of moisture may significantly affect sand movement. Reports in the literature indicate that, in the absence of other moisture sources, atmospheric humidity alone may generate surface moisture levels of 0.25%-0.60% by weight (Belly 1964, Bradley et al. 1992). At a moisture level of 0.60%, the seven models of moisture effects on threshold shear velocity reviewed in this paper predict increases in threshold shear velocity averaging about 48% of the dry threshold. The predicted increases range from 4% to 105%, indicating a substantive disagreement regarding the potential effects of low moisture levels. This disagreement is readily apparent when the available literature is considered as a whole. On the basis of laboratory investigations, several researchers have argued that moisture does not affect threshold shear velocity below contents of about 4% (Azizov 1977, Logie 1982), and Sarre (1987) places the lower limit of significant moisture effects at a content of about 14% from field

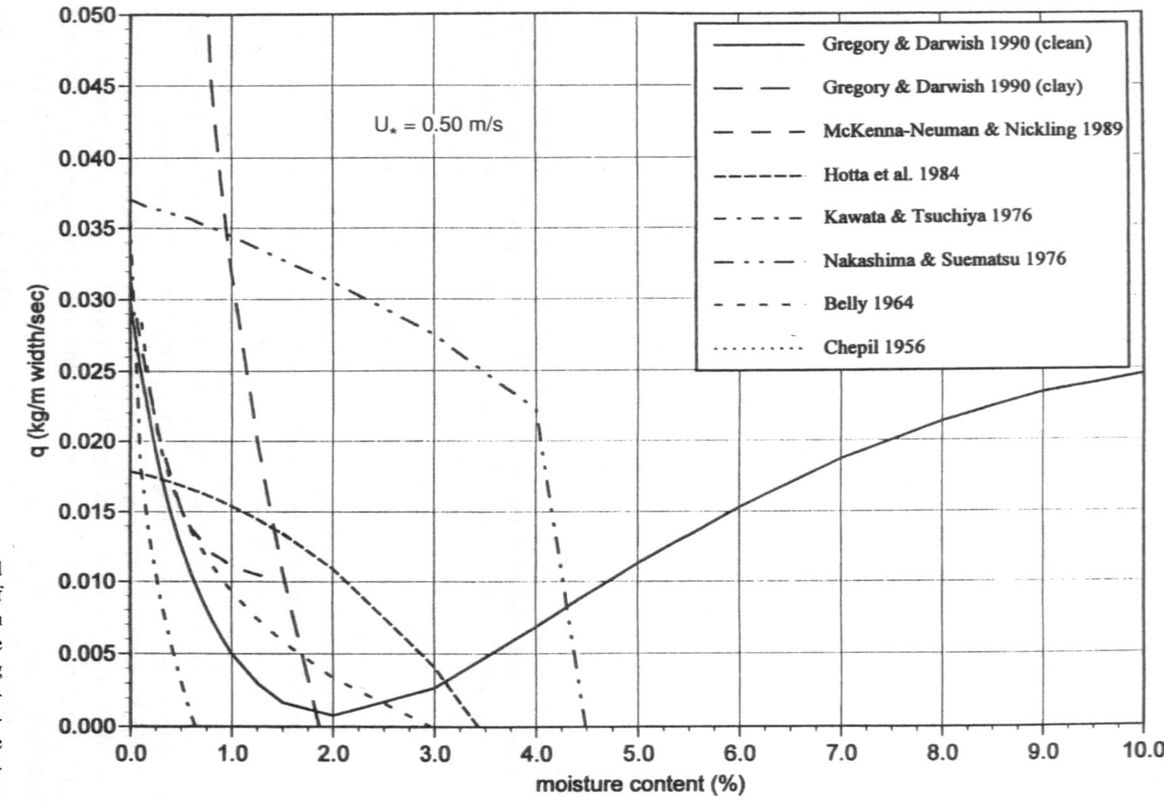

Figure 4. Model predictions of the change in transport rate with increasing moisture content, for 0.27-mm grain size and shear velocity of $0.50 m s^{-1}$.

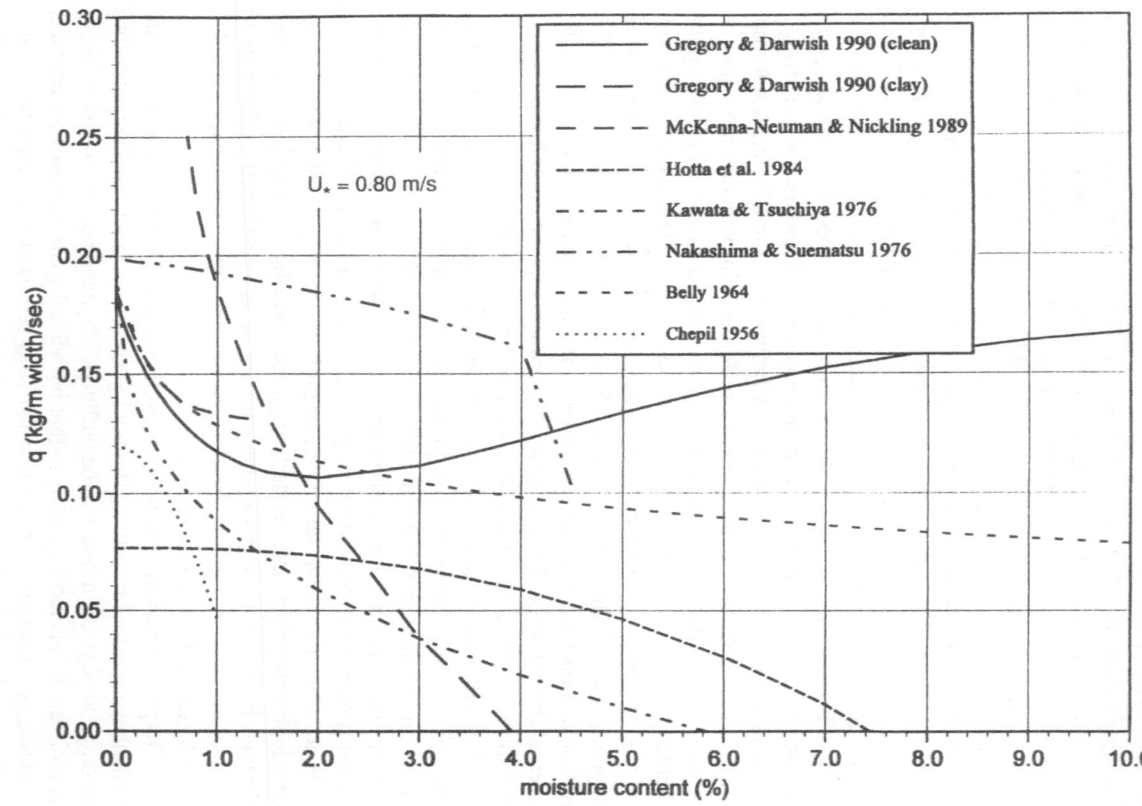

Figure 5. Model predictions of the change in transport rate with increasing moisture content, for 0.27-mm grain size and shear velocity of 0.80 m s^{-1}.

observations. A nearly opposite view is evident in McKenna-Neuman and Nickling (1989) and Gregory and Darwish (1990). The latter studies limit consideration to contents ranging up to 2%-3%, and argue that above such moisture levels the cohesion owing to moisture becomes so strong that the models can be considered to bracket the critical range of moisture contents.

It is difficult to reconcile the differences between the model results. However, several theoretical and empirical issues likely contribute to the disparities. First, measured moisture-content values are static representations of the moisture field. Evaporation rate is the dynamic control on how fast sediments may be released into the airstream for given flow conditions. The work of Logie (1982) illustrates the importance of evaporation over long time periods (i.e., hours), but virtually nothing is known about the short-term (i.e., seconds) effects. Part of the discrepancy between the models may result from variations in sediment release caused by differences in evaporation. Second, all of the studies focus on the representation of sediment grains using mean grain diameter. However, the presence of mixed sand sizes will introduce complexity into the geometry of intergrain pore spaces and thereby alter the magnitude of capillary forces. The approach of McKenna-Neuman and Nickling (1989) notably avoids this difficulty by considering moisture tension directly, however, at the not inconsiderable cost of introducing the need for knowledge of moisture tension in the uppermost layer or two of grains.

The empirical problems involve mainly the measurement of sediment moisture content. Some of these problems are alluded to earlier in the text. However, the primary problem comes in trying to isolate a surface sample only one or two grain diameters in thickness while trying to preserve the moisture content characteristics. None of the generally prescribed sampling methods are able to meet the ideal sample configuration. It may be that the further development of electronic methods (e.g., Jackson 1993) may address this issue through in-situ moisture measurement. Additionally, subjective elements in the identification of the initiation of motion in many of the studies may also introduce variability to the results.

Despite a recent focus on the theoretical aspects of the problem there are still many aspects of moisture effects which require further clarification before a complete physical model can be outlined. The nature of the cohesive forces at moisture levels large enough so that individual wedges at particle contacts coalesce is uncertain. The relation between sediment characteristics, water content, and moisture tension needs further investigation. Hanks (1992) has suggested that theoretical relations between these parameters cannot be established, and argues for the use of empirical measurement (using the tensiometer method, for example). Rogowski (1971) has proposed a model relating water content to moisture tension; however, determination of the various model inputs appears to require far more effort than directly measuring the characteristic curve when only a single, readily accessible site is concerned. While a general relation applicable to all soil types may be too complex to be developed analytically, the (relatively) simple structure of beach sands, for example, might allow for an analytical approach, or at least a generally applicable empirical model. An important part of the problem involves identification of the potential ranges in moisture tension that could be expected at given moisture levels as a result of hysteresis. A second issue, also largely unaddressed, involves assessing the effects of sediment-size distributions. Sorting effects may contribute substantially to differences in sediment response to moisture effects.

The body of work undertaken to date has focused on the influence of moisture on threshold velocity, and transport rate effects associated with altered thresholds, but a number of possible influences on the nature of transport remain unconsidered. Haff has observed "chunks of sand ripped or peeled off the surface when high winds impinge on wet dune sand" (P. Haff, pers. comm. 1994), a phenomena that cannot be accounted for by current threshold velocity models which assume grains move individually. Studies have also found that a significant amount of moisture may continue to adhere to individual grains during saltation (Hotta et al. 1984, Sarre 1990). The nature and potential effects of this retained moisture, including modification of the aerodynamic characteristics of the particles, resulting changes in saltation paths, rebounding, and transfer of momentum on landing, have not been examined. It is apparent that much work remains to be done.

ACKNOWLEDGMENTS

This research was partially supported by a grant from NATO (CRG900540) and by a contract from the California Department of Boating and Waterways. Peter Haff, William Nickling, and James Shulmeister read earlier drafts of the paper, and the authors gratefully acknowledge their valuable comments.

REFERENCES

Azizov, A. (1977) Influence of soil moisture on the resistance of soil to wind erosion. *Soviet Soil Science*, v. 9, p. 105-108.

Bagnold, R. A. (1936) The movement of desert sand. *Proceedings of the Royal Society of London*, v. A157, p. 594-620.

Belly, P. Y. (1964) *Sand Movement by Wind*. U.S. Army Corps of Engineers, Coastal Engineering Research Center, Technical Memorandum 1, Washington, D.C.

Bisal, F., and Hsieh, J. (1966) Influence of moisture on erodibility of soil. *Soil Science*, v. 102, p. 143-146.

Borøwka, R. K. (1980) Present day processes and dune morphology on the Leba barrier, Polish coast of the Baltic. *Geografiska Annaler*, v. 62A, p. 75-82.

Bradley, N. W., Gregory, J. M., and Wilson, G.R. (1992) Wet-bonding effects on threshold friction velocity. Paper presented at the Winter Meeting ofAmerican Society of Agricultural Engineers, Nashville, TN. Paper no. 922515.

Buckman, H. O., and Brady N. C. (1969) *The Nature and Properties of Soils*, 7th ed. The Macmillan Company, London.

Chepil, W. S. (1956) Influence of soil moisture on erodibility of soil by wind. *Soil Science Society*

of America Proceedings, v. 20, p. 288-292.
Childs, E. C. (1969) *Soil Water Phenomena*. Wiley-Interscience, New York.
Duncan, W. J., Thom, A. S., and Young, A. D. (1970) *Mechanics of Fluids*, 2nd ed. Edward Arnold, London.
Fisher, R. A. (1926) On the capillary forces in an ideal soil: correction of the formulae given by W.B. Haines. *Journal of Agricultural Science*, v. 16, p. 492-503.
Gares, P. (1987) *Eolian Sediment Transport and Dune Formation on Undeveloped and Developed Shorelines*. Ph.D. dissertation, Rutgers University, New Brunswick.
Gregory, J. M., and Darwish, M. M. (1990) Threshold friction velocity prediction considering water content. Paper presented at the Winter Meeting of American Society of Agricultural Engineers, Chicago, IL. Paper no. 902562.
Haines, W. B. (1925) Studies in the physical properties of soils: II. A note on the cohesion developed by capillary forces in an ideal soil. *Journal of Agricultural Science*, v. 15, p. 529-535.
Hanks, R. J. (1992) *Applied Soil Physics*, 2nd ed. Springer-Verlag, New York.
Hillel, D. (1971) *Soil and Water: Physical Principles and Processes*. Academic Press, New York.
Horikawa, K., Hotta, S., and Kubota, S. (1982) Experimental study of blown sand on a wetted sand surface. *Coastal Engineering in Japan*, v. 25, p. 177-195.
Hotta, S., Kubota, S., Katori, S., and Horikawa, K. (1984) Sand transport by wind on a wet sand surface. *Proceedings of the 19th Coastal Engineering Conference, American Society of Civil Engineers*, New York, p. 1263-1281.
Ismailov, M. I., Ismailov, M. M., and Mirzazhanov, K. M. (1991) Proneness of soils to wind erosion as a function of moisture content and wind speed. *Pochvovedeniye*, v. 4, p. 168-170.
Iwagaki, Y. (1950) On the effect of the sand-drift on the coast by wind for the filling up with sand in Ajiro-Harbor. *Journal of the Japanese Society of Civil Engineers*, v. 35, p. 265-271. (in Japanese)
Jackson, D.W.T. (1993) *Aeolian Entrainment of Surface Beach and Dune Sands*. Ph.D. dissertation, University of Ulster, U.K.
Johnson, J. W. (1965) Sand movement on coastal dunes. *Federal Inter-Agency Sedimentation Conference Proceedings*, U.S. Department of Agriculture, Miscellaneous Publication 970, Washington, p. 747-755.
Kadib, A. A. (1965) *A Function for Sand Movement by Wind*. Hydraulics Engineering Laboratory Report HEL-2-12, University of California, Berkeley.
Kawamura, R. (1951) *Study of Sand Movement by Wind*. Translated as Hydraulics Engineering Laboratory Report HEL 2-8, University of California, Berkeley.
Kawata, Y., and Tsuchiya, Y. (1976) Influence of water content on the threshold of sand movement and the rate of sand transport in blown sand. *Proceedings of the Japanese Society of Civil Engineers*, v. 249, p. 95-100. (in Japanese)
Kroon, A., and Hoekstra, P. (1990) Eolian sediment transport on a natural beach. *Journal of Coastal Research*, v. 6, p. 367-380.
Lettau, K., and Lettau, H. (1977) Experimental and micrometeorological field studies of dune migration. In K. Lettau and H. Lettau (eds.) *Exploring the World's Driest Climate*. IES Report 101, University of Wisconsin, Madison, p. 110-147.
Logie, M. (1982) Influence of roughness elements and soil moisture on the resistance of sand to wind erosion. *Catena*, Supplement 1, p. 161-174.
Lyles, L., and Woodruff, N. (1972) Boundary-layer flow structure: effects on detachment of non-cohesive particles. In H. Shen (ed.) *Sedimentation*. Department of Civil Engineering, Colorado State University, p. 2.1-2.16.
Marshall, T. J., and Holmes, J. W. (1988) *Soil Physics*, 2nd ed. Cambridge University Press, Cambridge.
McKenna-Neuman, C., and Nickling, W. G. (1989) A theoretical and wind tunnel investigation of the effect of capillary water on the entrainment of sediment. *Canadian Journal of Soil Science*, v. 69, p. 79-96.
Nakashima, Y., and Suematsu, T. (1976) Effect of moisture content of sand surface layer on blown sand (III): on the rate of sand movement, threshold velocity, and median diameter of blown sand. *Proceedings, 87th Conference of the Japanese Society of Forestry*, p. 361-362. (in

Japanese)

Nickling, W. G., and Davidson-Arnott, R.G.D. (1990) Aeolian sediment transport on beaches and coastal sand dunes. *Proceedings of the Symposium on Coastal Sand Dunes*, National Research Council of Canada, Ottawa, p. 1-36.

Nordstrom, K. F., and Jackson, N. L. (1992) Effect of source width and tidal elevation changes on aeolian transport on an estuarine beach. *Sedimentology*, v. 39, p. 769-778.

Pye, K., and Tsoar, H. (1990) *Aeolian Sand and Sand Dunes*. Unwin-Hyman, London.

Rogowski, A. S. (1971) Watershed physics: model of the soil moisture characteristic. *Water Resources Research*, v. 7, p. 1575-1582.

Sarre, R. D. (1987) Aeolian sand transport. *Progress in Physical Geography*, v. 11, p. 157-182.

Sarre, R. D. (1988) Evaluation of aeolian sand transport equations using intertidal zone measurements, Saunton Sands, England. *Sedimentology*, v. 35, p. 671-679.

Sarre, R. D. (1989) Aeolian sand drift from the intertidal zone on a temperate beach: potential and actual rates. *Earth Surface Processes and Landforms*, v. 14, p. 247-258.

Sarre, R. D. (1990) Reply to discussion: Evaluation of aeolian sand transport, Saunton Sands. *Sedimentology*, v. 37, p. 389-392.

Sherman, D. J. (1990) Discussion: Evaluation of aeolian sand transport equations using intertidal-zone measurements, Saunton Sands, England. *Sedimentology*, v. 37, p. 385-392.

Sherman, D. J., and Hotta, S. (1990) Aeolian sediment transport: theory and measurement. In K. F. Nordstrom, N. Psuty, and R.W.G. Carter (eds.) *Coastal Dunes: Form and Process*. Wiley, New York, p. 16-38.

Svasek, J. N., and Terwindt, J.H.J. (1974) Measurements of sand transport by wind on a natural beach. *Sedimentology*, v. 21, p. 311-322.

Tanaka, T., Sano, H., and Kakinuma, S. (1954) Study on the control of wind erosion. *Agricultural Meteorology*, v. 10, p. 24-30. (in Japanese)

Terzaghi, K., and Peck, R. B. (1967) *Soil Mechanics in Engineering Practice*, 2nd ed. John Wiley & Sons, New York.

14 WIND ABRASION AND VENTIFACT FORMATION IN CALIFORNIA

Julie E. Laity
Department of Geography
California State University, Northridge

ABSTRACT

Ventifacts are found in several physical settings in California: in formerly glaciated areas, in periglacial areas above or beyond glacier limits, in presently semiarid areas, along the coast, and in true deserts. In several localities, both active and fossil forms are found. Ventifacts and abraded surfaces develop wherever strong winds, laden with abundant sediment, erode resistant boulders or bedrock. In California, as elsewhere, the abrasive agent is most commonly a fine- to medium-grained aeolian sand. In arid areas, ventifacts occur near to Pleistocene lake shorelines, downwind of alluvial rivers, near dune fields, or in corridors of regional sand transit. They formed principally during a drier middle Holocene period from 8 to 5 ka. Staining and discoloration of the abraded face, patchy granular disintegration and spalling of abraded surfaces, and stabilized aeolian sand provide evidence for current inactivity of wind erosion. Ventifacts found on moraines beyond receding ice fronts are much older, dating from Pleistocene cold stages.

A long-term study in the Little Cowhole Mountains of the Mojave Desert seeks to further understand ventifact formation by studying active processes. Contemporary winds are monitored by a weather station and 11 anemometers. Strong winds blow from the north and south, and to date maximum gusts of 27 m s^{-1} have been recorded. Grooves are aligned with the strongest winds and ventifacts are abraded on the windward face: along the topographic ridge a sharp keel separates the facets. Polished ventifact surfaces that are very smooth to the eye or touch are quite rough at the microscale. Scanning electron microscopy shows that sand grains abrade by chipping, breaking along cleavage planes, rubbing, and microgouging. High-energy impacts near boulder tops causes cleavage to be the main process of abrasion, whereas near the boulder base rubbing and microgouging dominate. Future work will examine the formation of flutes and helices, abrasion maxima heights in hilly terrain, rates of abrasion, and the evolution of ventifact form.

INTRODUCTION

Ventifacts have formed in various environmental settings within California. Most are no longer active ("relict" or "fossil" forms). Ventifacts found on moraines along the eastern base of the Sierra Nevada developed in a periglacial environment (Blackwelder 1929) during early and late Wisconsin glacial stages (Bach, Dorn, this volume). These forms are presently inactive. Desert ventifacts, by contrast, appear to have formed principally during periods of

Desert Aeolian Processes. Edited by Vatche P. Tchakerian. Published in 1995 by Chapman & Hall, London. ISBN 978-94-010-6519-1

aeolian activity in the early- to mid-Holocene. However, active ventifact formation still occurs in some locales, including the Pisgah lava flow and the Little Cowhole Mountains of the east-central Mojave Desert (Figure 1).

The purpose of this chapter is first to briefly examine the nature and development of ventifacts, then to review the contributions of research in California, including the occurrence and age of ventifacts and their significance to paleoclimatic research, and lastly to provide preliminary observations of actively-forming ventifacts in the Little Cowhole Mountains.

VENTIFACTS: GENERAL OBSERVATIONS

The term ventifact is ill-defined and has been used to describe wind-eroded rocks of varying size, form, and material composition. Ventifacts range in size from small pebbles to large boulders. Bedrock outcrops may also be extensively abraded.

Erosional Forms

The astonishing variety of shapes and forms of ventifacts developed in a wide range of rock types is one of the factors that makes their study so fascinating. The size of the original material appears to be one control on the ultimate form of the ventifact. On the basis of studies in Death Valley, Maxson (1940) determined that the size of the rock is related to its form: if the rock is small, not exceeding a height of 8 cm, the ventifacts are of the classic faceted type (with planar faces and smooth surfaces) that lack lineations. Larger fragments are striated. Polish was observed to occur on both on smooth facets and within flutes and grooves.

The following categories of of erosional forms associated with ventifacts may be recognized:

Polishing

Smoothing and polishing of rock surfaces is perhaps the most common feature reported for ventifacts, and occurs both on smooth facets and within flutes and grooves. The degree of polish is a good indication of the relative age of the ventifact. Actively forming ventifacts on Pisgah lava flow and in marble outcrops of the Little Cowhole Mountains show considerable sheen and are very smooth to the touch (Figure 2). Weathering of carbonate rocks following the cessation of abrasion rapidly removes the polish. Clark and Wilson (1992) note that the retention of ventifact polish varies according to rock type.

Facets

The term facet describes a relatively plane surface which has been cut at right angles to the wind, regardless of the original shape of the stone. Facets commonly join along a sharp ridge or keel (Figure 3), and the number of keels (kante) is used to describe the stones as einkante, zweikanter, dreikanter (one-two-, three-ridged, etc.). These terms are particularly applied to small faceted ventifacts.

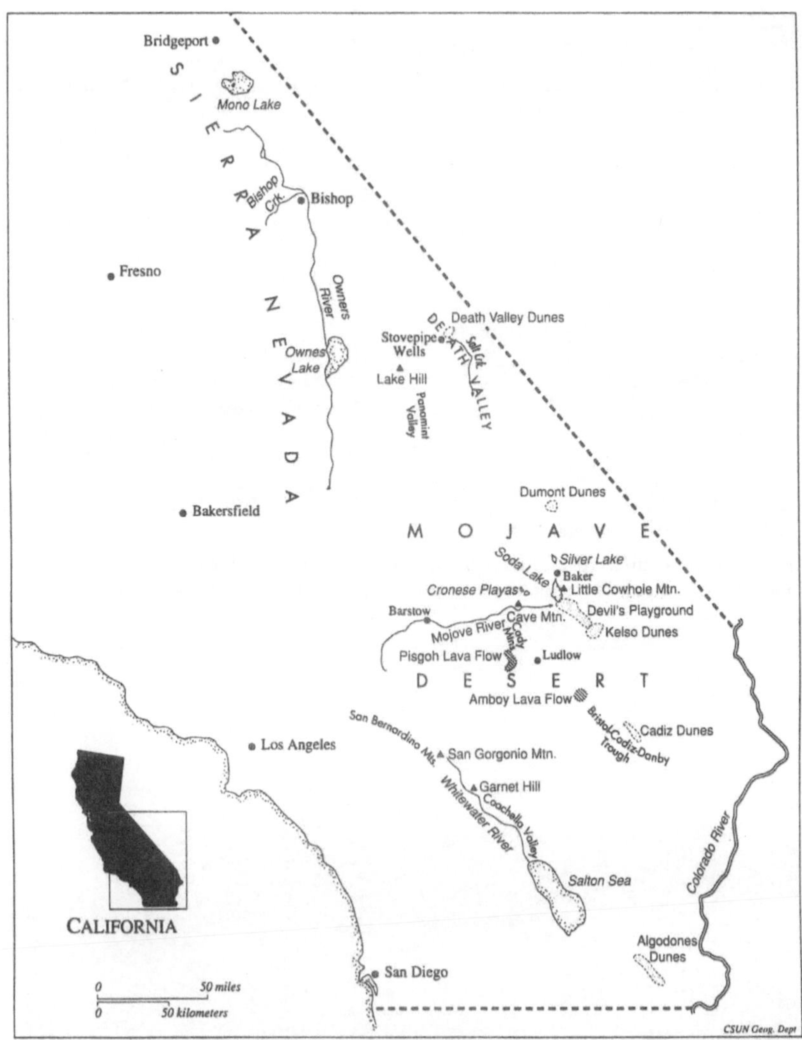

Figure 1. Ventifact sites in California discussed in the text.

Figure 2. Marble ventifacts of the Little Cowhole Mountains are grooved and polished, and exhibit a high degree of sheen. These ventifacts are periodically buried by dune sand.

Figure 3. Marble ventifacts along a topographic ridge of the Little Cowhole Mountains are abraded by opposing winds on both their north- and south-facing sides, the two faces separated by a sharp keel.

Figure 4. Pitted face of ventifact on moraine on the eastern side of the Sierra Nevada. Pits cover the boulder surface up to a height of several meters. Lichen growth (light spots) and spalling near the boulder base provide evidence that this feature is relict. Photo courtesy of Tim Boyle.

Figure 5. Closer view of pits shown in Figure 4. Lighter spots represent lichen growth. Photo courtesy of Tim Boyle.

Figure 6. Very large boulder on a high hill southwest of Soda Lake. The height of maximum erosion is about 1.5 m above the surface. The upper surface of the boulder is covered in flutes, the size of which increases toward the top of the boulder. Figures 7 and 16 show additional detail of this rock.

Figure 7. Closer view of the rock shown in Figures 6 and 16. As a consequence of increasing abrasion intensity with elevation above the surface, the upper face of the boulder has receded and the scale of fluting increased.

Figure 8. Large relict ventifact from the Little Cowhole Mountains. Features on the surface include flutes and long and deep grooves, up to 7 cm in width.

Pits

A pitted surface is one indented by closed depressions, often of irregular shape. They occur on surfaces that are inclined at high angles to the wind (55° to 90°) (Figures 4 and 5). These features are relatively common in basalt, and may result from enlargement and integration of preexisting vesicles. Granitic boulders may also be strongly pitted, and in some instances pit diameter increases with height up the boulder face.

Flutes

Flutes are scoop-shaped in plan, open at the downwind end and closed at the upwind end, and broadly U-shaped in cross-section. This "arrowhead" form is a useful indication of wind direction. Flutes become shorter and deeper as the inclination of the surface steepens (Maxson 1940, Sharp 1949). (Figures 6 and 7). Flute development appears to be independent of material hardness, composition, or rock structure. On large curved faces, the flutes or grooves radiate outward, often from a central pitted area.

Grooves

Grooves are longer than flutes and are best developed on surfaces gently inclined or parallel to the wind. Ventifact groove and flute trends are typically parallel on near-horizontal surfaces and reflect the flow direction of the highest velocity winds in an area (Laity 1987). An interesting example of groove formation was found on a large sheet of Styrofoam embedded in the Mojave River. Heavy rains in the San Bernardino Mountains caused flow of the river in the winter of 1992-1993, which removed a residual armor of coarse debris.

Figure 9. Etching occurs when the wind selectively erodes less resistant strata or foliation, such as in these volcanic rocks south of Soda Lake, near the Mojave River sink.

Figure 10. Basalt ventifact on a high ridge south of Owens Lake. Flutes and helical forms radiate away from the central area of the boulder. The face of the boulder is convex in both vertical and horizontal cross-sections through the central area. Photo courtesy of Tim Boyle.

Figure 11. Close-up of helical forms in basalt south of Owens Lake. Such forms appear to form in areas of very high wind intensity. Photo courtesy of Tim Boyle.

The desiccation of the channel was coincident with the spring peak of high wind velocity in the lower Mojave Valley, and blowing sand created near "white-out" conditions, with winds gusting in the channel to 16 m s^{-1} on March 14, 1993 and up to 22 m s^{-1} on May 6, 1993. The Styrofoam board was covered with grooves oriented at 260°, parallel to the unidirectional westerly flow, and perpendicular to wind ripples formed on the channel bed. These grooves had probably formed in less than three weeks.

The scale of ventifact grooves ranges from fine lineations, with a wavelength measured in millimeters, to long and deep grooves, up to 7 cm in width (Figure 8), in active and fossil ventifacts of the Little Cowhole Mountains.

Etching

Etching occurs when the composition of a rock mass is not homogeneous, and the wind selectively erodes less resistant strata or foliation. Excellent examples of foliation may be seen in volcanic rocks south of Soda Lake, near the Mojave River sink (Figure 9).

Helical forms

Helical forms begin as shallow grooves or flutes, deepen and spiral in a downwind direction, and terminate in a sharp point. Helices are relatively rare features. Excellent examples occur in marble in the Little Cowhole Mountains (see Laity 1994), but even there they are not common. Other examples may be seen in basalt exposed along high ridges south of Owens Lake (Figures 10 and

Figure 12. Boulder located in a low pass between hills, southwest of Soda Lake. Strong winds funneling through this pass (moving from right to left in the picture) have abraded the face of this large boulder (2.3 m long and about 1 m high). The height of maximum abrasion is about 1 m above the surface. As a consequence of the increase in abrasion up to the maximum level, the ventifact has developed a semi-planar face covered in flutes and grooves.

11). Their topographic position indicates that they form where wind velocities are very high.

Role of Topography on Wind Speed and Ventifact Formation

Topographic enhancement of wind speed is a major influence on ventifact distribution and ventifact form. Regional winds are sometimes strong enough to transport sand across relatively level surfaces and to abrade rocks, as on the Pisgah and Amboy lava flows. However, ventifacts are most common where winds are topographically enhanced, by either large- or small-scale topographic constrictions (causing a Venturi effect), or by the acceleration of wind flow up hills. The topographic gap immediately south of Ludlow in the Mojave Desert provides an example of a large-scale constriction. A series of eight 100-m transects was laid out at 2-km intervals immediately south of Ludlow in the Bristol-Cadiz-Danby trough. Faceting and grooving affected 70% to 90% of all exposed cobbles and boulders (Laity 1987). The ventifacts are relict and only small patches of sand remain in this area today. In the past, wind accelerated through this valley constriction, carrying sand that was deposited to the south in the area of the Cadiz dunes. At a smaller scale, dips and saddles along ridges and passes between hills also accelerate wind flow and result in enhanced abrasion (Figures 8 and 12).

Figure 13. Fossil ventifacts on hill crests to the south and west of Soda Lake show signs of present-day reactivation. Sand blast has freshened ventifacts (surfaces lighter gray in appearance), partially removing stained and weathered surface material (darker-toned in appearance). The boulders are about 1 m in height.

High wind speeds cause hilltops to be a particularly favorable ventifact locale. As wind passes over a hill there is a compression of streamlines in the boundary layer that causes winds to accelerate towards the crest of the slope, increasing sand transport and particle velocity. As a result, ventifacts form on windward upper slopes or summits, the most intensely abraded rocks occurring near the hill crest (Figures 6, 9, 10, and 11) (Laity 1987).

FORMATION OF VENTIFACTS

Agent of Abrasion

Research in California has attributed the formation of ventifacts to a sandblast action (Blake 1855, Blackwelder 1929, Maxson 1940, Sharp 1964, 1980, Smith 1967, Greeley and Iversen 1986, Smith 1984, Laity 1987, 1992, 1994). Although some researchers have attributed the formation of ventifacts to dust action (Whitney and Dietrich 1973, Breed et al. 1989), ventifacts in California do not appear to be found in areas subject to dust influx alone.

Sediment collected adjacent to marble ventifacts in the Little Cowhole Mountains and basalt ventifacts south of Owens Lake was composed of well-sorted fine sands: the grains are dominantly quartz in the Little Cowholes, but about 25% of the material from Owens Lake was basalt. Greeley et al. (1984) showed that there is little difference in abrasion by quartz and basalt particles.

Figure 14. The Little Cowhole Mountains ventifact site. The reversing dune along the ridge crest changes form and position in response to opposing winds, alternately burying and exposing ventifacts. The ventifacts are actively forming.

The Amboy lava field is also traversed and abraded by typical aeolian sands, consisting of fine- to medium-sized sand which is moderately well sorted (Greeley and Iversen 1986).

Rate of Abrasion

The time required to form a ventifact is difficult to determine. The original mass of the rock, the time of onset of abrasion, and the rates of abrasion are generally not known. Natural rates of abrasion are difficult to determine and are probably highly variable through time owing to the many different controlling factors that influence ventifact formation. Wind velocities are not constant, but vary according to season, time of day, and the passage of fronts. Most of the abrasion occurs during periods of high-velocity winds, which occur for only a small percentage of the time. Abrasion rates also change because, as erosion progresses, the rocks gradually wear, changing the angle of incidence of the impacting grain. If abrasion is episodic through time, weathering of the rock between erosional episodes may prepare it for abrasion and increase subsequent rates of surface wear. Weathered material may also be removed partly by deflation.

Many ventifacts have been subject to multidirectional winds, and are abraded on more than one face. The cessation of abrasion may have occurred at different times and considerable care must be taken in determining the ventifact "age," Furthermore, abrasion may cease for a period and later

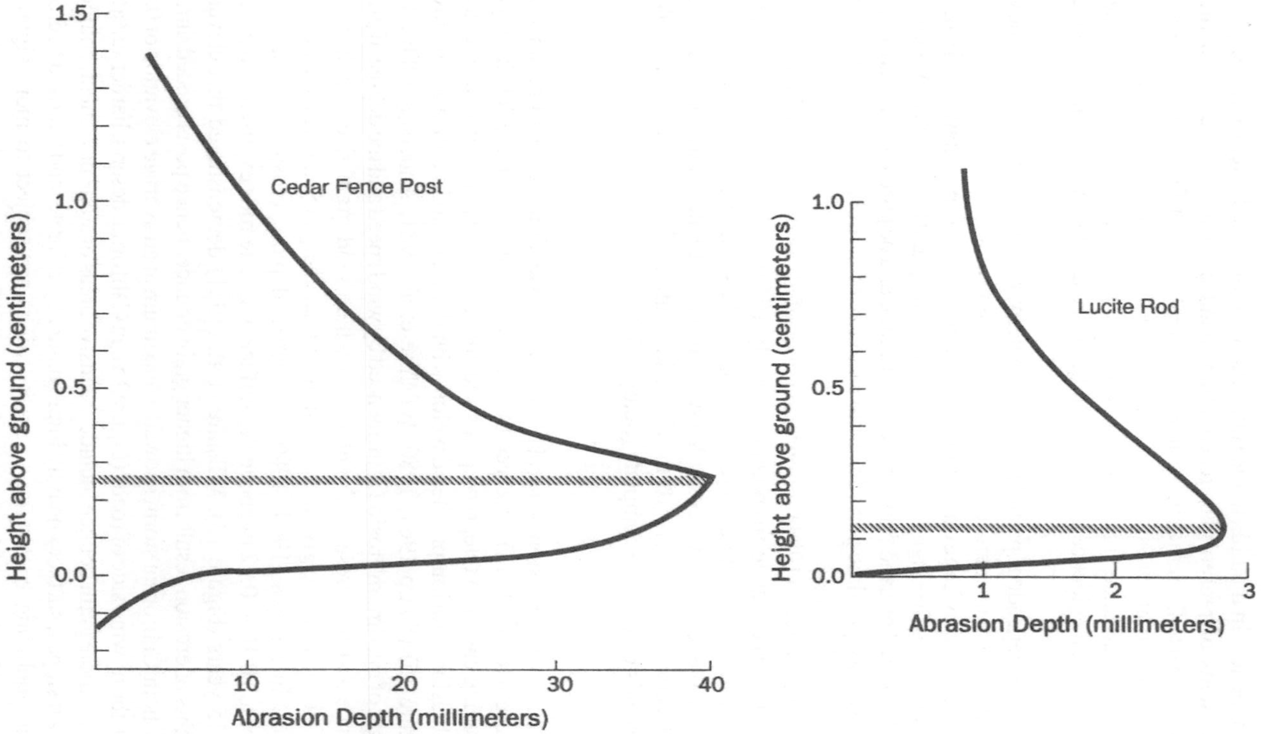

Figure 15. Within a general zone of sand abrasion, erosion profiles develop with distinct maxima of mass removal. Wilshire et al. (1981) demonstrated that elevated heights of erosion on fenceposts result from higher grain bounce on harder surfaces (roads). Posts eroded during a 24-hour California storm showed an average maxima of 0.28 m. In Sharp's experimental plots, erosion maxima on Lucite rods were recorded 0.10 to 0.12 m above the surface (Sharp 1964, 1980).

Cedar Fence Post

Height above ground (centimeters)

1.5
1.0
0.5
0.0

10
20
30
40

Abrasion Depth (millimeters)

Lucite Rod

Height above ground (centimeters)

1.0
0.5
0.0

1
2
3

Abrasion Depth (millimeters)

recommence. In the hills to the south and west of Soda Lake, fossil ventifacts show signs of present-day reactivation. Sand blast has freshened ventifact surfaces, partially removing stained and weathered surface material (Figure 13). The remobilization of sand may be a natural process or be promoted by surface disturbance from off-road vehicles. Regardless of the cause, these rocks provide evidence of the sometimes episodic nature of ventifact formation.

Time-dependent particle flux also determines abrasion rates. The movement of sand in aeolian corridors tends to be episodic, so that sites which have been traversed by large quantities of saltating grains may now harbor only small accumulations (Mainguet 1972, Sharp 1980, Sharp and Malin 1984). Ventifacts may be buried by sand for short periods of time and be protected (Figures 2 and 14). Over longer time periods, the availability of particles may decline owing to climate change.

Greeley et al. (1984) summarized rates of abrasion inferred from various sites. The rates range from approximately 10^{-5} to 10^{-1} cm yr^{-1}. Sharp (1964, 1980) concluded that ventifact formation in the Coachella Valley, California, would have required dozens or hundreds of years. In the central Namib Desert the faceting of boulders 10 to 50 cm in diameter is thought to have taken thousands of years to complete (Selby 1977).

Erosion Profile

Effective sand abrasion on level terrain rarely extends more than 1 m above the surface. Hobbs (1917) observed this limit on a variety of materials in Egypt, including cast-iron telegraph poles and adobe walls. Within this general zone of abrasion, erosion profiles develop with distinct maxima of mass removal (Figure 15) (Sharp 1964, 1980, Wilshire et al. 1981, Anderson 1986). The height of maximum erosion is influenced by wind speed and the degree of grain bounce and it represents the elevation where sand grain size, number, and velocity combine to give the maximum impact energy (Sharp 1964, Anderson 1986). In Sharp's (1964, 1980) experimental plots, erosion maxima were recorded 0.10 to 0.12 m above the surface in Lucite rods exposed to the wind for 15 years (Figure 15). Wilshire et al. (1981) demonstrated that elevated heights of erosion result from higher grain bounce. Fence posts eroded during a 24-hour California storm showed a maximum at an average elevation of 0.28 m on the upwind side of roads (Figure 15). In California deserts, harder surfaces that promote grain bounce include extensive areas of stone pavement, boulder-strewn slopes, surfaces heavily impregnated by calcrete, and exposed bedrock areas, including basalt flows. Such settings are subject to more vigorous erosion (Greeley and Iversen 1986, Laity 1987).

Field observations in hilly terrain suggest that the height of maximum abrasion can be higher than hitherto recorded where winds are accelerated upslope and funneled through passes or saddles. Figures 12 and 16 show abrasion maxima at heights of 1 m and 1.5 m above the surface.

As a consequence of the increase in abrasion up to the maximum level,

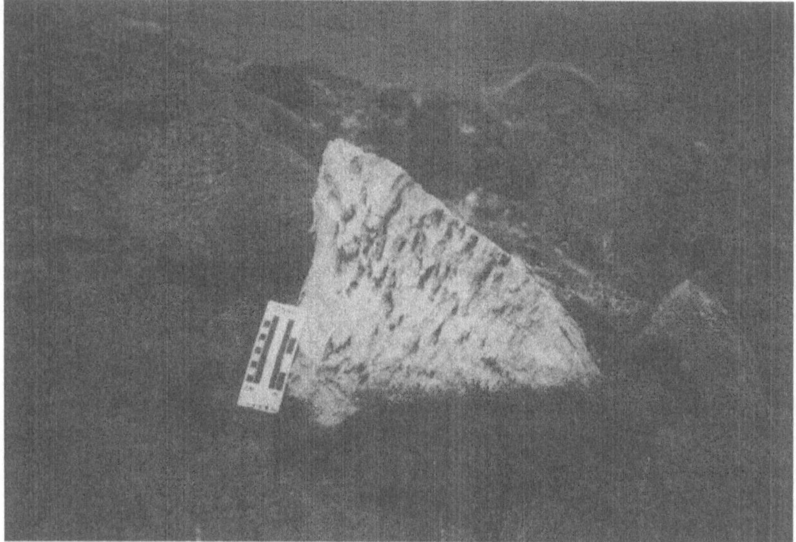

Figure 16. Fossil ventifacts on a hill to the north of the Dumont Dunes. The relict nature of the site is indicated by the lack of active sand at the surface and the weathered nature of the boulders. The light carbonate rock in the center of the image is intensely fluted.

many ventifacts ultimately develop semi-planar faces, with the upper part of the abrasion face receding more rapidly than the lower part. Flutes and grooves may cover the receding face (Figure 12). Some boulders develop convex faces (Figure 10 and 16), wherein the upper face and receding limbs of the boulder are eroded more rapidly than the central surface of the rock.

Preservation of Ventifacts

Ventifacts develop in a wide range of rock types including basalt, granite, aplite, andesite, marble, dolomite, and limestone. Polishing and smoothing of rock surfaces have been observed on all lithologies. The occurrence and long-term preservation of ventifacts in different rock types are affected by physical and chemical weathering processes. In general, coarse-grained rock such as granite does not preserve grooving well, owing to granular disintegration and spalling. In the Cady Mountains evidence of abrasion on granite ventifacts is patchy and discontinuous, whereas on adjacent basalt ventifacts grooves and flutes are very well preserved. Andesite also appears to weather rapidly in the desert, and evidence of wind abrasion is very patchy, with little preservation. Ventifacts formed of limestone, such as those at Lake Hill in the Panamint Valley, are prone to chemical weathering and etching, and such secondary erosion may cause initial problems in directional interpretation. Fine-grained materials that are less sensitive to physical and chemical weathering, such as marble and basalt, best preserve evidence of abrasion. Marble produces beautiful ventifacts, and appears relatively resistant to chemical weathering. The surface dulls and discolors, but there is no loss of directional information.

VENTIFACTS IN CALIFORNIA

Early Observations of Ventifact Formation: 1855-1940

Published works dealing with ventifacts in the southwestern United States date back to the mid-19th century. The first recognition of aeolian erosion was by Blake (1855), who described ventifacts on a projecting spur of San Gorgonio Mountain, in the San Bernardino Mountains of southern California (Figure 1). The granite was "cut into long and perfectly parallel grooves and little furrows, . . . was beautifully smoothed, and though very uneven, had a fine polish." While Blake contemplated whether such forms could have been caused by glacial erosion, the solution to the problem presented itself—the rocks were enveloped in moving sand that was deposited in deep banks and drifts on the lee side of his point of observation. Garnet or quartz crystals embedded in feldspar stood out in relief, resulting in projecting "fingers" of rock, such that the ventifacts "form in reality a perfect index of the winds' direction." Blake noted that the effects of driving sand were seen in "all parts of the Desert where there are any hard rocks or minerals to be acted upon." He pondered whether a ventifact covered for a long period with moist earth would suffer surface decomposition, removing the polish from the furrows and leaving doubt as to the origin of the feature. He also observed that whereas the action of a single grain was essentially immeasurable, the combined action of many grains operating over long time periods had resulted in "cubic yards of granite . . . cut into dust and driven before the wind over the expanse of the Desert."

Numerous other authors have contributed to our knowledge of ventifacts based on research conducted in the southwestern United States. Gilbert (1875) echoed the observations of Blake, attaching considerable importance to sand-blast action in arid climates where it was "not merely appreciable, but even important." Working in the Colorado Plateau area, he observed that topographic controls on abrasion were significant: erosion was greatest in "passes and contracting valleys, where the wind is focused, and its velocity augmented...but no inconsiderable work is accomplished on broad plains, where its normal force only is felt." Gilbert described the erosion of small pebbles, large boulders, pedestal rocks, and the etched bases of cliffs. Polish, erosional form, and ventifact preservation were observed to vary according to material type. Abrasion of sandstone, for example, was seen infrequently owing to poor preservation of the friable material.

Blackwelder (1929) examined fossil ventifacts along the eastern base of the Sierra Nevada. The rocks exceed a meter in height and are abraded extensively across their windward surface with pits 5-10 mm in diameter and 10-25 mm in depth, and long grooves 5-15 mm in width. On the leeward face "no such markings were found, but on the contrary the rocks are merely cracked and exfoliated in the ordinary manner." Blackwelder's study helped to demonstrate that most ventifacts show significant erosion on their windward faces and no erosion on their lee surfaces, a point of contention in the present-day literature (see Breed et al. 1989).

Maxson (1940) studied ventifacts near Stovepipe Wells and Salt Creek in the northern part of Death Valley. The area was characterized by strong winds blowing from two opposing directions. Abraded fragments of limestone on alluvial fans developed facets in response to each wind, separated by a ridge running along the spine at right angles to the wind.

The work of Blake (1855), Gilbert (1875), Blackwelder (1929), and Maxson (1940) in California included the following observations that are a mainstay of modern day research: (1) ventifacts are eroded by moving sand; (2) sand is responsible for surface microfeatures, including grooves and polish; (3) such grooves, as well as projecting fingers of rock and inter-facet ridges, may be used to ascertain wind direction; (4) the surface texture and morphology of ventifacts is altered following the cessation of abrasion; (5) significant mass removal of rock may result from prolonged abrasion; (6) topography enhances the wind, and thus sand-blast action; (7) ventifact preservation varies according to rock type; and (8) ventifacts are abraded on their windward faces, and show no erosion to the lee.

Recent Ventifact Studies: 1941-Present

Sharp (1964, 1980) extended an interest in ventifacts initially gained in Wyoming (Sharp 1949), and conducted the first experimental field research on ventifacts in the Coachella Valley, California. Lucite rods, common red bricks, and cubes of commercial gypsum cement were placed in an experimental plot and the amount of wear measured. The 15-year study of abrasion (Sharp 1964, 1980) demonstrated that abrasion rates are in part determined by time-dependent particle flux. Saltating sand driven by a strong unidirectional wind was derived from the nearby Whitewater River channel. Occasional floods in the river provided a fresh source of abradant. Sharp (1980) showed that the annual rate of wear was 15 times greater during the final three-year interval than in preceding years owing to an increased flux of material derived from recent flooding. Polish, pitting, and incipient fluting developed within 10 months during periods of intense abrasion.

Ventifact Occurrence and Age

Occurrence and distribution of ventifacts in California

Ventifact formation is influenced by factors similar to those affecting dunes: wind frequency, magnitude, and persistence, as well as sediment supply, basin geomorphology, and vegetation cover. Sediment supply is of particular significance to an understanding of the occurrence of ventifacts in the southwest United States. Present-day winds often exceed the threshold velocity for sand transport, particularly when topographically enhanced, but actively forming ventifacts are relatively rare. Thus, erosion must be limited principally by the availability of abradant.

Figure 17. The same boulder as shown in Figure 6. This rock is developing a convex face, wherein the upper face and receding limbs are eroding more rapidly than the central surface.

Ventifacts are commonly found in areas where strong winds are combined with abundant moving sand. These include sites near lake shorelines, downwind of alluvial rivers, adjacent to dune sands, and in corridors of former sand transit.

Along former lakes, sand was blown from the shore lines as water level dropped. Resulting ventifacts include those found in hills to the northwest of Silver Lake, south of Owens Lake, and west of East Cronese playa.

Sand blown out of alluvial rivers following flood events abraded downwind boulders. Sand from the Mojave River, in transit through the Lower Mojave Valley, abraded boulders in the Cady Mountains. The Whitewater River was the source of abrasive material which formed ventifacts at Garnet Hill (Sharp 1964, 1980).

Ridges and hilltops near dune fields are good areas to locate ventifacts. Examples include ventifacts near the Dumont Dunes, Death Valley Dunes, Algodones Dunes, Panamint Dunes, the Devils Playground, and the Kelso Dunes. Such ventifacts are of particular value to the understanding of aeolian geomorphology as they commonly record wind directions associated with dune formation.

Aeolian corridors or plains, which have been traversed by large quantities of saltating grains in the past, may also show signs of abrasion. Ventifacts to the south of Ludlow provide an example of these conditions.

Finally, ventifacts are found on moraines along the eastern margin of the

Sierra, Nevada (Blackwelder 1929; Bach, this volume). Sediment derived from the glacial environment is transported by strong katabatic winds to abrade till at Bishop Creek and near Bridgeport, California.

Ventifact ages

The age of ventifacts and the time necessary for their formation are difficult to ascertain. The relict nature of ventifacts is often indicated by the weathered, dulled, fretted, or partly exfoliated condition of the rock surfaces (Blackwelder 1929, Smith 1967, Smith 1984, Laity 1992) and by the presence of rock varnish and rock coatings (Dorn 1986, this volume). Grooves and fluting may not cover the entire surface of the boulder or outcrop, but rather occur in patches where weathering has failed to remove them. The growth of vegetation surrounding ventifacts, the lack of any apparent wind-blown material, or the stabilization of and incipient soil development on aeolian deposits also indicate the fossil nature of ventifacts (Figure 8 and 17).

Weathering which is not visible macroscopically may be visible microscopically. Scanning electron micrographs of rocks undergoing active abrasion show sharp impact structures with fresh cleavage facets and no loss in definition at high magnifications. In addition, the surrounding abradant sands are mobile. By contrast, 6000-year-old ventifacts south of Ludlow show considerable post-erosional smoothing of the surface (Laity 1992).

Numerical ages of ventifacts have been determined from rock varnish that forms within grooves following the cessation of abrasion. Dating has been by either radiocarbon or cation-ratio techniques, or a combination of the two. The small amounts of organic matter incorporated during varnish formation require accelerator mass spectrometry (AMS) ^{14}C dating in order to ensure adequate precision. Such ages provide minimum limits for the subaerial exposure of the rock underlying the varnish. The initial results suggest that abrasion in the Mojave Desert was probably greatest in the early- to mid-Holocene and greatly diminished in the late Holocene (Dorn et al. 1989, Laity 1992). This work confirms the suggestions of H.T.U. Smith (1967). As pluvial conditions diminished in the late Pleistocene to early Holocene, the deflation of exposed stream-channel and lakeshore sediments resulted in the widespread movement of sand and dust in the east-central Mojave Desert and promoted the development of ventifacts.

In contrast to the Holocene ages of desert ventifacts, cation-ratio ages for varnishes formed on ventifacts of glacial moraines at Bishop Creek, eastern California, suggest aeolian abrasion during the early and late Wisconsin glacial stages at about 60-65 ka and 15-22 ka (Bach, this volume). The sites of abrasion on the moraines changed through time in response to changing glacier positions.

Ventifact Significance: Reconstructing Paleocirculation Using Ventifacts

Ventifacts provide an excellent record of near-surface circulation. Their morphology (the position of the sharpest bounding edge of a facet, the pitting on a face, or the direction of grooving and pitting) allows determination of wind direction. For faceted ventifacts, the keel is oriented in a large majority of cases at right angles to the wind (Maxson 1940). Fossil ventifacts have been used to reconstruct paleocirculation and to interpret climate change. Blackwelder (1929), H.T.U. Smith (1967), R.S.U. Smith (1984), Dorn (1986), and Laity (1992) have noted the relict nature of many California ventifacts. H.T.U. Smith (1967) attributed ventifact formation to a time of more intense and protracted wind action during the Altithermal. Fossil ventifacts in coarse-grained crystalline rock at Cave Mountain showed ubiquitous weathering at ground level, with the marks of aeolian abrasion best preserved on the upper surfaces (H.T.U. Smith 1967).

Both Groat (1967) and Anders (1974) used ventifacts as a proxy for detailed surface wind data in the Mojave Desert east of Barstow. Present-day winds are unidirectional (westerly). Ventifacts in the Cady Mountains were fluted in a west to east direction, and were severely pitted on the west sides and unpitted on the east (Groat 1967). Anders (1974) used pitting and fluting to indicate the direction of sand transport in a study of topographic influences on wind patterns and sand deposition. A comparison of ventifact orientation with modern wind directions determined from vegetation sand tails (shadows) suggested that the wind regime had shown little change in this area since the interval of more arid conditions in the mid-Holocene.

R.S.U. Smith (1984) examined the aeolian geomorphology within a 50-km radius of Silver Lake, north of Baker, California, and concluded that ventifacts provide paleocirculation data which enhance our understanding of dune field formation. Ventifacts record all directions of sand transport, unlike climbing and falling dunes, regional sand streaks, and lee dunes which indicate a resultant azimuth of sand transport (Smith 1984). The multidirectional wind regime recorded by inactive Holocene ventifacts near the Kelso Dunes (Smith 1984) is consistent with Sharp's (1963) conclusion that the dunes are localized at a crossroads of winds whose net effect on sediment transport is near zero.

Most fossil ventifacts in California appear to record wind directions similar to those of today. However, ventifacts in the Lower Mojave Valley east of Barstow record both northerly and southerly flow in an area presently dominated by unidirectional westerly flow (Laity 1992). Ventifacts showing north- and south-trending grooves are best preserved on slopes sheltered from the prevailing Westerlies. However, in some cases, cross-grooving of the boulders is observed. Accelerator mass spectrometry (AMS) ^{14}C and cation-ratio dates of organic matter at the base of varnish formed on ventifacts in the Cady and Cronese Mountains are in close accord and suggest that a pulse of erosion associated with northerly and southerly flow ceased about 5100 to 6700 yr B.P. The fossil grooves may have been formed owing to a westward incursion of an enhanced monsoonal system during the intensification of summer solar radiation values in the Northern Hemisphere at 6000-9000 yr B.P (Laity 1992).

Case Study: Ventifacts in the Little Cowhole Mountains

To date, fossil ventifacts have provided the main basis for the study of ventifact formation. Studies of active wind abrasion are uncommon, and were conducted on relatively flat ground (Wilshire et al. 1981) or in areas of limited sand supply (Sharp 1964, 1980). Thus, a sand-blasted ridge covered by marble ventifacts in the Little Cowhole Mountains provides a unique opportunity to study ongoing processes of erosion in an area of hilly terrain and abundant sand (Figures 2, 3, and 14).

The Little Cowhole Mountains are located east of Soda Lake and north of the Devil's Playground and Kelso Dunes (Figure 1). Both actively-forming and fossil ventifacts occur here. Fossil ventifacts are found in association with stabilized sand, vegetation, and lag deposits. Deep grooves on large ventifacts attest to the intensity of past abrasion (Figure 8). Wind speeds at the sites of fossil ventifacts equal or exceed those recorded where active ventifacts are forming. Thus, in the present environment, it appears that the supply of abradant determines whether forms are active or fossil.

Active ventifact formation is being examined on a low eastern ridge of the mountains. An automatic weather station monitors wind flow speed and direction and other meteorological variables, and an array of 11 additional anemometers around the study site provide data on the role of topography on sand transport and ventifact abrasion. During the first year of operation (1993-1994) the anemometer recorded wind gusts exceeding the threshold velocity for sand transport during every month of the year, and recorded maximum gusts of 27 m s^{-1}. The prevailing wind is southerly: however, when winds exceeding 10 m s^{-1} are analyzed, the significance of bidirectional flow (northerly and southerly) is apparent. A partially stabilized lee dune, with a north-south axis extending from the south-facing slope, suggests that long-term sand transport has been southward, towards the Devils Playground and Kelso Dunes. An additional geomorphic indicator of the wind regime is a small reversing dune near the summit which migrates in response to the opposing winds, but shows no apparent net movement. Moving dune sand periodically buries boulders and bedrock, protecting the underlying surfaces from abrasion (Figures 2 and 14).

The majority of erosion involves individual boulders, but in some areas, planation of the marble bedrock has leveled areas of tens of square meters. Abrasion occurs on exposed rock of the windward slope during periods of strong wind action (10 m s^{-1} or greater). Ventifacts along the topographic ridge are abraded on both their north- and south-facing sides, the two faces separated by a sharp keel (Figure 3). Rocks on the lower slopes show abrasion only on the surface which faces the impinging wind. In addition to the direction of the wind, the effect of local ridgeline topography is important in determining groove and flute orientation. Notches or passes along the ridge strongly funnel wind flow, and ventifact grooves parallel the notch axis (Laity 1992).

As previously discussed, natural rates of ventifact abrasion are difficult to determine and rates change as rocks wear, changing the angle of incidence of

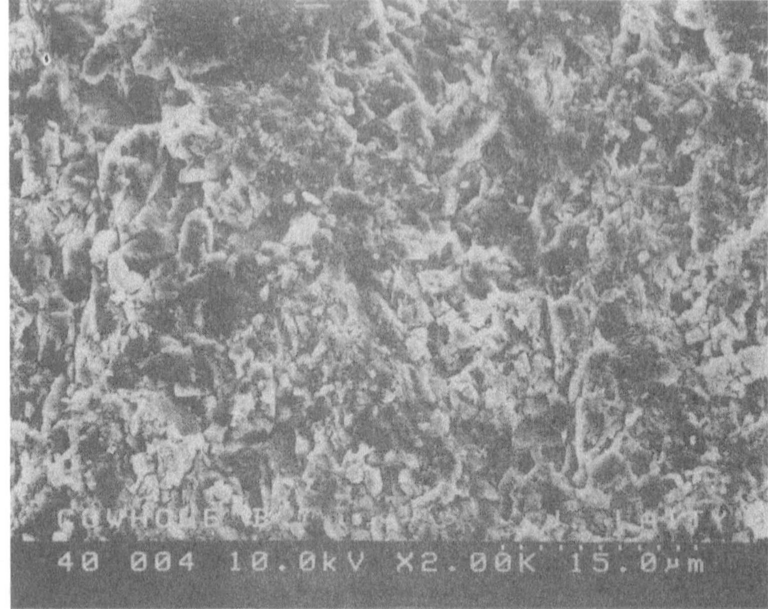

Figure 18a, b, c. Scanning electron micrographs of the top (a) and below (b) and right (c) of a 30 cm high marble ventifact from the Little Cowhole Mountains. The scale is 2000X. At the top of the rock cleavage fracture is the major abrasion mechanism. Figure 18a shows a groove surface covered entirely by sharp cleavage features. The ridge at the top of the boulder (Figure 18b) show patchy cleavage, but rock smoothing by rubbing and microgouging is also apparent. Small microgouging marks from grains skidding up the face of the rock are prevalent.

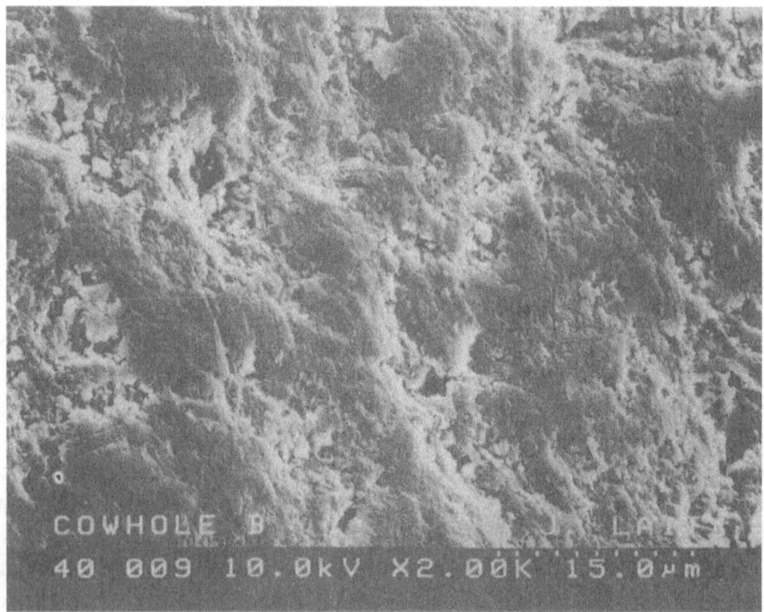

Figure 18b.

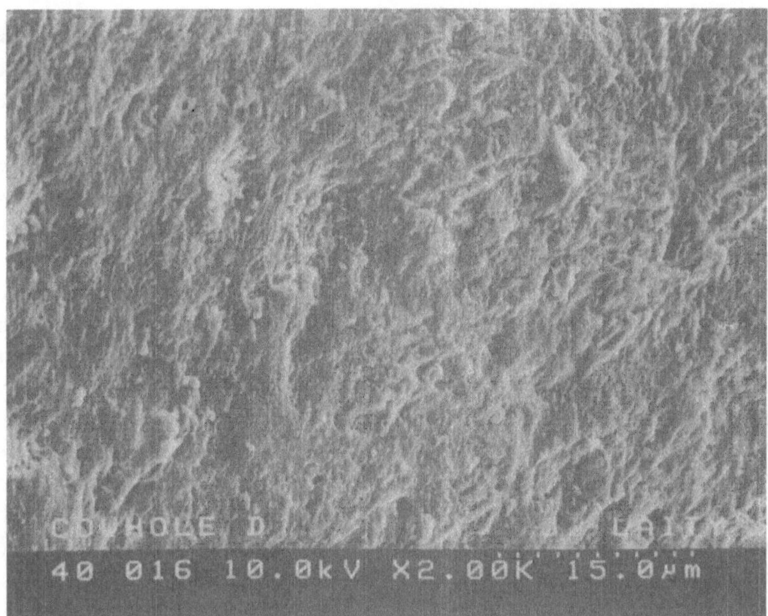

Figure 18c. At the base of the rock, there is no evidence of cleavage and material appears to have been rubbed and gouged away, enhancing the visibility of differential hardness in the marble. By contrast, at the top of the boulder high impact abrasion and cleavage obliterate effects owing to variations in hardness.

impacting grains. Six rigid, closed-cell foam blocks of different densities are being field tested to determine the most suitable for long-term experimentation on rates and patterns of erosion. The foams have great resistance to water absorption and to UV degradation and in other contexts are commercially sand-blasted to make sculptured sign letters, shapes, and figures. "Boulders" are placed near the anemometers to examine mass removal as a function of wind velocity, time, and rock shape. The role of changing face angle on rates of abrasion, and the role of high-velocity winds on the long-term evolution of ventifact form and surface features will be studied.

Nature of Abrasion

In addition to studying macroscopic effects of abrasion, the microscopic evolution of ventifacts is being examined using scanning electron microscopy. For saltating grains, particles traveling at greater heights have higher velocities, owing to an increase in wind speed above the ground and the longer saltation paths that allow more time for them to be accelerated by the wind (Greeley et al. 1984). As the mass removed per impact on most rocks scales roughly with the kinetic energy of the impact, it is likely that microscopic differences in abrasion occur with height up a boulder face. In order to test this hypothesis, the top and bottom of a 30-cm-high ventifact from the Little Cowhole

Mountains was examined using a scanning electron microscope (Figure 18a, b, and c). Both grooves and ridges were examined. Megascopically, there is little difference between the top and base of the rock: both areas have small-scale grooving, and the rocks are polished and very smooth to the touch. However, microscopically there is considerable variation with height. At the top of the rock, cleavage fracture of crystalline material is widespread. It is the only mechanism observed to remove material from the upper grooves (Figure 18a). The ridges show patchy cleavage, but rock smoothing by rubbing and microgouging is also apparent (Figure 18b). At the base of the rock, there is no evidence of cleavage, either in the grooves or on the ridges. The rocks appear to have been rubbed and gouged away (Figure 18c). Small microgouging marks from grains skidding up the face of the rock are prevalent. These lower energy forms of abrasion enhance the visibility of differential hardness in the marble, whereas at the top of the boulder high-impact abrasion and cleavage obliterate effects caused by variation in hardness. Both the Little Cowhole marbles and basalt ventifacts from the Pisgah flow (Laity 1991) show that surfaces which appear smooth to the eye and touch can be very rough at the microscale. These two sites suggest that dust need not be invoked to explain polish or fine features.

CONCLUSIONS

Ventifacts are relatively common in many regions of California, occurring where strong winds have driven available abrasive material against resistant rocks that weather slowly. Climate change influences the development and longevity of ventifacts by affecting wind speed and direction, the nature of the surface over which the wind blows (rocky or sandy conditions, vegetation type and density), the supply of abrasive sand, and the rates and nature of rock weathering. Thus, there are spatial variations in the age and intensity of abrasion in different environmental settings.

Most ventifacts are relict. In desert areas they formed during a drier, middle Holocene period from 8 to 5 ka. However, it appears that selected areas remained active into the late Holocene, and a few areas remain active today. Ventifacts occur proximal to Pleistocene lake shorelines, downwind of alluvial rivers, near dune fields, or in areas of regional sand transit. They are more common in the central and eastern Mojave Desert than the western Mojave owing to the long-term emplacement of sand in this region by westerly winds. Their occurrence in the Colorado Desert has received somewhat less attention (Sharp 1964, 1980, Dorn, personal communication).

Mapping and dating ventifacts provides one approach to understanding the nature of past climate change. Rocks and outcrops abraded by wind-transported sand are often faceted, grooved, fluted, and pitted. Numerous investigators have shown that wind direction may be determined by reference to the position of the sharpest bounding edge of a facet, by pitting on a face, or by the orientation of grooves and flutes. Paleocirculation patterns may be deduced when ventifact mapping is conducted over large geographic regions, providing detailed information on surface flow that is not attainable by other methods of paleoclimatic reconstruction. Determining the ages of ventifacts is an essential element of the research. Such work provides an important complement to existing research that infers paleowind regimes by examining sand dunes (Tchakerian 1994).

The study of past climates using fossil ventifacts is enhanced by research on the formation of contemporary ventifacts. A long-term study of ventifact formation in the Little Cowhole Mountains seeks to relate contemporary ventifact formation to the modern wind regime. A weather station and an array of anemometers placed around the site indicate that erosion occurs on the windward face of boulders, causing grooves and flutes that are aligned with the highest velocity winds. Processes of abrasion are examined using the scanning electron microscope. Preliminary studies indicate that the development of the erosion profile in ventifacts may be related to different mechanics of abrasion, with higher energy impacts inducing cleavage fracture near the tops of marble boulders, and lower energy impact near the boulder base rubbing or gouging away material.

Future research in the Little Cowhole Mountains will use micrography to examine how abrasion varies spatially in flutes and helices, in order to shed light on these poorly understood forms. The difference in abrasion between ridges and grooves in grooved rocks, and type of abrasion in low- and high-angle impacts will also be examined. The study will investigate the pattern of wind velocity around the hill and its relation to ventifact erosion and form. Foam "boulders" will be placed adjacent to each of the anemometers to examine changing rates of abrasion and the long-term evolution of ventifacts. Finally, the height of abrasion maxima in hilly topography will be related to microtopographic influences on wind speed.

ACKNOWLEDGMENTS

The work described in this paper benefited from the generous assistance of several colleagues. Saverio D'Agostino and Tim Boyle helped in the field. Kevin Mulligan provided the anemometer array, and Tim Boyle and Mark Kuhlman helped to establish and maintain the weather station. Scanning electron micrographs were obtained with the assistance of Ken Evans and Saverio D'Agostino of the Jet Propulsion Laboratory. Reviews of this paper by Antony Orme and Amalie Orme are gratefully acknowledged.

REFERENCES

Anders, F. J. (1974) *Sand Deposits as Related to Interactions of Wind and Topography in the Mojave Desert, near Barstow, California.* M.S. thesis, Department of Environmental Sciences, University of Virginia.

Anderson, R. S. (1986) Erosion profiles due to particles entrained by wind: Application of an aeolian sediment-transport model. *Geological Society of America Bulletin*, v. 97, p. 1270-1278.

Blackwelder, E. (1929) Sandblast action in relation to the glaciers of the Sierra Nevada. *Journal of Geology*, v. 37, p. 256-260.
Blake, W. P. (1855) On the grooving and polishing of hard rocks and minerals by dry sand. *American Journal of Science*, v. 20, p. 178-181.
Breed, C. S., McCauley, J. F., and Whitney, M. I. (1989) Wind erosion forms. In D.S.G. Thomas (ed.) *Arid Zone Geomorphology*. John Wiley Sons, New York, p. 284-307.
Clark, R. and Wilson, P. (1992) Occurrence and significance of ventifacts in the Falkland Islands, South Atlantic. *Geografiska Annaler*, v. 74A, p. 35-46.
Dorn, R. I. (1986) Rock varnish as an indicator of aeolian environmental change. In W. G. Nickling (ed.) *Aeolian Geomorphology*. Allen & Unwin, Boston, p. 291-307.
Dorn, R. I., Jull, A.J.T., Donahue, D. B., Linick, T. W., and Toolin, L. (1989). Accelerator mass spectrometry radiocarbon dating of rock varnish. *Geological Society of America Bulletin*, v. 101, p. 1363-1372.
Gilbert, G. K. (1875) *Report on the Geology of Portions of Nevada, Utah, California, and Arizona*. Geographical and Geological Surveys West of the 100th Meridian, Vol. III.
Greeley, R., and Iversen, J. D. (1986) Aeolian processes and features at Amboy Lava field, California. In F. El-Baz and M. Hassan (eds.) *Physics of Desertification*. Martinus Nijhoff Publishers, Dordrecht, p. 210-240.
Greeley, R., Williams, S. H., White, B. R., Pollack, J. B., and Marshall, J. R. (1984) Wind abrasion on Earth and Mars. In M. J. Woldenberg (ed.) *Models in Geomorphology*. Allen & Unwin, Boston, p. 373-422.
Groat, C. G. (1967) *Geology and Hydrology of the Troy Playa Area, San Bernardino County, California*. M.S. thesis, University of Massachusetts, Amherst.
Hobbs, W. H. (1917) The erosional and degradational processes of deserts, with especial reference to the origin of desert depressions. *Annals of the Association of American Geographers*, v. 7, p. 25-60.
Laity, J. E. (1987) Topographic effects on ventifact development, Mojave Desert, California. *Physical Geography*, v. 8, p. 113-132.
Laity, J. E. (1992) Ventifact evidence for Holocene wind patterns in the east-central Mojave Desert. *Zeitschrift für Geomorphologie, Supplement Band*, v. 84, p. 1-16.
Laity, J. E. (1994) Landforms of aeolian erosion. In A.D. Abrahams and A.J. Parsons (eds.) *Geomorphology of Desert Environments*. Chapman & Hall, London, p. 506-535.
Mainguet, M. (1972) *Le Modelé des Grès*. Institute Geographie National: Paris.
Maxson, J. H. (1940) Fluting and faceting of rock fragments. *Journal of Geology*, v. 48, p. 717-751.
Selby, M. J. (1977) Palaeowind directions in the central Namib Desert, as indicated by ventifacts. *Madoqua*, v. 10, p. 195-198.
Sharp, R. P. (1949) Pleistocene ventifacts east of the Big Horn Mountains, Wyoming. *Journal of Geology*, v. 57, p. 173-195.
Sharp, R. P. (1963) Kelso Dunes, Mohave Desert, California. *Bulletin of the Geological Society of America*, v. 77, p. 1045-1074.
Sharp, R. P. (1964) Wind-driven sand in Coachella Valley, California. *Geological Society of America Bulletin*, v. 75, p. 785-804.
Sharp, R. P. (1980) Wind-driven sand in Coachella Valley, California: Further data. *Geological Society of America Bulletin*, v. 91, p. 724-730.
Sharp, R. P., and Malin, M. C. (1984) Surface geology from Viking landers on Mars: A second look. *Geological Society of America Bulletin*, v. 95, p. 1398-1412.
Smith, H.T.U. (1967) Past versus present wind action in the Mojave Desert region, California. *Air Force Cambridge Research Laboratories*, 67 - 0683, p. 1-26.
Smith, R.S.U. (1984) Eolian geomorphology of the Devils Playground, Kelso Dunes and Silurian Valley, California. In J.Lintz (ed.)*Western Geological Excursions*, v. 1, Geological Society of America 97th Annual Meeting Field Trip Guidebook, Reno, Nevada, p. 239-251.
Tchakerian, V. P. (1994) Paleoclimatic interpretations from desert dunes and sediments. In A. D. Abrahams and A. J. Parsons (eds.) *Geomorphology of Desert Environments*. Chapman & Hall, London, p. 631-643.

Whitney, M. I. and Dietrich, R. V. (1973) Ventifact sculpture by windblown dust. *Geological Society of America Bulletin*, v. 84, p. 2361-2582.

Wilshire, H. G., Nakata, J.D, and Hallet, B. (1981) Field observations of the December 1977 wind storm, San Joaquin Valley, California. In T.J. Péwé (ed.) *Desert Dust*. Geological Society of America, Special Paper 186, p. 233-251.

Index